Risk Management in the Oil and Gas Industry

Risk Management in the Oil and Gas Industry

Offshore and Onshore Concepts and Case Studies

GERARDO PORTELA DA PONTE JR

Doctor of Risk and Safety Management from the Department of Naval and Oceanic Engineering at COPPE - Federal University of Rio de Janeiro, Brazil, with specializations in Human Factors and Safety Engineering, Charles W. Davidson College of Engineering, The California State University, San Jose, CA, United States and in Maritime and Offshore Safety, Kelvin Hydrodynamics Laboratory, University of Strathclyde, Scotland United Kingdom

ELSEVIER

Gulf Professional Publishing
An imprint of Elsevier

Gulf Professional Publishing is an imprint of Elsevier
50 Hampshire Street, 5th Floor, Cambridge, MA 02139, United States
The Boulevard, Langford Lane, Kidlington, Oxford, OX5 1GB, United Kingdom

British Library Cataloguing-in-Publication Data
A catalogue record for this book is available from the British Library

Library of Congress Cataloging-in-Publication Data
A catalog record for this book is available from the Library of Congress

ISBN: 978-0-12-823533-1

For Information on all Gulf Professional Publishing publications
visit our website at https://www.elsevier.com/books-and-journals

Publisher: Joe Hayton
Senior Acquisitions Editor: Katie Hammon
Editorial Project Manager: Naomi Robertson
Production Project Manager: Sojan P. Pazhayattil
Designer: Christian J. Bilbow

Typeset by MPS Limited, Chennai, India

It is always possible to cheat.
Even if we place a watchman next to each person,
the watchman can also cheat.
It is cultivating good VALUES that we will be righteously protected.

—Gerardo Portela da Ponte Junior

Dedication

God

Angelica

Alana

Micaela

PARENTS

Family

Friends

Teachers Students

And to the professionals who prioritize the benefit of people and society in technological enterprises of all kinds.

Contents

Special acknowledgment xvii
Editorial acknowledgment xix
Complementary sources xxi
Declaration xxiii
About the author xxv
Foreword xxvii
Acknowledgments xxix

1. Introduction and reading guide 1

2. Fundamentals of risk management 5

2.1 Nonquantifiable risk 7
2.2 Safety culture and risk acceptance 9
2.2.1 What is right attention at the right time? 12
2.2.2 Safety pendulum 14
2.2.3 Seven principles of the safety culture 14
2.3 Human factors and the error-inducing environment 18
2.3.1 Seven principles of human factors 20
2.4 Efficiency and strategic risk management line 23
2.4.1 Efficiency 23
2.4.2 Risk management strategic line 29
2.4.3 Technical and operational knowledge 31
2.4.4 Hazard reduction 35
2.4.5 Removal of agents (people) 39
2.4.6 Emergency control 41
2.4.7 Design-basis accident 42
2.4.8 Beyond design-basis accident 42
2.4.9 Reducing unpredictability 43
2.5 Lessons learned 46
2.5.1 The theory specialist 46
2.5.2 The "best gas sensor in the world" 50
2.6 Exercise 52
2.7 Answers 54
2.8 Review questions 54

3. Technical and operational knowledge **55**

3.1 Oil industry 56
3.1.1 John Davison Rockefeller and risk management 56
3.1.2 Components of the oil and gas productive chain 58
3.1.3 Onshore and offshore facilities 58
3.1.4 Accidents in the oil and gas industry 60
3.2 Getting to know upstream facilities 62
3.2.1 Drilling rig and completion 64
3.2.2 Primary processing equipment 66
3.2.3 Fixed offshore platforms 67
3.2.4 Semisubmersible offshore platforms 70
3.2.5 Floating production, storage, and offloading system platforms 71
3.2.6 Special offshore platforms 74
3.3 Getting to know downstream facilities 76
3.3.1 Refining facilities and petrochemical plants 77
3.3.2 Transportation and distribution 79
3.3.3 Marine terminals (inshore or at shore) 81
3.4 Knowing process safety 82
3.4.1 Loss of containment (liquid and gas leaks) 82
3.4.2 Stable or explosive burning combustion 84
3.4.3 Safety in physical and chemical operations with hydrocarbons 86
3.5 Knowing operational practice (field experience) 90
3.5.1 Safety barrier 94
3.5.2 Professional work in operational activities and in the field 94
3.6 Knowing the project routine 96
3.6.1 Project routines 97
3.6.2 Professional work in project activities 98
3.6.3 Safety systems design documents 102
3.7 Lessons learned 102
3.7.1 Avatar for "experts" without operational experience 102
3.8 Exercises 108
3.9 Answers 108
3.10 Review questions 109

4. Hazards reduction **111**

4.1 Segmentation of the hydrocarbon inventory 113
4.1.1 Layout techniques 113
4.1.2 Blocking segmentation technique 115
4.2 Disposal of the hydrocarbon inventory during an emergency 117

4.2.1 Pressure relief and depressurization 118
4.2.2 Controlled burning and dispersion 118
4.3 Automatic emergency shutdown 125
4.3.1 ESD level 1 125
4.3.2 ESD level 2 126
4.3.3 ESD level 3 126
4.3.4 ESD level 4 127
4.3.5 Example of an emergency shutdown sequence 127
4.3.6 Shutdown requires caution 129
4.4 Lessons learned 129
4.4.1 Piper alpha hazards reduction failure 129
4.4.2 Lessons learned from piper alpha 132
4.5 Exercises 134
4.6 Answers 135
4.7 Review questions 137

5. Agents (people) evacuation 139

5.1 Importance of the systems of escape and abandonment 141
5.2 Accidents in facilities with hydrocarbon inventories and survival 142
5.3 Human–system interaction during escape and abandonment 143
5.4 Escape and abandonment operation 145
5.5 Technical recommendations for escape and abandonment system 149
5.5.1 Possible operational sequences 149
5.5.2 Basic dimensions and recommendations for escape routes 150
5.5.3 Evacuation, escape, and rescue analysis 152
5.5.4 Spaces with limited access and machine rooms 152
5.5.5 Applicable materials in escape and abandonment systems 153
5.5.6 Meeting points (muster stations) and abandonment points 153
5.6 Sea survival equipment 154
5.6.1 Lifeboats 155
5.6.2 Life rafts 161
5.6.3 Rescue boat 162
5.6.4 Salvage equipment 163
5.7 Lessons learned 163
5.7.1 SOS: emergency in FPSO 163
5.8 Exercise 169
5.8.1 Crisis scenario simulator 169
5.8.2 General instructions 170
5.8.3 Instructions about scenario evolution 171

5.9 Answer 174
5.10 Review questions 175

6. Emergency control 179

6.1 Power generation systems 180
6.1.1 Essential consumers 182
6.1.2 Safety consumers 184
6.1.3 Special requirements for cables and lighting 184
6.1.4 Area classification 185
6.2 Heating, ventilation, and air conditioning systems 196
6.3 Flushing, purging, and inerting systems 199
6.4 Gas detection system 200
6.4.1 Flammable gas detection 200
6.4.2 Toxic gases detection (H_2S) 205
6.4.3 Monitoring gas contamination (H_2S/CH_4) 207
6.4.4 Specification and location of gas detectors 208
6.5 Fire detection systems 209
6.5.1 Flame detection 210
6.5.2 Heat detection (fusible plug) 212
6.5.3 Smoke detection 214
6.5.4 Thermovelocimetric detection 216
6.5.5 Fixed temperature heat detection 216
6.5.6 Specification and positioning of fire detectors 216
6.6 Automatic fire-fighting systems 218
6.6.1 Water spray fixed systems (deluge) 220
6.6.2 Foam-water spray systems 224
6.6.3 Fire-fighting water pumps 225
6.6.4 Fire-fighting water distribution system 236
6.6.5 Carbon dioxide fire extinguishing system 239
6.6.6 Water mist fire suppression system 242
6.7 Additional fire protection systems 244
6.7.1 Fire hydrants 245
6.7.2 Mobile foam generating equipment 245
6.7.3 Fire-fighting monitor cannons 246
6.7.4 Fire extinguisher 246
6.7.5 Auxiliary equipment 246
6.8 Passive fire protection 246
6.8.1 Determination of the type of partitions 249
6.8.2 Observations cited in the tables 250

6.8.3 Interference of classified bulkhead and penetrations 253
6.8.4 Interference between classified bulkheads and doors and windows 253
6.8.5 Structural protection 253
6.8.6 Materials for passive protection 254
6.9 Protection systems for confined equipment 254
6.10 Accidents with cryogenic products (LNG) 254
6.10.1 Knowing the cryogenic characteristics of liquefied natural gas 255
6.10.2 Basic accidental scenarios and liquefied natural gas cryogenics 257
6.10.3 Emergency control and liquefied natural gas cryogenics 260
6.10.4 Rapid phase transition and liquefied natural gas cryogenics 264
6.11 Subsea safety equipment 265
6.12 Fire brigade and rescue crew performance 265
6.13 Crisis management and decision making 266
6.14 Selecting and identifying accidental scenarios 268
6.14.1 Design basis accident 271
6.14.2 External origin accidents 271
6.14.3 Beyond design accident 275
6.15 Special safety strategies applied to automation 276
6.16 Conception of redundancies and ways to start up safety systems 277
6.16.1 Types of redundant configurations 278
6.16.2 Classical failures 279
6.17 Understanding explosion phenomena 281
6.17.1 Types of explosion involving flammable products 282
6.17.2 Formation of explosive atmospheres in open space 282
6.17.3 Formation of explosive atmospheres in closed space 283
6.17.4 Formation of explosive atmospheres by BLEVE 286
6.17.5 Shock waves and factors that influence explosions 287
6.18 Lessons learned 289
6.18.1 Correction of conceptual error results in 50 million dollar savings 289
6.18.2 Strategy 292
6.18.3 Interactivity, arrangement, and risk management 293
6.18.4 Criteria and results 297
6.19 Conclusions 299
6.20 Exercises 300
6.21 Answers 301
6.22 Review questions 303

7. Reducing unpredictability **307**

7.1 Risk analysis techniques 310
7.1.1 Quantitative and qualitative risk analyses 312

7.1.2 Preliminary risk analysis 314
7.1.3 Preliminary hazard analysis 319
7.1.4 Hazards and operability analysis 322
7.1.5 Other risk analysis techniques 331
7.2 Studies and consequence analyses 338
7.2.1 Fire propagation study 340
7.2.2 Study of dispersion of gases and smoke 341
7.2.3 Explosion study 342
7.2.4 Escape, abandonment, and rescue study 343
7.2.5 Analysis of loss of liquid containment and environmental control 346
7.2.6 Studies of stability and naval damage condition 351
7.3 Full safety analysis 352
7.3.1 Features of the analysis of offshore rig 355
7.3.2 Importing documents to build the 3D model 356
7.3.3 Building the 3D model 357
7.3.4 Adaptation of the process plant area 357
7.3.5 Adaptation of FPSO hull internal areas 358
7.3.6 Adaptation of the superstructure internal area 359
7.3.7 Definition of agents on board and their behavioral parameters 360
7.3.8 People on board definition 360
7.3.9 Operational experience of agents on board 360
7.3.10 Gender of agents on board 361
7.3.11 Age of agents on board 362
7.3.12 Travel speeds of agents onboard 362
7.3.13 Reaction times for agents on board 363
7.3.14 Physical positioning of agents in the rig 363
7.3.15 Special tasks for specific agents during the emergency 365
7.3.16 Measuring the effects of emergency on people's integrity 366
7.3.17 Conceptual definition of accidental scenarios 367
7.3.18 Standard and gas leakage scenarios 368
7.3.19 Fire scenarios 368
7.3.20 Naval damage condition scenarios 369
7.3.21 Theoretical scenarios for comparative purposes 371
7.3.22 Representative simulations for offshore rigs 371
7.4 Lessons learned 372
7.4.1 Risk analysis and team work 372
7.4.2 HAZOP chaos 376
7.5 Exercises 378
7.6 Answers 380
7.7 Review questions 381

8. Human–system interaction **385**

8.1 Human error 386
8.2 Human factors 388
8.2.1 Main influences related to human factors 393
8.2.2 Human factors analysis 402
8.2.3 Programs for consideration of human factors 408
8.2.4 Human factors in the life cycle of technological enterprises 410
8.2.5 Intelligent identification of systems and equipment 415
8.3 Limitations of quantification techniques related to human reliability 416
8.4 Rapid Entire Body Assessment 418
8.4.1 Example of the application of the REBA technique 419
8.4.2 Human body mechanics during execution of tasks 419
8.4.3 Anthropometry 420
8.4.4 Static work 421
8.4.5 Repetitive work, cumulative trauma, and use of hand tools 421
8.4.6 Rapid Entire Body Assessment evaluation 422
8.4.7 Rapid Entire Body Assessment recommendations 422
8.5 Lessons learned 422
8.5.1 Influence of Global Positioning System on the driver 423
8.5.2 Global Positioning System position in the dashboard 423
8.5.3 Audio information 424
8.5.4 Visual information 424
8.5.5 Software and configurations 424
8.5.6 Driver's GPS knowledge 425
8.5.7 Definition of the GPS position in the dashboard 425
8.5.8 Limiting the level of audio information 425
8.5.9 Visual information overload 425
8.5.10 Human–system interface 426
8.5.11 Conclusion 426
8.7 Exercises 427
8.8 Answers 428
8.9 Review questions 429

9. Risk management systems **431**

9.1 Risk management in the corporate environment 432
9.2 Centralization and decentralization of risk management 433
9.3 Association of different technical fields 434
9.4 Historical data records and management by indicators 435
9.5 Risk management, occupational safety and safety engineering 436

9.6 Risk-based design 438
9.7 Safety peer review 439
9.8 Accident investigations 441
9.8.1 Systematic cause analysis technique 442
9.8.2 5 Whys technique 443
9.8.3 SHELL method (Software, Hardware, Environment, Liveware 1, Liveware 2) 444
9.8.4 Causal tree or fault tree technique 444
9.9 Surveillance system 446
9.10 Capillarity of concepts and principles of risk management 447
9.11 Risk and safety management in the energy industry postpandemic COVID-19 448
9.11.1 "World energy outlook": International Energy Agency 450
9.11.2 Risk management in the postpandemic world energy future 461
9.12 Risk and safety management and the potential of the new digital tools 464
9.12.1 New risk analysis methods 465
9.12.2 Risk analysis including the human behavior 467
9.13 Applicable technical standards 469
9.13.1 Reference standards 470
9.14 Lessons learned 471
9.14.1 False safety improvement plans 471
9.15 Exercises 473
9.16 Answers 474
9.17 Review questions 476

10. Synthesis **479**

Bibliography *481*
Index *485*

Special acknowledgment

Special thanks to two families that have been indispensable for carrying out this work. Firstly the Mike, Kim, and Phillip Kirouac's family who received us, a Brazilian couple with a two-month old baby, in the small town of Campbell in California. Our work in the United States would have not been possible without the continued support of the Kirouac family from settling in the city to the issuance of certificates of completion of the course. In addition, after all of their dedication in California, the Kirouac family has also referred us to another Scottish Christian family to continue support in Glasgow, Scotland, United Kingdom. To our dear Roddy and Moira Shaw, who also welcomed us wonderfully in Glasgow, we are also grateful to have been by our side, helping us to overcome the typical difficulties of a Brazilian family, alone, in such a distant country. At no time did we feel alone while away from our land. We will always have a special affection for the Americans and Scots who will always be remembered by us, symbolized by these extraordinary friends.

Editorial acknowledgment

Writing a book is a team effort. This work was born under the guidance of the publisher Andrea Rodrigues (Elsevier Brazil). I thank you Andrea for your technical competence and dedication, enabling the publication of two books that were prepared during some rather difficult times. A few months later, an opportunity arose to publish one of the books internationally. It took 5 years for the Senior Acquisitions Editor Katie Hammon (Elsevier Inc., USA) to put together the complete publication proposal in the United States and submit it for approval in the midst of the pandemic that paralyzed the planet. Thank you very much Katie, you guided me with your competence and now a professional dream is coming true with the help of so many other professionals, editors, diagrammers, proofreaders, graphic designers, whom I hereby represent in the names of editorial project manager Naomi Robertson and translator Luiz Souza. Everyone added value to my work. Without you all, it would be impossible. Thank you very much for this technical scientific work done as a team!

Complementary sources

Complementary technical information and about the author and institutions related to this subject matter, videos, and interviews in the media are available in the following links:

www.gerardoportela.com.br
https://www.youtube.com/channel/UC8R-9vvefegkd-krihbNlgA
www.risksafety.com.br

Risk & Safety

The success of a technological enterprise is associated with respect for human, environmental, economic, and social factors that are under its influence. Good values establish good safety culture.

Declaration

Although the cases presented in each chapter under the subtitle "Lessons Learned" have similarities with real world situations, they have been included for didactic purposes. If the reader identifies the correspondence of the narratives with people or companies, that correspondence may be a pure coincidence.

About the author

Gerardo Portela da Ponte Junior has a doctorate degree from the Department of Ocean Engineering at COPPE—Federal University of Rio de Janeiro, Brazil, with a doctoral thesis on Risk Management and Offshore Safety. He has a specialization in Safety Engineering from Charles W. Davidson College of Engineering, The California State University at San Jose, CA, United States (Silicon Valley) where he also worked as a researcher in the area of Human Factors.

In the experimental part of his doctoral program, Portela conducted research on computer simulations of escape and abandonment in offshore rigs, at the Kelvin Hydrodynamics Laboratory, University of Strathclyde, in Glasgow, Scotland, United Kingdom. Portela also has a master's degree in Technological Management and also a BSc in Mechanical and Industrial Engineering from the Federal Center for Technological Education RJ, Brazil. He has more than 40 years of professional experience in the engineering sector, having worked in the areas of design, construction & assembly, and operation of safety systems. He has worked in the fields of shipbuilding, steel, construction & assembly, airport infrastructure, oil, electricity generation, and nuclear power plants for large companies such as Petrobras, Eletrobras Thermonuclear, Furnas Power Stations, Infraero Airports, Ishikawajima, and Shell, among others. He is a professor of graduate courses in Offshore Safety at COPPEAD/COPPE/UFRJ, Brazil, and in the courses on Human Factors, Risk Analysis and Offshore Safety at Universidade Corporativa Petrobras, the company where he currently works in the area of risk management and safety. Gerardo Portela is a recognized authority on the topic of risk management, acting as a guest commentator for several press agencies in Brazil and abroad.

Foreword

I was very happy to write the foreword of Gerardo Portela's new book entitled *Risk Management in the Oil and Gas Industry: Offshore and Onshore Concepts and Case Studies*. I consider this book to be an additional element that will contribute to the safety of teams that work in the oil and gas sector. I remember that back in 2005 I was invited to organize a graduate course at COPPE/UFRJ together with engineers from Petrobras, called Safety Applied to Oil Exploration and Production Projects. We had seven groups of students graduating in that year, and as the result of this work, three theses were developed at COPPE, with Gerardo Portela's doctoral dissertation being one of them. I had great pleasure being the supervisor of his thesis and I must acknowledge how much I learned during the professional and personal interaction. Gerardo with this book goes one step further to offer knowledge to all those involved in offshore and onshore operations that are directly or indirectly related to the safety of people, the environment, processes, and equipment used in oil exploration and production. This book is both a gift and a challenge for all of us involved with the subject matter of safety in the oil and gas sector.

José Márcio de A. Vasconcellos, Dsc Engenharia Naval
Professor at the Department of Naval Architecture and Engineering—COPPE/UFRJ, Brazil

Acknowledgments

Thanks for the privilege of being able to share in this book the knowledge I have acquired. Thanks for the health, for the support of my family, for the teachers, and colleagues who taught us so much. Thanks for the sustenance through work, for Your protection from the dangers that surround us. May the content of this work be blessed. May this book serve to enlighten professionals who work in high-risk activities, protecting lives and preventing accidents. Help us to be humble when facing danger. Help us to have courage so that we will never be cowards in the struggle for life. Bless us so that we can endure the difficulties of our work, that we can never be discouraged for lack of knowledge, never be discouraged by lack of recognition, never be discouraged in the face of political pressures, never be discouraged in the face of economic pressures and in the face of personal interests that try to divert us from our true professional objective, which is to protect life, the environment, our society, and our property. Thanks because, above all things, our greatest safety is deposited in You, who govern with wisdom and love in a way that exceeds our comprehension. Accept our thanks, with the request of forgiveness for the failures that we unfortunately commit, in the name of our Lord and Savior Jesus Christ, Amen.

CHAPTER 1

Introduction and reading guide

The knowledge of risk and safety management presented in this book can be applied to different fields of industrial activities and nonindustrial alike.

Although all the examples mentioned in this work are focused on the oil and gas industry, risk management concepts and principles can be useful for professionals and students in other fields.

This book is based on the results obtained during more than 18 years of work and research on design of safety systems for the oil and gas industry, and almost 40 years of effective professional work experience in engineering, technology, and risk management. The research work conducted as part of the author's doctoral studies in Risk Management and master's in Technology Management also contributed to the development of its content. The studies and research were conducted at the Federal University of Rio de Janeiro, Coppe, Cefet RJ, Brazil, The California State University at San Jose, USA, and University of Strathclyde, Glasgow, UK. Aiming at facilitating the reading of this book, we present in this chapter a summary that serves as a guide for its full or partial reading.

Chapter 2, Fundamentals of Risk Management, introduces the fundamentals of risk management, its principles, and main concepts. It also presents the risk management strategic line, which is essential to organize all the breadth of multidisciplinary knowledge that is scattered in different fields of human knowledge. The risk management strategic line is composed of five elements, which are presented in detail in the subsequent chapters.

Chapter 3, Technical and Operational Knowledge, shows the importance of "technical and operational knowledge" for the risk management professional to achieve the technical authority required to provide solutions that aim for the reduction of hazards and risks in facilities in the oil and gas industry. Chapter 3, Technical and Operational Knowledge, also provides a minimum information base on upstream facilities, downstream facilities, process safety, operational practice, and design routine.

Chapter 4, Hazards Reduction, presents the "hazards reduction" element, which in the case of the oil and gas industry is centered on

Risk Management in the Oil and Gas Industry.
DOI: https://doi.org/10.1016/B978-0-12-823533-1.00014-9

emergency shutdown systems (Emergency Shutdown). Chapter 4, Hazards Reduction, also shows the main techniques for segmentation of the hydrocarbon inventory, how to properly dispose of this inventory during an emergency, and the sequences of operational actions for the protection of the facilities.

Chapter 5, Agents (People) Evacuation, describes the "removal of agents" element and shows the importance of escape and abandonment systems for the protection of personnel and, consequently, to maintain emergencies at levels where there are no losses of people's integrity. It describes the survival conditions of people in accidental scenarios in oil and gas facilities and also shows: the influence of human X system interaction factors during escape and abandonment operations; the main recommendations for the effectiveness of escape and abandonment systems; the types of rescue boats (offshore rigs); and other equipment associated with escape, abandonment, and rescue operations.

Chapter 6, Emergency Control, gathers the technical description and recommendations related to the main "emergency control" systems in oil and gas facilities, such as power generation, ventilation, heating, air conditioning, flushing, purging, inerting, gas detection, fire detection, fire fighting water systems, deluge systems, passive protection, cryogenic systems, subsea equipment, and explosion.

Chapter 7, Reducing Unpredictability, presents the "reduction of unpredictability" element whose content is related to safety studies and risk analyses. It shows the characteristics and applications of quantitative and qualitative risk analysis; preliminary risk analyses; preliminary hazard analysis (PHA and HAZID); analysis of operational hazards (HAZOP); consequences analysis; and the concept of Full Safety Analysis as a technique with great potential in risk management.

Chapter 8, Human–System Interaction, addresses the relevant aspects related to human X system interaction in oil and gas facilities and shows the technical approach to human error, human reliability, human factors, and the Rapid Entire Body Assessment technique.

Chapter 9, Risk Managements Systems, presents concepts and benchmarks for experts to develop efficient risk management systems. In addition, it conceptually describes risk management techniques and methodologies such as: Risk-Based Design, Safety Peer Review technique and methodologie and tests Surveillance System. Chapter 9, Risk Managements Systems, also shows information about accident investigations, their techniques, and methods.

Finally, Chapter 10, Synthesis, shows a superficial synthesis of the book, which contains the general description of its content so that the reader can quickly compose an "overall vision" of the book.

There are several valid ways to use nomenclatures and technical terms in the area of risk management and safety. It would be exhaustive to survey all the technical terms applicable to the topic and the several definitions attributed to each of them. As an instrument for facilitating technical communication, we indicate as an important general reference: the Center for Chemical Process Safety—"CCPS Process Safety Glossary." It can be accessed through the electronic address: https://www.aiche.org/ccps/resources/glossary and also the very general "GLOSSARY FEMA" of the Federal Emergency Management Agency (FEMA), the official body of the United States Government specialized in risk management. It can be accessed through the electronic address: https://www.fema.gov/about/glossary

International standards and references are extremely important for those who wish to work in the area of risk management and safety. It is necessary to know how to deal with the difficulties in finding the most appropriate technical standard when we know that certain themes are mentioned in various standards, from different countries and continents. Therefore the author chose to concentrate in Chapter 9, Risk Managements Systems, some of the most important references of technical regulations applicable to the content of the book. Professionals in the area of risk and safety management need to know how to access these standards in their official sources, as these standards are constantly evolving and updated. The author's proposal is that the reader searches the internet portals of each source of technical standards, those that are adapted to the problems faced in daily life. The names and identification numbers of these standards change frequently, often surprisingly. That is why we chose to avoid frequent citations of standards and indicate in Chapter 9, Risk Managements Systems, the main sources of standards for professionals, by their own means, to seek the most up-to-date version of the standards. The author understands that this is the safest way for each professional, and the book itself, to keep up-to-date.

CHAPTER 2

Fundamentals of risk management

The accelerated technological progress creates increasingly complex challenges for engineers. But this progress faces limitations such as unavailability of complete scientific solutions, high costs, and risks to the safety of people, the environment, and society. Scientific limitations require time and research to be overcome thus allowing the consolidation of new technology. The term technology may have several interpretations, but in this work we consider the following to be the most suitable definition: technology is productive science. Science generates knowledge, but the knowledge does not always have an immediate practical application. In some cases, a scientific discovery may remain without any apparent use for years after being announced until someone finds a practical application for it. When it happens, the scientific knowledge starts producing practical results in people's lives both in terms of benefits for their daily routine as well as economic results for the society.

Even when technology is available, associated costs require willingness to pay for it. But risks related to safety will always need to be managed as the final limiting factors, which restrict technological progress with objective facts and evidence, regardless of the availability of economic and scientific resources. In other words, we can have financial capital and technology available but if the risks of accidents and the damage caused by them fall outside all "acceptance" criteria there will be no one willing to promote such technological progress. Risk can always be represented by a number or by a percentage value. Once a hazard and an accident associated with it are identified, we can evaluate the frequency of occurrence of this scenario and come up with a number, a percentage of chances of the accident happening. For instance, when we say that for every 100 drunk drivers 95 are unable to complete a lap in a driving circuit without colliding with the traffic cones, we can say that "the risk of a drunk driver colliding with a cone in this circuit is 95%." In the previous example, for some drivers the risk of 95% chance of colliding with the cones is unacceptable. So these drivers can refuse taking the risk of trying

Risk Management in the Oil and Gas Industry.
DOI: https://doi.org/10.1016/B978-0-12-823533-1.00010-1

to do the circuit in the drunk state. But other drivers, for their own reasons, may consider the same 95% risk acceptable since they believe to be part of the 5% with capacity or be lucky to complete the circuit lap even under the influence of alcohol. The most important point in managing risks is knowing when to accept them and when to reject them. There is always a subjectivity aspect in a decision. Despite efforts by experts to create scientific means to provide support for such a decision, there is still no perfect statistical model capable of objectively ensuring 100% protection against an accident. Only refusing to take a risk assures full guarantee against it. That is what happens when one does not travel by plane so there is no chance of becoming a victim of a plane crash or when nuclear power plants are not built so as to not be subjected to the risk of an accident at a nuclear power plant. It is very important to realize that if we eliminate certain risks many others will always be there because being at risk is part of human existence and the nature itself. The difference is in the value of the risk, the probability and frequency of occurrence of the accident, which can be lower or higher, that makes it more or less "acceptable." Without traveling by plane and without building power plants we eliminate (with 100% assurance) some specific risks. But we remain subjected to others. The fundamental point is to choose wisely the risks that we must accept, because the more risks we accept the greater will be the work required to manage them. So, if we have the ability to select the absolutely necessary risks, discarding the greatest possible amount of unnecessary ones, we will have an efficient risk management.

But what is a "absolutely necessary" risk? That is exactly where the subjective component lies which we mentioned earlier. Both in our personal life when, for instance, we decide to practice an extreme sport as well as in the corporate world, when we choose a certain technology, by accepting such risks we are also sending signals that a certain risk, under these circumstances, is for us "absolutely required." We can decide by accepting or not accepting a risk independently or with the help of arguments of others that justify an opinion contrary to ours. Even with figures and statistics on accidents presented in risk analysis reports, some societies may reject the risk of a nuclear power plant while others accept it. Elements of subjective influence are present in decisions whether or not to accept risks. Sometimes explicitly like when we accept to practice an extreme sport without scientific basis but just for pleasure and other times it is hidden among the premises of introductory text of quantitative risk analyses. Even the numerical values of frequency of occurrence of accidents are obtained through data acquisitions associated with statistical calculations, but are always based on some premises. The premises are part of the scientific method, but as the scientific risk analysis evolves and its final results are quantified the premises get weaker in the process and it is perceived as less influential than the real one. For a better understanding of the subjectivity component, it is necessary to deal comfortably both with mathematical models and with subjective aspects associated with the safety culture and with the human factors that influence risk acceptance.

2.1 Nonquantifiable risk

The subjectivity present in the decision whether or not to accept a risk can be better understood by the concept of nonquantifiable risk. There are risks for which, in order to be calculated, many variables need to be taken into consideration that their calculations become unfeasible or intractable. We call such risks nonquantifiable risks. The "nonquantifiable risk" factor represents a set of influences sometimes explicit and sometimes hidden that cannot be measured within the context of technical feasibility. When someone, after a scientific risk analysis, despite obtaining results that recommend acceptance of the risk, nevertheless decides not to accept it, that person is probably assigning more weight to the "unquantifiable risk" than the "quantifiable risk" portion. Conversely, if the decision is to accept the risk the manager is minimizing the weight associated with the "nonquantifiable risk" factor. The "nonquantifiable risk" refers also to the risks related to the scenarios not considered in the analyses. To every quantified risk there is a component of unquantified risk which is associated with it. This second, not quantifiable, component is always treated in a subjective manner according to the criterion adopted and the experience of each organization, professional, and person. This might be disturbing to those who do not accept the weight of nonquantifiable risks. However, when scientific rigor is applied there is no quantitative risk analysis that can ensure that an accident will not occur. Even if such an analysis indicates extremely low probability of an accident occurring, it does not ensure that the accident will in fact not going to happen.

A case where the "nonquantifiable risk" factor contributes to the explanation of a catastrophic accident is that of the nuclear power plant disaster of Fukushima, Japan, where one of its reactors exploded on March 12, 2011, as a result of a tsunami that had occurred the day before. The original design of the Fukushima nuclear power plant included studies and safety analysis that took into consideration the influence of natural phenomena such as earthquakes and tsunamis. These analyses influenced the design calculations and response strategies and nuclear safety of the nuclear power plant. In the majority of nuclear power plants, starting from the design phase, consequences analysis is performed which serves as a basis for the development of protection against accidents of external origin—nomenclature used for this type of accident (caused by tsunami, earthquake, etc.) in the field of nuclear energy. Additionally, Fukushima was built before Chernobyl and as a result it had been retrofitted to incorporate supplemental modifications regarding its adequacy after the occurrence of the accident at the nuclear power plant of the former Soviet Union, in 1986. Despite the good engineering tradition and the Japanese operational discipline, Fukushima plant did not withstand the effects of the extremely strong tsunami due, in part, to unexpected failures at the nuclear plant and also for the scale of the natural phenomenon that originated the accident which was beyond the expected. In both cases, the risk associated with the accidental event can be considered as "nonquantifiable risk" with respect to the standards and premises adopted in safety studies in the original design.

"Nonquantifiable risk" is the term that represents the uncertainty associated with every risk management process. There is no complete assurance against accidents. Risk management is a multidisciplinary activity which tries to gather as much technical information as possible to assess the risks and contribute to the decision of their acceptance or rejection. Sports attract people all over the world despite many cultural differences. One of the probable justifications for the worldwide fascination with sports is that they reproduce a relatively well controlled risk environment in real life. When there is less control on the sports environment than usual, then the term "extreme sport" is used as reference to those modalities where the risks are very similar to those most critical ones that are part of people's daily lives (Fig. 2.1). An interesting experience recommended for those whose work is devoted to professional risk management is the practice of extreme sports. Evidently, practicing an extreme sport is far from being a typical academic recommendation for the professional development of risk management experts. Nevertheless, sport practice provides opportunities to exercise decision-making, the understanding of

Figure 2.1 Surfing is considered a "radical" sport. Surfers need to deal with natural forces that are totally beyond human reach and control. Fast decisions do not leave margin for errors, and should they occur, they can result in serious accidents. However, it is possible to practice surfing safely based on a good assessment of the sea conditions, the surfer own physical conditions, the conditions and adequacy of the necessary equipment. The practice of "extreme" sports can be considered a playful form of risk management exercise.

accident dynamics and the reflections on many lessons learned from victories and losses some of which can be traumatic. We recommend the practice of sports for those who plan to dedicate themselves to risk management so that they are able to deal with issues related to uncertainty, human failure, calculation errors, and "nonquantifiable risks" in a relatively safe environment.

2.2 Safety culture and risk acceptance

On April 26, 1986, one of the most important accidents of all time occurred: the nuclear accident at the Chernobyl power plant, in the Soviet Union. Many lessons were learned from the investigations and studies in the wake of the accident. It was a high-cost learning because in addition to the victims and immediate fatalities, the consequences of the damage to the environment and to the population remain to this date, decades after the accident happened, and will continue for a long time to come.

After the Chernobyl accident many changes took place in nuclear power plants around the world. These changes were not limited to the field of nuclear engineering but rather they influenced changes and created new concepts applicable to the safety of all types of technological enterprises. The nuclear energy sector, which is considered a benchmark for high technology and safety, also leaves as a legacy the history of the generalized cascading failures that culminated in a catastrophe that became a "symbol" of technological failure—Chernobyl.

From the studies and lessons learned from Chernobyl, what has become one of the most positive concepts for the increase of the safety level of technological enterprises is the concept of "Safety Culture." Experts and researchers from all over the world have studied and continue ongoing studies of the events of that dawn in April 1986, in Ukraine. One of the most important conclusions is that the set of factors and conditions that resulted in the Chernobyl nuclear accident is so complex and unexpected that it extends beyond technical and operational problems. The set of factors and conditions that allowed the escalation of the accident constituted a problem of cultural scope.

The former Soviet Union had, at the Chernobyl Nuclear Power Plant, not only a power generation asset of its energy grid but also an additional objective which was to reprocess the used fuel elements as raw material for nuclear warheads for military purposes. We need to recognize that all nuclear reactors, after the burning cycle of the fuel elements, provide radioactive material that can be used for such purposes. However, the problem at Chernobyl was that this objective was overvalued in the structure of the time in the former Soviet Union, and it influenced the design of the Chernobyl Power Plant, and also its operation and procedures.

There was a productivity culture of power generation associated with a military strategic culture.

To a great extent these cultural characteristics influenced the methods, the operational actions, and the design. Using graphite as moderator, one of the important technical factors that increased the accident magnitude, may have been adopted as a suitable solution within that context but it certainly would not have been accepted if the dominant culture were truly "Safety Culture" which effectively prioritized the protection of lives and the environment. After the accident, experts identified the need for developing a specific approach where proper attention was given at the right time to matters related to safety. Thus "Safety Culture" has reached a broader meaning, which today is also applicable to other technological enterprises.

Safety culture is a very broad topic, and it involves different technical and social aspects. There is a wealth of literature and research on the subject matter. Several ideas are widely accepted and well received by managers and risk management professionals. However, there is some difficulty in bringing the safety culture concepts from theory to practice through scientific methodologies that are also compatible with the engineering dynamics and corporate routine. But in the nuclear industry the safety culture has already built a solid pathway. Experiences with accident investigations over the years and most importantly the need for a high level of risk management have developed the safety culture in the Western nuclear industry. Today, the concepts are applied in a practical way in the daily operation of nuclear power plants. For this reason, we can also find the general concepts for the formation of a solid safety culture in its standards and procedures.

The original and practical concept of safety culture defined by the International Atomic Energy Agency (IAEA), Safety Series number 75-INSAG-4, defines: "Safety culture is that assembly of characteristics and attitudes in organizations and individuals which establishes that, as an overriding priority, nuclear plant safety issues receive the attention warranted by their significance." This is the concept of safety culture successfully adopted internationally by the nuclear industry.

Nuclear safety is recognized as a technical reference, given the operational rigor required by this industry. We have made some changes to the original text to make it more comprehensive and applicable to other industries and facilities, such as offshore platforms, refineries, and so on. The modified text reads as follows: "Safety Culture is the combination of commitments and attitudes, by organizations and individuals, which establish as an absolute priority that safety-related issues receive the right attention at the right time."

As much as possible, we take advantage of the technical content related to the safety culture established in the Safety Series number 75-INSAG-4—International Nuclear Safety Advisory Group. These principles cannot always be utilized by engineers with the same objectivity as in the usage of the mathematical models. This is one of the biggest challenges for experts. It is not possible to develop a complete model that includes all the subjective aspects that make up a safety culture. The objective of engineers must be to include means to treat some of these aspects as a contribution to the improvement of the safety culture. In objective terms, we can make it simpler and consider that the procedures and standards of each group reflect the main aspects of the safety culture of that group.

So, addressing the quality of standards and procedures, written or not, can be a good starting point towards building a strong safety culture.

2.2.1 What is right attention at the right time?

The definition of safety culture that we have presented is focused on searching for "the right attention at the right time." But how can we understand what this means in practical terms? To facilitate understanding, we will use the following illustration based on the unfortunate skiing accident of the seven-time world Formula One champion Michael Schumacher. Michael Schumacher was one of the greatest Formula One champion of all times. He suffered some accidents during his career, but none of them as serious as the accident he suffered after retirement, while skiing on a public ski area, where recreational skiers practice the sport regularly. Even though he drove cars in the fastest car racing series in the world for most of his life and under high risks, reaching speeds greater than 350 km/h, he suffered his biggest accident in an apparently lower risk scenario having only his own muscles and the force of gravity as the driving forces.

Schumacher has always been recognized for his mental ability to create strategies to face risks with total safety and making very well calculated passing maneuvers with the lowest possible risk. His racing skills and technical knowledge provided him with the ability to give "the right attention at the right time" to achieve his goal of performing a passing or risky maneuver.

A very impressive episode happened during his career at the 2003 Austrian Grand Prix when he overcame a fire in his car during a refueling stop. Formula One teams train exhaustively for such refueling stops, besides creating devices and equipment to make the operation faster and more accurate. But even in the Formula One high-tech environment there is no way to prevent something unexpected causing a serious accident and that was the case during the refueling stop by the seven-time champion at the Austrian Grand Prix in 2003. After changing tires and refueling, the refueling hose coupling did not disengage as expected, despite all the technology adopted in its design and all the team training for such an operation. As a result, the hose was in an intermediate position, neither connected nor disconnected, creating an opening through which flammable gases were dangerously leaking near the hot parts of the car.

While the mechanic in charge of refueling made desperate attempts to remove the fuel nozzle from the Formula One car fuel tank, fuel dripped out of the nozzle of the car and at that moment the gas mixture ignited, starting a fire involving an already fueled car, a fuel hose, a team of more than 10 mechanics on the track, and a Formula One car driver who buckled up in their tiny survival cell right in the center of the emergency scene. Michael Schumacher's reaction was impressive. The video available on the internet clearly shows him moving his head through the flames with great accuracy, alternatively looking in the rear view mirrors, monitoring the fire fighting by mechanics and the crew. He performed this practical work of crisis management and risk analysis under all the psychological pressure of someone who is at the center of the accident as the main operator and possible victim. The fire was extinguished despite critical fire fighting conditions. Michael Schumacher then quickly understood that he was in the condition to continue the race normally, mainly because he had not wasted time trying to get out of the vehicle before the technical limit to attempt controlling the fire. Not only did Michael Schumacher overcome the accident but he also won the 2003 Austrian Grand Prix and, at the end of the season, he won for the fourth time the title of world champion in the car race series. His Ferrari's refueling stop at the Austrian Grand Prix made history as an example of human ability to manage crises, even being in the center of them, using the accumulated technical knowledge and some well-developed personal skills.

However, unfortunately, the same Michael Schumacher, with his indisputable technical strength to manage risks, suffered an extremely serious accident at the end of 2013 while practicing recreational skiing when he crossed the boundary of the ski area authorized for this practice leaving the original ski run. He did it without any well-founded reason.

Living with professional risk management activities also generates an obligation for zero tolerance for errors in circumstances where human lives are at risk, including one's own. This creates a high-pressure psychological environment, and it can produce side effects such as the need to "be able to make small mistakes" within well-calculated limits outside of professional life. It happens because of the desire for freedom and to be able to relax the strict limits that are routinely imposed by the nature of those who work directly in risk management. This "self-licensing" to take risks can be of great significance for professionals who live constantly under a strict safety limit regime. It can work as a compensation that represents some kind of sense of extra freedom. We do not know exactly the rationale for the seven-time champion to leave the ski trail to take risks in places where possibly stones could meet him along the way. But we do know that, on that tragic day, he failed to pay, even if only for a few few moments, the "right attention at the right time" to his safety-related actions. It takes only a single moment like this for the consequences of an accident, that had been avoided for a lifetime, to happen. Even for an expert on the subject matter, like Michael Schumacher.

2.2.2 Safety pendulum

Risk management involves the acceptance and rejection of risks based on technical and scientific arguments under the influence of safety culture and human factors. This establishes a dynamic management process that can be compared to the motion of a pendulum moved by all these forces of influence and that oscillates between risk rejection and risk acceptance. Risk rejection is a tendency following recent accidents, corporate, and social traumas. Risk acceptance is a tendency associated with too much self-confidence, excessive costs, and competitiveness. Risk management is about the ability to avoid accidents while maintaining the dynamics of the safety pendulum between maximum and minimum rigor without exceeding the limits that lead respectively to complete halt due to the rejection of all the risks and to the catastrophic accident due to inappropriate acceptance of risks. Fig. 2.2 shows the representation of the safety pendulum. In the risk management of technological enterprises, in order for the concept of safety culture to become even clearer and more practical we subdivide it into seven principles. These principles need to be considered as values to be cultivated with the objective of developing a solid and sustainable safety culture.

2.2.3 Seven principles of the safety culture

2.2.3.1 Principle 1 of multidisciplinarity

The development of a safety culture requires a multidisciplinary vision of the accidents. Accidental scenarios present themselves as adverse situations

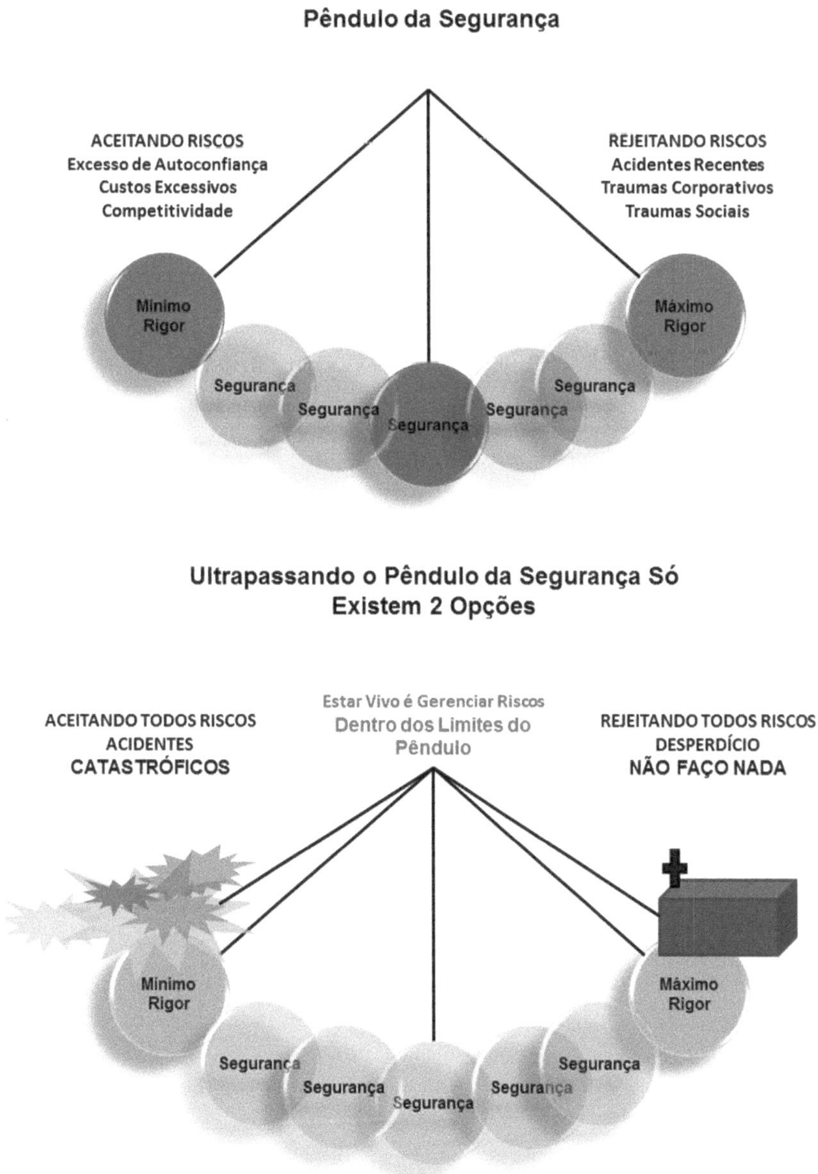

Figure 2.2 Safety pendulum: it represents the dynamic risks management process that oscillates between acceptance or rejection of risk decisions, corresponding to the minimum and maximum rigor with respect to safety.

with multidisciplinary characteristics related to consequences of the unpredictability of certain facts, natural phenomena, equipment failures, procedural failures, behavioral failures, management failures, among others.

In summary, accidents are problems in need of a multidisciplinary solution. And the multidisciplinary solution depends both on typical engineering knowledge and knowledge about natural phenomena and failures arising from human behavior deficiencies under the greater influence of the safety culture.

2.2.3.2 Principle 2 of subjectivity

The development of a safety culture requires the inclusion of subjectivity in the set of objective themes that make up the scope of work for risk management and safety engineering. Being able to relate subjective themes to objective themes coherently and efficiently justifies the development of a safety culture. As an example, the (subjective) commitment to the concepts acquired in technical development leads to the right (objective) attitude.

2.2.3.3 Principle 3 of prioritization

The development of a safety culture requires prioritization of safety-related matters.

It is not possible to develop a safety culture when it is allowed that other matters reduce the attention that should be given to safety-related topics.

2.2.3.4 Principle 4 of right attention

The development of a safety culture requires the ability to provide the right attention to matters related to safety. It is not sufficient to provide attention but right attention is required.

Implementing various safety and prevention measures, plans and safety design, redundancies of safety systems, advertisement and dissemination, courses, training, and qualification, all mean attention. The right attention is the one that is sufficient and effective to avoid a specific accident. A driver can be very skillful behind the wheel, respect all traffic rules, and keep their vehicle in perfect condition and care. But still, the driver can hit a power line pole head-on. In this case, despite driving skills, compliance with traffic laws, and the proper car conditions, in order to avoid this specific accident, the "right attention" required is to see the pole and avoid hitting it. It does not matter whether all other aspects have been dealt with with properly.

2.2.3.5 Principle 5 of right time

The development of a safety culture requires the ability to identify the right time to act. It is not sufficient to act nearly all the time, but it is required to act at the right time when the action is effective to avoid the accident.

Continuously performing preventive and systematic safety actions does not ensure the perception of the right time to have the attitude to avoid the accident. Formal safety routine is not a guarantee against an accident. Realizing the right time and acting that is what does it. Let us consider the driver's example, it is useless to have dodged the pole all the days prior. In order to avoid the accident, the "Right Time" to deflect the pole is the day of the accident.

2.2.3.6 Principle 6 of inclusion of human factors project

The development of a safety culture requires a design of human factors capable of controlling the extent of the consequences of unavoidable human errors.

Human error is unavoidable. To avoid accidents due to human error we can change everything except the human beings as they will not lose their traits of making mistakes even with the best possible training. All factors capable of influencing the extent of the consequences of the unavoidable human errors to avoid accidents due to human error need to be considered so that these consequences can lie within acceptable limits defined by a human factors design. Training may reduce human errors but it cannot avoid the errors altogether.

2.2.3.7 Principle 7 of technical intelligence

The development of the safety culture requires technical intelligence to provide engineering solutions. These solutions should be free from biases such as legalism, heroism, and most importantly, free from mechanistic behavior. These can reduce or be an impediment to the ability to analyze and provide multidisciplinary solutions in accidental scenarios where the unpredictable and unexpected elements are always present.

2.3 Human factors and the error-inducing environment

Human factors is a subject matter that has become increasingly important for the understanding and avoidance of accidents. All technological enterprises originate in people's minds and are intended to produce some kind of consequence on people's interest. But the technical rite of engineering is influenced by the values of those who participate in it and by the values of the society in which they are introduced. Pressures associated with deadlines and economic interests as well as strategies to achieve personal or corporate goals can interfere with the level of importance given to the people's interests, which somehow does or will interact with the technological enterprise. Be it the design of an industrial plant or unique equipment, be its construction or operation, or even while conducting scientific research, every technological enterprise generates an associated "human factors" design. The "human factors" are those related to human–system interaction and hence exert direct influence, by increasing or decreasing the error-inducing effect on people.

The "human factors design" can be developed consciously by engineering professionals, under the technical care with the objective of reducing the error-inducing environment. Even when the professionals involved with the technical rite do not care specifically about these human factors, these designers are still creating a human–system interaction environment, that is, an unattended "human factors design" which will certainly create an increased error-inducing environment in comparison to designs where the human factors are carefully dealt with.

The way to approach aspects of human factors in technological enterprises is a challenge for engineers who need to combine objective technical aspects with the subjectivity of human behavior. There is an important distinction between human error and the term human factors which causes some confusion. Human error is unavoidable and an aspect of human nature. Human factors are related with the error-inducing environment created by technological enterprises. In other words, humans make mistakes and despite training and technical qualification of professionals, sooner or later, this human characteristic will manifest itself with few options for engineers to avoid it. On the other hand, dealing with human factors more specifically with the human–system interaction provided by technological enterprise projects, engineers can significantly reduce the error-inducing environment and also reduce the consequences of human errors so that they will not result in catastrophes. Fig. 2.3 shows the three components that have strong influence on the likelihood of human error to occur. The natural vulnerabilities resulting from external phenomena that are intrinsic to the enterprise technology are beyond the control of engineers. The human limitations that lead to human error are also not part of direct actions by the engineers. Nevertheless, with regards to the error-inducing environment generated by the technological enterprise it is of total responsibility of engineers and should be the focus of efforts to reduce human error and its consequences.

The human factors subject matter is multidisciplinary and very complex. It is a broad field of ongoing engineering research, mainly related to the ability to put into practice ideas and strategies that can effectively reduce the error-inducing environment. But the risk reduction requirements that characterize our times lead us to the first steps in this field. We do not intend to explore this topic exhaustively, but we want to give

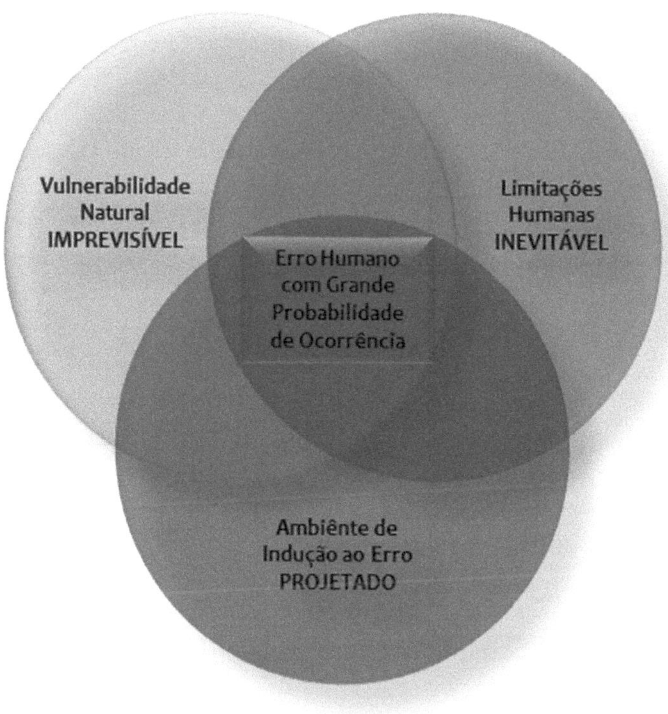

Figure 2.3 Influences of human error with a high probability of occurrence.

minimal treatment to technical matters related to human error induced by the designed environment. To that end, we have established seven basic principles for engineers, designers and other professionals to consider in their activities in technological ventures. These principles should be considered as values to be cultivated in practical engineering activities, so that the elements related to human factors receive at least a minimal attention.

2.3.1 Seven principles of human factors

2.3.1.1 Principle 1 of centralizing objectives in people

The objective of any technological enterprise should be centered on the benefit to human being, both as an individual and as a society, including the safety necessary for protection related to the greatest extent of consequences possible, arising from technological enterprise, which may affect individuals and the society.

Although prioritizing benefit to people may seem an obvious requirement, the objectives of technological enterprises end up naturally changing their priorities, for example, to economic performance and to meet deadlines.

A simple example of this natural departure, including in the form of engineering documents, is the lack of representation of people in engineering and architectural plans and drawings. In general, the human being appears represented only in marketing materials. In terms of engineering and architectural drawings, the consideration of the profile of the people who will interact with the equipment and the diversity of the associated population should be represented in all documents, given that the equipment is being designed to interact with people, and this interaction needs to be represented as it is the case with the other components of the project.

The design documents for an offshore platform, a building, a vehicle should include the representation of all people who will need to interact with the equipment depicted in the drawing. The representation of people should only be removed from the drawings in views where it is essential to enable complete visualization. By simply inserting the representation of the totality of agents assumed to interact with the equipment or facility in the future, many improvement opportunities may be identified.

2.3.1.2 Principle 2 of adaptation of the design to humans

The technological enterprise needs to be designed to interact safely with the greatest diversity of human beings possible, regardless of anthropometric, behavioral or cultural characteristics. Whenever possible, the work needs to be designed to be adaptable to as many people as possible, rather than people adapting themselves to the work.

2.3.1.3 Principle 3 of control of human–system interaction

Every technological enterprise generates, consciously or not, a human factors design that defines the form of interaction of this enterprise with people (human–system interface). This human factors design has a direct influence on the occurrence of failures, errors, and accidents. The human factors design should act on the error-inducing environment enabling a positive influence on the human–system interaction and limiting the consequences of human errors, so that they do not cause catastrophic accidents.

2.3.1.4 Principle 4 of protection against human error

Human error is influenced by natural vulnerabilities (unpredictable), human limitations (unavoidable), and the error-inducing environment (designed for). The control of the consequences of human error within the limits of acceptable levels is only possible through a design of human factors that works by limiting the error-inducing environment. Natural vulnerabilities and human limitations are not within the grasp of engineering. To protect against human error is to recognize that human errors are unavoidable and it is up to the human factors design to develop engineering solutions that limit the consequences of these errors to acceptable risk levels.

2.3.1.5 Principle 5 of human decision superiority

No type of automation, interlock or process computer offers a better decision than the technical professional adequately trained to perform emergency mitigation measures. Accidents always include unpredictable or unexpected aspects whether due to equipment failures, procedural failures, human errors, or unexpected natural phenomena. The combination of all these factors added to the perception of the impact of the accident escalation both in its technical effects and environmental and social effects generate a degree of complexity that is magnified by subjective aspects that limit the automation systems' ability to provide the best decision. There is a greater chance of positive results through decision-making by a adequately trained professional at the forefront of crisis management.

Automation is indispensable as a supporting tool for fast, simultaneous actions in complex processes. Nevertheless, the main objective of automation should be to reduce the workload demand on the technical professional regarding operational activity so that his or her information processing capacity can be redirected to the most critical and complex decisions and thus be spared from receiving demand shares beyond human processing capacity. It is part of the automation design to consider the human factors involved with the operational task, so as to prevent the automation design from assigning variables and sending alarm signal related to the operators without due consideration on the limits of human processing capacity. This capacity, despite its biological limitations, needs to be increased and developed through specific training for the feasibility of efficient processing of information during a possible crisis.

2.3.1.6 Principle 6 of nonmechanization of human labor

The technological enterprise should provide engineering solutions that prevent the mechanization of human work at all levels through a comprehensive human factors design. The mechanization of any human activity increases the risk of catastrophic accidents by leading to a reduction, albeit momentary, of the analytical ability and the ability to deliver solutions in accidental scenarios where unpredictable and unexpected elements are always present.

Human–machine interaction systems are included as mechanization of human. They limit the interaction related to the adoption of standards, rules, and procedures without any possibility of being questioned, evaluated, and, if necessary, disregarded at any time as a means to avoid an accident.

Standards and procedures, even if specifically designed for safety, should be emphatically adopted merely as reference considering that in theory their contents are based on the best of engineering experience and best practices applicable to the activity in progress. Standards and procedures may contain errors or be unsuitable for assessing the actual accidental scenario, which is unique and might never have been anticipated, even by the experts who write the standards. Therefore legalism bias as well as the mechanical behavior of engineers and technician needs to be eliminated to avoid as well as to deal with accidents. It is essential the existence of an environment that allows operators to perform their work with intelligent freedom, rich in technical skills and operational experience, and at the same time, backed by a solid base of theoretical technical knowledge.

2.3.1.7 Principle 7 of inclusion of anthropometric and psychological project

Engineering designs, in order to achieve higher level of safety, need to include an anthropometric and psychological approach for an adequate human factors design. Based on anthropometric data, analyses of biomechanical risks, risks of static and repetitive work, and risks of manual work should be considered as well as the influence of temperature, visual environment, hearing, vibration, among others.

With respect to psychology, matters such as stress and individual fatigue, environmental stress, work overload, human information processing, and mental workload should be considered.

2.4 Efficiency and strategic risk management line

2.4.1 Efficiency

The risk management field is multidisciplinary and crosses the boundaries established by exact sciences. Despite that, though, engineers are usually the professionals who receive most requests to provide solutions to risk management problems. Even for experienced professionals some accidental scenarios pose difficulties in identifying the best strategy for approaching the problem due to the various tools applicable to risk analysis. Those are accidents with numerous possibilities for investigative approach. Some are so complex and investigative value not clearly evident that the decision of where the investigation should start can be difficult to make.

At the same time, the challenges for engineers who conduct safety studies early in the design phase involve so many multidisciplinary problems and so many possibilities for interrelationships among them that any premise adopted at the beginning of the risk analysis discards a slew of possibilities upfront. This can generate fear that something more relevant than first thought may be inappropriately left behind. How to start approaching risk management problems without getting lost in the organization of ideas? Should we establish detailed premises to narrow down the extent of the accidental scenario under analysis or should we instead avoid the premises because they can lead to an unrealistic and excessive simplification? How to organize objective facts and subjective perceptions in the formulation of a risk management problem?

For a good risk management work, it is necessary to have an initial approach strategy that leaves room to accommodate the whole complexity of the problem without impoverishing or oversimplifying it. Excessive simplification compromises the analysis drifting it away from reality. A common mistake in the quest for an adequate approach to risk scenarios is to focus on details of one particular aspect under study at the expense of the overall vision. This mistake can produce the illusion that the approach is being carried out at a deep level whereas in reality the depth of the analysis is only related to a certain aspect or to one of the several possible subscenarios that make up the problem. The importance of the big-picture vision can never be underestimated and therefore should be revisited in parallel while an aspect or subscenario of a risk management problem is being progressively analyzed in details.

A methodology for a balanced approach to the problems of risk management can be obtained through what we call "risk management strategic line" which will be presented later in this chapter. But first, some general concepts need to be outlined as values to be cultivated in the efficient practice of risk management. Without a well-established set of minimum values, the multiple influences throughout the risk analysis work can lead even the best expert to get lost.

Similarly to what we did to organize safety culture and human factors, we will also present a summary in seven principles, which should be cultivated as important values for the efficiency of the management of technological enterprises, as we will show next

2.4.1.1 Seven principles of efficiency in risk management

2.4.1.1.1 Principle 1 of rejecting unnecessary risks

Accept only the absolutely necessary risks. All activities include risks, more than that, life includes risks. The more unnecessary risks are rejected, the more attention will be paid to managing indispensable risks.

2.4.1.1.2 Principle 2 of respect for natural laws

The more the intervention caused by the technological enterprise is opposed to natural phenomena, physical, chemical, and biological laws, the greater the risk. Human interventions through technological enterprises should explore the sense of evolution of the natural phenomena associated with them.

For example, during the design of a lifeboat equipment for abandonment of an offshore rig, designers can choose between the predominant systems in the market: descent into the sea by cables or descent into the sea by free fall. The equipment that is conceptually more aligned with the principle of respect for natural laws is the one that performs the descent into the sea by free fall. During the descent operation, the cable rescue boat wrestles with the most imperative force known in the physical universe: the force of gravity. The purpose of the cables is to counterbalance the action of gravity as conventional lifeboats are not designed to withstand the forces from the impact with water resulting from the free fall. Conversely, the lifeboat that can be launched into the sea by free fall uses gravity as an ally when it is released to execute the intended motion. Thus this extremely important natural force works in the same direction as the intervention promoted by the technological enterprise which in this case is a lifeboat for abandonment of an offshore platform towards the sea.

Obviously there are other components that need to be taken into account by the designer which may, depending on the design, make the use of a free fall lifeboat unfeasible. But the concept of descending into the sea in an emergency situation due to the effect of free fall is undoubtedly a concept more aligned with natural forces than the one based on the use of conventional lifeboats whose descent into the sea is assisted by cables. The use of the force of gravity to set the lifeboat in free motion conceptually reduces the theoretical malfunction risks. This means more efficiency in the management of the risks involved.

2.4.1.1.3 Principle 3 of simplicity

Regarding efficiency in risk and safety management, minimum is maximum. The fewer moving parts, the fewer people involved, the less automation, the fewer variations, the fewer procedures, the fewer words to communicate an idea, the less sophistication, the less complexity there are in a technological enterprise the more efficient it will be in terms of risk management compared to other projects that accomplish the same final result.

Certain results cannot be achieved without sophisticated and complex technological developments, including widespread use of automation systems. But the risk management of these enterprises will be as efficient as the designers can reduce such a sophistication, automation, and complexity to the minimum necessary. If for the same result, other technological enterprises appear simpler they will achieve better efficiency in managing their risks. For risk management purposes, "minimum is maximum."

2.4.1.1.4 Principle 4 of conciseness of rules

Rules, standards, procedures, signage, technical specifications, guidelines, manuals, alarms, panels, consoles, screens, and any safety-related texts and media materials should be as concise and simple as possible. The fewer signs, letters, and words used to convey an idea in safety-related texts the better the efficacy of communicating important technical safety information. In colloquial language, we can say that texts and signs related to safety should be as efficient as a "bikini": big enough to cover the essential parts and small enough to attract attention.

2.4.1.1.5 Principle 5 of combating legalism

This is an especially controversial concept. Therefore it is important to clarify that the author does not propose any kind of "rebellion" to the rules and norms. On the contrary, although they are not "laws" as with penal and civil codes, technical rules and standards are the best references to avoid accidents. Unfortunately, many professionals make this good truth a subterfuge to justify errors that cause accidents and deaths. Some professionals use "compliance with the rules" as a justification for deliberately making mistakes against safety in scenarios not yet foreseen in the rules as if they could not make a critical analysis of the rules and standards and realize that an application of rules without a critical analysis would result in an error against safety. The author is in favor of applying technical rules and standards intelligently, under professional review. Risk management experts must apply the rules and standards with a "thinking head" and not as "robots." If they identify an inadequacy in the rules for the application scenario, they should suggest revising these rules and not simply move on and cause an accident. In the case of criminal and civil codes this is not possible, but in the case of technical rules this is necessary. That is why we describe this concept as "Combating Legalism." That is, technical rules need to be applied as technical rules and not as criminal and civil codes.

Texts, procedures and formal safety rules should be treated as the best technical references for safety-related actions and should be, as much as possible, respected, widely questioned, and revised as frequently as possible. They should never be treated as absolute and definitive truths or its application be taken for legal rites imposed by the formally constituted legislation, which cannot be challenged in court. Engineers and experts are solely responsible for the scientific analysis of the phenomena associated with the scenarios of application of texts, procedures, and formal safety rules and, if necessary, adjust or even in extreme cases ignore such texts, procedures, and rules so as to avoid an accident. Consequently, this act of noncompliance requires sound technical, operational, and phenomenological knowledge of the developing scenario and those who decide to do so accept all consequences of their decisions.

In some situations, the accidental scenario formation (which always includes unpredictability as one of its components) may require from both the designer and the final operator to close the gap between rules and reality. This lapse can be greater or smaller, depending on the quality and conciseness of the rules. When the rules fail or are omitted, for distancing themselves from the operational reality, emergency solution is sought in the elements of safety culture, operational experience, and conceptual technical base. When the safety culture is poor, the only solution left is to adhere to the rules even if, due to some failure, these rules are deemed insufficient to provide the right attention at the right time to avoid the accident. Under these conditions, the operator who should act as a "thinking head" starts acting mechanically complying with standards without a technical conscience on their suitability for the developing emergency scenario.

Often times, given the difference between the ideal designed world and the real world, it is necessary to break the rules and procedures in order to meet the requirements imposed by natural laws (physics, chemistry, biology, etc.) involved in the escalation of the accidental scenario. Safety is not stopping at the red signal light. Safety is moving ahead in the green light in the same way as one would in the red light if this were necessary. In order to avoid accidents, it is necessary to use the established rules as a reference for instance, the colors of traffic signal lights, but at no time should they be considered a guarantee to avoid the accident, for it being necessary to assess the real developing phenomena. In the driving example, parameters such as speed of the vehicles involved, space and time available, in order to actually make the right decision, whether to move forward or not regardless of the signal light color or condition, that is, green, yellow, red or not working.

2.4.1.1.6 Principle 6 of fighting heroism

The contribution of engineering to risk management is the technology capable of promoting the right attention at the right time in order to avoid an accident. Seeking the safety of people, the environment, and property may be a natural desire but the means to achieve it is not. It requires technique.

The same is applicable in the case of keeping the production of a facility heroically for the fulfillment of corporate objectives while at the same time it is subjected to risks. It is very important to know the various technologies and the various physical phenomena that characterize the possible accidental scenarios. Through this technical knowledge, the productive activity can be developed with the minimum risk of human, environmental and property losses.

Heroism may represent commitment and altruism. But it does not always translate into efficiency in risk management. Deciding on a heroic act is everyone's right, but for engineers and risk management professionals, the heroic act can only be employed after the intelligent use of analysis technologies and response to the accidental scenario. The risk manager has an obligation to provide the solutions based on the most effective technique in reducing the number of victims. It makes no logical sense to increase the number of victims by exposing more people to the risk, driven only by a sense of heroism. The engineer and risk management professionals can also exercise the right to act like a hero, but first they must fulfill their professional obligation to provide solutions that reduce the number of victims and losses to the absolute minimum possible.

Heroism may also be understood as behavior aimed at maintaining the production or the availability of a plant despite conditions of unacceptable risks. In this case, in the name of sustained productivity, operators may keep a plant in operation heroically albeit under unacceptable risks. This is a mistake. This type of behavior is still present in organizations, groups, individuals, and society and it is a malpractice that can impair the efficiency of risk and safety management.

Unfortunately, some people and even organizations abuse certain noble characteristics of human behavior and demand unbalanced attitudes between risk and objective from collaborators. For all these reasons, we must combat heroism and it must be considered risk and safety management malpractice. During a crisis, only technique, training, and professional experience may adequately guide decisions and actions in order not to take any unnecessary risk. Therefore it is possible to save the greatest number of human lives, instead of increasing the number of victims.

2.4.1.1.7 Principle 7 of humility

Technological enterprises are human interventions in the natural world, which has its own imperative rules and that are beyond the grasp of absolute control by human kind. What is the worst accident that can happen to any human being? The one where one becomes a fatal victim. However, despite all the science and technology developed over centuries, an accident will one day happen to us all whether due to behavioral failures, equipment failures, natural forces, or biological accidents (diseases) that develop in our organisms. Due to our complete inability to control these accidents, we simply qualify them as natural.

Science and technology seem to have advanced significantly, and the intelligence they provide fascinates us, captivates us, and, unfortunately, also temporarily deceives us with their impermanent spells. But maybe, for the most important matter related to risk management, they add nothing to avoid the fatal accident that nature imposes on every human being which can at most be only postponed. In a significant number of cases where fatalities could be avoided, the right attention at the right time is not efficiently given to avoid the accident caused by the lack of recognition of engineering limitations due to lack of humility.

2.4.2 Risk management strategic line

From a well-constructed set of minimum values and with characteristics similar to the values that constitute a culture we will then have a "compass" or a "GPS." These values serve to guide us through the countless bifurcations that represent the various decisions that make up the work of risk management. Each bifurcation means a decision that will lead to other bifurcations. Without the vision of the whole, the best solution to an immediate problem may not result in the best solution overall to reduce the risks of the enterprise. This disparity occurs when the decision on a immediate problem is obtained by analyses that overvalue the specific scenario of a part of the problem, exaggerate in the detailing of the partial solution without paying attention to the consequences of the adoption of the solution of the broader and more complete scenario.

Fig. 2.4 illustrates graphically how risk management is carried out through risk acceptance or risk rejection decisions. When a technological enterprise is subjected to a risk analysis, this analysis can be done correctly using different approaches. In Fig. 2.4, the letters "X," "Y," and "Z" represent distinct approaches to risk analysis. For example, they may represent three different accidental scenarios resulting from three different approaches to the problem. For each risk (represented by the bifurcations) there is a decision to be made: to accept it or not. It is as if for each risk there is a bifurcation path for those who accept it and another for those

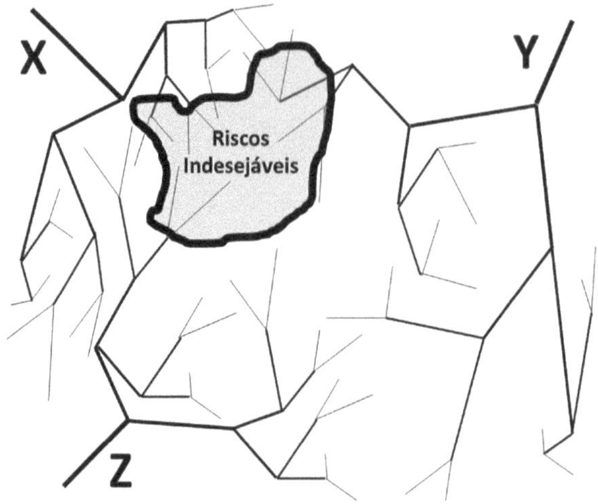

Figure 2.4 The entries "X," "Y," and "Z" represent the possibilities for addressing a risk management problem. Each branch point represents a decision between accepting a risk or not. The sequences of decisions of acceptance/rejection of risks can take the technological enterprise to bifurcation-limit in which any of the subsequent paths results in reaching a region of undesirable risks where the technological enterprise necessarily meets its limits.

who reject it. Each decision drives the advancement of the technological enterprise through risk management. After a few bifurcations, depending on the decisions made, the technological enterprise may end up in completely distinct regions of risks with scenarios and problems pertinent to each risk region. If the technological enterprise gets stuck in a region of undesirable risks, reverting this situation will require going back up along the same path, or alternatively, the invention of a new path, as risky as the first, to try the adjustment. We can define limit bifurcation as the one beyond which any of the paths takes the enterprise to a region of undesirable risk. At the limit bifurcation point the decision to accept or not the risk of the immediate problem is irrelevant. All that matters is to realize the limit, take a step back, and revise the strategies so as to get out of the region of undesirable risk.

Finally, Fig. 2.5 illustrates the representation of what we call the risk management strategic line. On the left side, we have the five risk management components, in their order of importance, from top to bottom. We believe that managing risks is, above all, to know how to decide whether or not to accept risks along a sequence of problems that makes up the history of the technological enterprise. In this context, the most important

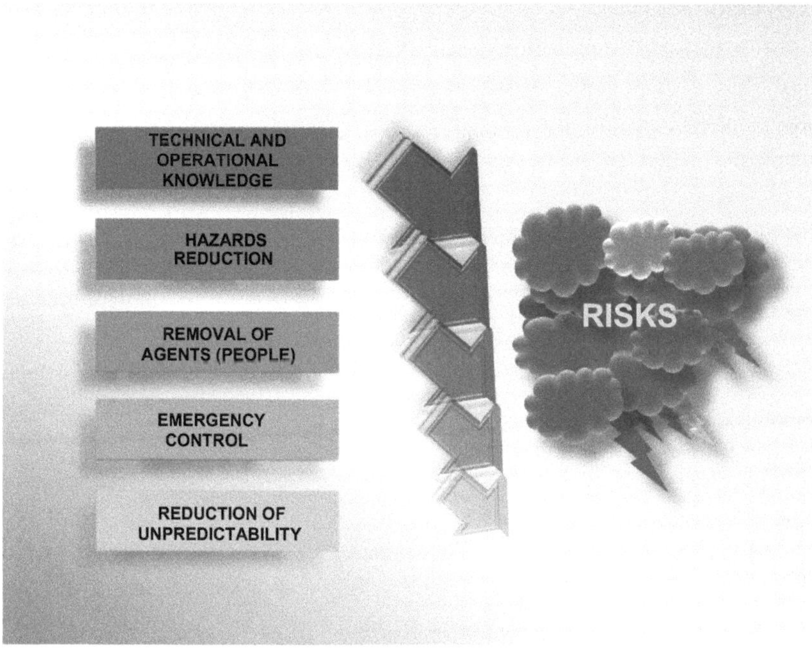

Figure 2.5 Risk management strategic line. The five main components are shown on the left, in order of importance, from top to bottom.

component is technical and operational knowledge, followed by hazard reduction, removal of agents, emergency control, and finally, the reduction, as much as possible, of the unpredictability related to accidents.

A summary of each of the five main components of the risk management strategic line is presented below. In the subsequent chapters of this book, these components will be presented in detail. Together they form a strategic action line that guides all technical content of risk and safety management.

We call this problem-solving concept related to safety the risk management strategic line.

2.4.3 Technical and operational knowledge

Nothing is more important than the knowledge to avoid an accident or minimize its consequences. Deep understanding of the natural phenomena that exert influence on the accidental scenarios avoids incomplete solutions. A solid basic foundation in physics, chemistry, mechanics, electricity, thermodynamics, and biology is essential in understanding the evolution of accidental scenarios. Previous experience in as many possible routine and emergency operational situations as possible avoids naive solutions that do not have correlation with the operational reality, both for design problems and for problems related to the routines of users and operators. Effective knowledge involves practical field experience. Academic experience and study of technical standards are not sufficient. Purely theoretical knowledge requires successful experiments that validate its efficacy, therefore operational experience is the most important and valuable source of technical knowledge for efficient risk management.

We can say that all good theoretical documents such as the technical standards, guidelines, literature, and technical articles are only *records made by those who at least tried to do it so that those who have not yet tried do not make or repeat previous mistakes.*

Therefore the true source of knowledge is in operational practice or in scientific laboratory experiments and not in the records which are simply the best attempts to preserve this precious knowledge acquired through practice. Designing, without technical field experience, based solely on standards and rules is like trying to paint a canvas by capturing the scene through a small mirror. Viewing through a mirror, despite the distortions, everything that one sees has some correspondence with reality. But having just a mirror as the image source, the final painting will contain a limited view of the real scenario. It will be poor in relation to the vision of the whole, and it will be only a simple partial view of the real scenario that is infinitely bigger.

Unfortunately, there are some distortions in the training of risk management professionals that make it possible for unprepared professionals from the standpoint of operational experience to enter the labor market. We are not only talking about recent graduates, but even professionals with decades of theoretical experience for example in design, work in activities related to risk management without proper operational experience. This has generated rather unrealistic safety designs since many designers have never participated as operators or even as end users of the technological enterprise they are meant to design. Many of these professionals justify their decisions with good academic argumentation, or standards-based argumentation, or even with years of experience exclusively in projects. But risk management decisions, as we have said earlier, work as successive bifurcations that lead to regions of acceptable and unacceptable risks. In this context, good decisions require a vision of the whole and need to be made based on some minimum experience requirements which often is only possible with decades of operational experience.

It is very common to find conflicts and road blocks between operators and designers who lack field experience. The operational experience gap in the training of designers limits the design vision. These classical designers are very skillful at using the scientific method in an academic context but nevertheless produce bad projects for being unable to see the operational reality directly. The vision that these inexperienced designers have related to operational reality is indirect, acquired through standards and rules or previous trial-and-error projects. For this reason, this "classical blindness" is often associated also with the interference of legalism bias due to the total dependence on the rules that these designers have without operational experience. This dependency occurs not only due to zeal or prioritization of safety, as some may try to argue, but mostly because for these designers the true source of technical knowledge—operational experience—is inaccessible.

As for the profile and technical background of a risk management professional due to the fact that subject matter is multidisciplinary, a solid academic background in the field of exact sciences is necessary, but so is the ability to connect other relevant knowledge sources to the subject matter. This includes the ability to deal with a certain amount of subjectivity and to understand the influence of biological and social sciences on more complete risk and safety management solutions. Academic training only is not sufficient, a certain amount of time dedicated to technical work is also required which allows an operational experience without devaluing academic development such as specialization and graduate courses. This complementary training should be done progressively and in parallel with practical operational activities. Practical experience will allow the development of the perception of multidisciplinary sources of knowledge associated with risk management. After a solid basic academic foundation in exact sciences, the professional will need to acquire some operational experience and then complement one's original academic background with specific graduate and extension courses. Experts in risk management need to build a multidisciplinary curriculum. In Brazil, a "safety engineering" course is available at academic institutions but the training is tailored for an engineering market where the "safety engineer's" main role is to prevent companies from being fined by the Ministry of Labor. The value of the "safety engineering" course does not deserve to be questioned but such a training certainly does not enable a professional to engage in risk management activities to their fullest. At most, it can only serve as one more training to be added to the multidisciplinary mosaic of the professional trainee's curriculum. However, it is worth mentioning that this type of training, in some cases, can be deemed nonessential.

Some "safety engineering" courses offered in the United States have a broader scope, less restricted to the presentation of standards, placing priority on the transfer of operational experience. In many countries, "safety engineering" course as established in Brazil is not even available, engineering professionals being individually responsible for safety in the respective system related to each one's area of specialization. This is the best and most efficient way: first engineering education, then operational experience, and finally, complementary multidisciplinary courses at graduate level. Of course, in addition to knowledge, it is necessary that prioritizing safety becomes a "value " for each person. In this way, we form a solid safety culture.

Ideally, risk management professionals would be trained after years or, if possible, decades of previous operational experience. This long period of hands-on learning should include all major engineering areas such as design, construction, maintenance and operation.

One of the biggest credibility issues for risk management and safety professionals is the perfect solution in terms of theory and poor in terms of operational practice. After all, as we have already explained, the source of greatest technical knowledge in this field of engineering is to be a seasoned professional with operational experience. Rules, standards, and literature only amount for the minimum record of that knowledge. After years of dealing with practical maintenance problems, design failures, construction and assembly shortcomings, operational difficulties, emergencies, and quasiemergencies engineers will be much better prepared to provide multidisciplinary solutions to risk management problems which are much more compatible with the operational reality and much more efficient for risk reduction.

California State University at San Jose, located in the heart of Silicon Valley, offers the Safety for Engineers course. As part of a basic safety engineering training, it has broader content than traditional courses in Brazil. Classes are focused on the identification of hazards and prevention measures. They promote approaches to management problems of risks in engineering design of equipment, installation project, and process routines. The course also prepares the students to get familiar with various safety systems, understanding of the types of safety analyses, and the forms of studies and alternative solutions for the establishment of countermeasures in emergencies. Also included is an update on regulations and agencies responsible for their enforcement. The course, offered by the Department of Industrial Engineering at Charles W. Davidson College of Engineering, also includes the historical description of the industrial safety movement and the role of safety engineering in technological enterprises.

The classes cover computational techniques for safety analysis, accident databases, fire and explosion risk analyses, and accident investigation as well as risk identification, assessment, and management. In parallel to the activities throughout the course, students present a final seminar on the worst fires in history. Each group of students, during the entire duration of the course, studies a particular historical fire as well as all the dynamics of the events related to it, its consequences and losses, and most importantly, the lessons learned and incorporated by the engineers through good designs of safety systems.

Chapter 3, Technical and Operational Knowledge, of this book is entirely dedicated to providing the minimum foundation of technical and operational references so that professionals can achieve more efficiency in their practical field activities necessary for their training as specialists in risk and safety management. As we have already mentioned, the source of technical knowledge is in the operational experience, that is, it is necessary to work in the field in operation, designs, and in the various fields of engineering although this work can be temporary like an internship. Chapter 3, Technical and Operational Knowledge, presents an overview of main equipment and facilities related to the activities of the oil and gas industry, as a starting point, but without the objective of replacing the practical experience which only the design routines and field operational routines can provide.

2.4.4 Hazard reduction

Following the risk management strategic line model, the second item to be studied is the hazard reduction. Hazard is the source of risk. It is important to have the concepts of risk and hazard very well distinguished.

2.4.4.1 Hazard

Hazard is a threat to the technological enterprise that can generate losses.

Hazard may exist, be identified, and never actually generate losses. We consider losses as damages to life, society, property, and the environment. Hazard is not expressed by a number but rather it requires a description.

It can be a material or a substance. For example, hydrocarbons are considered hazards in the oil and gas industry and radioactive materials are hazards in the nuclear industry.

Hazard can also be a behavior, a habit, or cultural characteristics. It may also be an amount of energy or a particular situation. So, illustratively, we can compare hazard to a scenario prepared for a play. Like at an operations theater, even with the scenario ready, the play that represents the sequence of facts that results in losses may or may not take place.

2.4.4.2 Risk

Risk is the probability that a given hazard will effectively generate losses.

Risk is a prognosis which may or may not be the result of a calculation, higher or lower possibility of a hazard generating losses. Risk can be expressed by a number, and even if it is small it does not mean a guarantee that there will be no losses.

Risk, even when quantified, always contains a margin of uncertainty about the possibility of occurrence of losses. For each quantified risk, a second component of nonquantifiable risks is always associated with it, either due to the infeasibility of the calculation or due to the impossibility of performing the calculation in the lack of representative historical data. Fig. 2.6 shows an on—off switch, a disconnected wire, and a fuel storage barrel. The scenario indicates a hazard. In this case, the hazard is that a person may turn on the switch button with the wire still disconnected thus generating a spark and fire in the fuel barrel. The risk of a person setting the on—off switch in the wrong position and cause a fire is 50%.

2.4.4.3 Reduction of the hazardous scenario

Technological progress and the engineering evolution cannot fail to follow its natural path in the face of hazardous scenarios that present themselves throughout the challenges of each new enterprise. Hazardous substances and hazardous situations are challenges that need to be overcome for a technological enterprise to become viable. It is exactly at this point that the risk management engineering activity enters the scene. For each hazard presented risk management experts need to study its associated risks, that is, the chances of that hazard actually generating losses.

This management includes risk analysis and, wherever possible, mathematical calculations to assess and keep risk values within an acceptable range by the organization and society in where it operates.

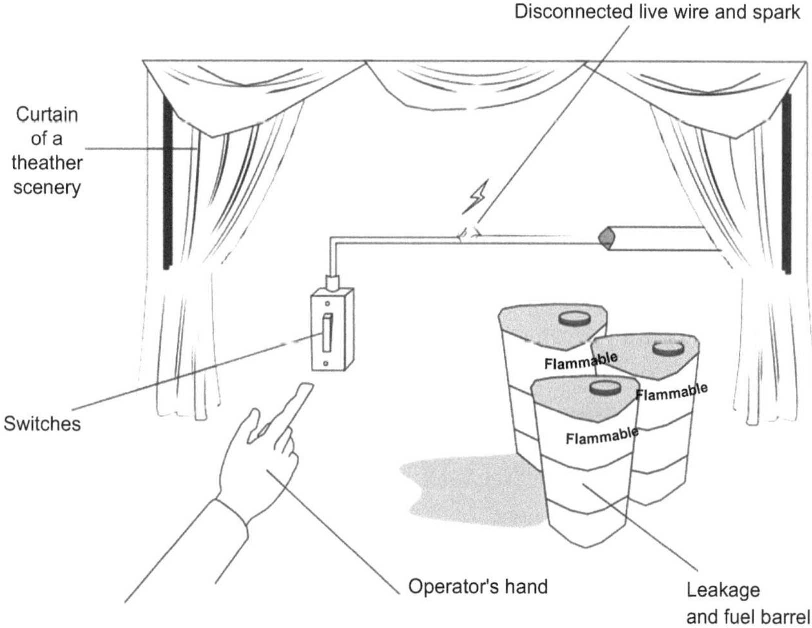

Figure 2.6 Hazard can be compared to an operations theater and its characters. In this case, there is a leaking fuel barrel, an on–off switch and a disconnected wire in the electrical installation. The sequence of facts within this scenario can cause losses. The hazard is the scenario itself, while the risk can be associated with a numerical value. The risk of an operator energizing the loose wire and generating a spark through the wrong choice of a switch position is 50%.

However, throughout the operational history of the technological enterprise unforeseen situations and operational deviations may occur that will modify the original hazard scenarios, turning them into abnormal, with greater than acceptable risks. In these cases, the immediate action to be taken by safety engineering of the technological enterprise is to reduce the hazard itself as much as possible.

Consequently, risks will also be more easily reduced and approach the maximum acceptable values. This must be done even at the expense of the operational result, always prioritizing the protection of life, society, property, and the environment.

Hazard reduction should be an objective to be pursued starting in the design phase. For instance, fuel inventories should be segmented as much as possible within the limits that allow the operation of the technological enterprise. Controlling fire in a large fuel storage tank is much more difficult than if the fire occurs in just one of several smaller tanks spaced apart during construction to prevent the fire from spreading more quickly. But costs, operational requirements, technological limits, and market competitiveness impose limitations to designers who in turn need to optimize these factors by balancing benefits and their associated risks. In other words, in terms of engineering, there would be no limits for studying the inclusion of new equipment and the adoption of additional strategies that can reduce the hazards and risks of a design or activity. The limit for making improvements in safety is economic viability. We must always seek to improve the safety of engineering processes. Beyond a certain point, the inclusion of additional protection can lead to higher costs that the enterprise might not able to absorb. At this stage, the risk of the enterprise needs to be assessed to verify whether it is acceptable thus ensuring the feasibility of its continuity. Otherwise, if the risk is unacceptable, the technological enterprise can be considered unfeasible.

This feasibility assessment of whether or not to include each new safety item may compel designers to establish limits for the ideal operational scenarios. That is, in order for the project to remain viable, it may be necessary to establish some premises considering that certain hazardous conditions cannot occur. Then the project may proceed without overly conservative design of the safety systems that would be justifiable only in scenarios that, as a premise, were excluded. But there is obviously a compensation to be adopted for the completion of this strategy. As we know that a premise is not an assurance in itself, the design needs to provide solutions to reduce the risks if the scenarios discarded in the premises should unfortunately occur during the operation of the technological enterprise. These emergency measures aim to reduce the hazard at the expense of operational results in the presence of a degraded safety scenario that exceeds the limits of response capabilities of the safety systems.

This happens, for example, when a nuclear power plant reduces its power output during an emergency after the detection of a major failure. By reducing the power of the nuclear reaction, the danger of overheating the reactor is also reduced.

Under normal operation, technological enterprises can live with a certain level of hazard. However, during emergencies one of the first actions to be taken is the reduction, as much as possible, of the hazards that affect the enterprise so that the event escalation can be avoided. Even if it means loss of operational continuity or economic loss.

Another example of hazard reduction is what happens on an offshore platform for oil and gas exploration and production. During normal operation the platform has pipes, tanks, and vessels storing a considerable hydrocarbon inventory needed to make production feasible. In the event of a major emergency, the hydrocarbon inventory will, as much as possible, be reduced or at least segmented to minimize the risk of fire propagation and explosion. These hazard reduction measures will cause many operational delays in the eventual return to production and a significant impact on economic and productivity results. Nevertheless, with a good risk reduction strategy, such inconveniences can be managed to reduce losses to a minimum with respect to the catastrophic accidents. In terms of risk management engineering, reducing hazards is a task that requires significant strategy and operational experience. In the example of the offshore platform, we attend to hazard reduction by reducing the basic hazard agent which in this case is hydrocarbon. But a comprehensive strategy shall also include reducing or eliminating sources of ignition, converging with the ultimate objective of preventing fires and explosions.

The basic hazard in the oil and gas industry is hydrocarbon. This is due to its characteristic of high energy concentration and its associated risks of fire and explosion. In Chapter 4, Hazard Reduction, we present several strategies and techniques used by engineers (design and operation) to reduce hazards in response to emergencies in the oil and gas industry facilities.

2.4.5 Removal of agents (people)

The sequence presented in Fig. 2.5 shows the risk management strategic line, in which five components are displayed in blocks, on the left side of the figure. From top to bottom, the components are ordered in priority, forming a strategic line to manage risks. The first component is Technical and Operational Knowledge, the second is Hazard Reduction, and right afterwards, the Removal of Agents (People) component appears. What has been described so far shows that good risk management starts with solid technical knowledge associated with operational experience followed by the reduction, as much as possible, of the hazards associated with technological enterprise. Agent Removal is the next component of the strategy, and can be understood basically as the escape and abandonment system.

No other safety system is totally dedicated to saving lives like the escape and abandonment system. The fire and gas detection systems, water and foam firefighting systems, passive protection systems, and all others are also very important for saving lives. But these systems' objectives are split between protecting facilities and agents. In order to achieve that, they act by avoiding the escalation of emergencies, maintaining control over the accidental scenario, and protecting the facility including its escape and abandonment routes. On the other hand, the escape and abandonment system is exclusively dedicated to providing the means to remove the agents from an accidental scenario which can evolve and lead to losses of human lives. This direct link between the escape and abandonment system and the reduction of loss of human lives is what characterizes this system as the most important to avoid fatalities in any type of emergency and any technological enterprise.

It is important to note that the escape and abandonment system promotes the removal of agents from one high-risk scenario to another of lower risks and criticality. In the event of a fire in a marine facility, be it a platform or a ship, the escape and abandonment system promotes the removal of the agents from a scenario of risk of physical damage resulting from fire to a second scenario of survival at sea. This second scenario also poses considerable risks, but which become relatively smaller in comparison with the risks of a large fire scenario on a ship or offshore platform. It is important to know exactly when and how to activate the escape and abandonment process in any emergency situation.

For the layperson, it may seem that activating the abandonment process as quickly as possible is the best option. However, depending on the case, the emergency situation may have been poorly assessed and the emergency may even be aggravated by rushing into the decision on the abandonment order. It is a difficult decision because if there is a delay, lives can be compromised and if there is a hasty and disorganized start it can also cause problems and victims. Once again, risk management problems guide us to the need for professionals with technical knowledge and solid operational experience. Only a high level of technical knowledge, combined with broad operational experience, can provide the best decision on whether or not to order escape and abandonment.

Risk management and safety strategies need to pay special attention to the escape and abandonment system. It must be available and preserved at all times because, in the event of emergency escalation, the removal of agents from a scenario will mean the greatest contribution to the preservation of human lives. However bad the material losses of a serious accident may be, when there are no victims, at least from this standpoint, the risk management strategy can be considered successful. When there are victims, even if property is preserved, the risk management strategy will always be considered a major failure. Avoiding human losses to the minimum can be expected of a risk management strategy. And the escape and abandonment system is the most important part of the strategy.

Chapter 5, Agents (People) Evacuation, is dedicated to the study of escape and abandonment systems. In addition to the main rules adopted by designers, techniques will be presented which improve the means of escape and abandonment making them more efficient. The standards and regulations only provide general guidelines, which are very important, but that do not address the risks involved in escape and abandonment matters in all types of projects. Each technological enterprise has some inherent risks, due to its specific nature. That is why it is very important to develop a technical conscience related to the relevance of conducting studies related to the escape and abandonment system in a specific way in each project considering its particularities. Such study needs to consider the engineering technical knowledge about the equipment, the safety culture, and human factors associated with the specific project. It is the only way for the most efficient life-saving safety system (escape and abandonment) to receive the deserved attention.

2.4.6 Emergency control

The fourth component of the risk management strategic line is emergency control. During the operational dynamics of an emergency, all the accumulated technical knowledge and operational experience can be utilized with the purpose of reducing hazards and removing people protecting them from the consequences of the accidental scenario. Besides that, though, a direct response to the accident will be necessary in order to keep control of the emergency thus reducing the escalation of its consequences.

The response to an emergency starts long before it occurs. When risk management experts study accident scenarios early in the design phase of the technological enterprise, they also initiate demands for the designers of the safety system that will respond in the event of an accident. The associated hazards and risks, which are identified in these studies, indicate how the safety instrumentation system needs to be designed regarding detection of fire and gas, for example. Preliminary water demand values used in fire fighting can also be estimated through comparisons with reference projects. Thus a strategy for emergency response and control starts taking shape based on the accident scenarios postulated in the conceptual and basic design phases.

With technical and operational knowledge, risk management strategies are developed throughout the entire design and operation of technological enterprises. Thus a well-located gas detector, based on operational experience and gas dispersion studies, can initiate the response to an accident even before a gas cloud from a leak gets ignited. This is an accident response and emergency control already in place even without any fire, explosion, or casualties.

The emergency response capability of any technological enterprise is not infinite. It always has limits. Its capability is usually only sufficient to exercise control over an accident up to a certain degree of escalation. Beyond a certain point, safety systems are unable to provide response and control. For better understanding, we need to adopt two important concepts.

2.4.7 Design-basis accident

It is the worst and most serious accidental scenario for which the safety systems of the technological enterprise have the ability to offer some response and control, albeit in a limited way.

2.4.8 Beyond design-basis accident

These are accidental scenarios for which safety systems, due to their technical and economic feasibility limitations, are unable to provide any further means of response and control.

The safety systems are designed to provide a response, albeit limited, up to the maximum degree of degradation considered in the description of the "Design Basis Accident." Beyond that point, the evolution of the accident leads the technological enterprise to the emergency condition of "Beyond Design-Basis Accident." Although in this condition the safety systems are no longer able to provide responses compatible with the extent of aggravation of the accident, risk management strategies may still indicate some actions for operators and agents who at this point should be focused primarily on avoiding human losses, including their own lives, in the case of fire brigade members. For this reason, technological enterprises designed in environments with a strong safety culture have specific strategies to be adopted when an accident reaches the condition of "Beyond Design-Basis Accident."

This happens, for example, in the nuclear industry. The "Design Basis Accident" of a nuclear power plant is, in general, defined as a Loss Of Coolant Accident (LOCA). All safety systems are designed to respond to a small LOCA, a medium LOCA, or even to a large LOCA. But there are degradation conditions that cannot be reached from a LOCA that can change the accident framework to "Beyond Design-Basis Accident." Engineers and nuclear plant operators, past this condition, start to focus their attention to monitoring a small number of functions called "Critical Safety Functions," previously defined in the design. With this strategy, regardless of the severity of the accident and the difference between the emergency scenario and all the postulated scenarios related to previous studies and thoughts, there will always be a strategic line to provide minimal guidance to operators and engineers, with the objective of reducing losses, mainly human losses.

In Chapter 6, Emergency Control, we will introduce the main safety systems used in most installation projects for the oil and gas industry. We will identify the most important technical aspects of each safety system, highlighting the correspondence between such aspects and the overall risk management strategy of the technological enterprise. Each one of the safety systems contributes its share to the overall risk management strategy of the enterprise. For each safety system, different technologies are employed, involving experts from various disciplines such as instrumentation, industrial pipelines, electricity, architecture, shipbuilding, and mechanical and electrical equipment. Each safety system requires one or more types of specialists for its design and operation.

But the risk management professional should have multidisciplinary knowledge related to all systems. Evidently, each specialist will know more about the safety system associated with their work but those responsible for risk management need to have a minimum general knowledge in order to understand the systems, their functions, interrelationships, and influences on the general strategy of risk management of the technological enterprise. Providing an overview of the safety systems responsible for the response and control of emergencies is the objective of Chapter 6, Emergency Control.

2.4.9 Reducing unpredictability

Finally, the fifth and final component of the strategic line of risk management is "Reducing Unpredictability" (Fig. 2.5). Every accident or emergency includes some aspect of unpredictability in the description of your scenario. In a sequence of facts that result in an accident, there is always something that happens differently than expected or desired. In some cases, it may seem as though the previous facts were signaling, or practically "saying" that the accident would happen. But always, at least from one of the possible points of view of observation of the phenomena associated with the accident, something will appear to have happened differently from what had been anticipated.

This ever-present unpredictability disrupts the technical scientific work of risk management. This characteristic involves a broader approach which includes subjective questions that cannot be modeled mathematically. One of these subjective aspects is human behavior which can even be the root cause of accidents. Scientific methods have been created as an attempt to address this matter and extract as much objective evidence as possible to allow a scientific analysis of the risks of a technological enterprise. Such methods range from qualitative techniques that study accidental scenarios to meetings of specialists to quantitative techniques using computational tools. Qualitative analyses attempt to organize the hazards, scenarios, and risks in order to better deal with them. Quantitative risk analyses use sophisticated computational tools as attempts of theoretical simulation of events related to novel scenarios on which there is insufficient previous operational experience to provide an effective contribution to the decision-making of acceptance or rejection of risks (bifurcations of risk acceptance decisions, Fig. 2.4).

Often those interested in entering the risk management and safety field show special interest in the risk analysis subfield. This interest has been justified in part by the latest generation of computational tools and the sophisticated computer simulations that these tools produce. Another attractive aspect is that after the phase of establishment of premises and definition of scenarios the risk analysis work, either quantitative or qualitative, starts to follow a logical sequence similar to those adopted by other engineering disciplines. Beginners may have an erroneous understanding that risk analyses have similar scientific accuracy for example, those of structural calculations based on the strength of materials. But it is a big mistake. When risk analysis reaches this perceived level of mathematical precision, it has already gone through a phase in which subjective arguments served as the basis for decisive orientation. These arguments shaped the logical sequence of the risk analysis. Thus hidden behind all risk analyses using computer simulations rigorously performed within a scientific mathematical modeling framework, there is a minimum set of assumptions and decisions regarding the acceptance or rejection of risks (Fig. 2.4), often obtained in a subjective manner, which has much greater weight and influence on the final results than the processing of mathematical calculations itself and the computer simulations.

Even when quantitative risk analyses use equipment failure frequencies from international databases still there is enormous subjectivity in the choice of these databases and in the assessment of the validity of matching database information with respect to the actual local conditions of the technological enterprise under study. The use of international databases of this kind has as its objective the reduction of subjectivity in the establishment of the risk assessment premises but their efficiency is questionable due to the disconnection between the environments where data are collected and where they are applied. International databases contain large amount of information about possible failures of various types of equipment. From the use of these data it is possible to estimate frequencies for events and thus in theory, the subjectivity of the risk analysis process could be reduced.

But the data acquisition for construction of these databases is widely dependent on the operators of countless companies involving institutions from different countries and cultures. As much as standards and guidelines help to standardize the interpretation of equipment failures, there may still be differences in the criteria adopted by the collaborators who feed these databases. Another aspect is the collaborators' environment, which can be similar or rather completely different in terms of external influences including cultural ones. When making queries to international databases for obtaining equipment failure frequencies, specialists find values obtained statistically for use in their mathematical models. Since these are numerical values, they appear to be more accurate than they actually are. It is not unusual for equipment failure frequency data obtained in international databases to be totally inconsistent with actual internal data from large companies. A quantitative risk analysis could achieve greater accuracy if the sources of the frequency values were internal databases obtained in each technological enterprise's own environment.

At this point, the strategic cycle already presented as risk management strategic line (Fig. 2.5) gets closed. After all, internal databases and the technical and operational capacity to maintain them with quality scientific research, in a certain way, convey the meaning of the priority component of the risk management strategic line: operational technical knowledge. International databases are nothing more than the attempt to record technical knowledge and, most importantly, the record of operational experience. Therefore the greatest relevance of the technical and operational knowledge component gets confirmed which is obtained through the study of phenomena related to the technologies required by the technological enterprise and mainly obtained in practice through the accumulated operational experience associated with it.

The important contribution of risk analysis is limited to reducing the margin of unpredictability. The more hypothetical accidental scenarios are postulated and analyzed in the design phase the greater will be the reduction of the unpredictability of scenarios to be considered by safety system designers and operators. That is the consistent contribution of risk analyses. But we need to remember that in the risk management strategic line the Unpredictability Reduction component is positioned at the fifth level of importance. It is not that it is of little importance, quite the contrary, it is one of the five most important components that make up the risk management strategies. But this relative position is due to the fact that, in terms of effectiveness against accidents, technical operational knowledge, the reduction of hazards, the removal of people, and emergency control will be higher than risk analysis.

In order for the results of risk analyses to be at least minimally realistic, it is necessary that those who conduct them have sound technical operational knowledge, ability to understand and to know the specific hazards of the technological enterprise under study as well as the associated physical and chemical phenomena, ability to understand and to know the means for removal of people, in addition to, obviously, understanding and knowing each safety system to be employed in response to the accident.

Chapter 7, Reducing Unpredictability, presents well-established risk analysis techniques and also some of the most recent and innovative techniques. Each analysis method requires courses and training in order to be performed efficiently. Although some techniques seem easy to apply, only experienced professionals will know how to fully explore them avoiding distortions and results contradictory to the operational reality.

2.5 Lessons learned

2.5.1 The theory specialist

The lack of operational experience may compromise the complete technical training of professionals even if they have a solid academic background.

Technical (theoretical) knowledge is the basis for the best risk management decisions. The standards and guidelines are the records of cumulative learning from previous experiences. But when a professional has no operational experience, he or she thinks that all problems can be solved only with theoretical technical knowledge and in compliance with standards and guidelines.

The operational dynamics of a technological enterprise is very complex, influenced by elements of unpredictability, behavioral aspects, and human and cultural factors. Through the acquisition of operational experience, one can realize that the main role of the operational team is to promote the adjustment between the theoretical design and reality. That is because the influences on the technological enterprise generate the continuous need for corrections. Operators sense inconsistencies between the what was postulated in the design and what is actually happening. Within technical safety margins, they make the necessary adjustments to keep the enterprise in operation. Over time, operators learn that certain functions that work perfectly in theory do not work in practice. The reasons for this may seem disconcerting to designers but the operational experience is confirmed decade after decade: not even the best design in the world works without adjustments and operational monitoring. For this reason, after effectively acquiring operational experience, the designer shall be much better prepared to create good designs compatible with the operational dynamics. When this experience is lacking, there is a tendency for the designer to push theoretically perfect solutions, based on the best standards, without realizing that in practice these designs do not consider the numerous factors that are not part of standards or technical books but that are only perceived by the professionals who have gone through the refinement promoted by the operational experience.

Some professionals spend decades designing facilities that create operational difficulties and safety risks due to the lack of operational experience. But these designers are unable to find errors in their designs because, in fact, in theory they do not really exist. As a result, they only point the finger at engineers and operating professionals as the ones responsible for operational deficiencies. This creates an old conflict between designers and operators. On one hand, the designer who delivers a project without any theoretical errors, on the other hand, an operator who cannot make it work as designed.

Almost always the operator is right in his arguments more times than the designer, unfortunately. The fact that a design works perfectly in theory does not mean that it will work in practice. There is a gap between theory and practice. It will always exist. The more the designers admit it, the less naive their designs will be.

A very experienced design engineer was called in to provide technical assistance in one of his own old designs. It was about the fixed system for fire protection (this type of system will be described in Chapter 6, Emergency Control) of an oil platform. The complaint of the operators was that the system had gone automatically into operation several times without any fire, generating a spurious fire alarm, resulting in interruption of operation and financial and operational losses besides causing a climate of distrust among almost two hundred platform operators. After all, when another fire alarm goes off, will it be another spurious alarm? Will it be worth it to act like a real fire? And if a real fire happened, how credible would the alarm be for this team which is already used to spurious alarms?

The designer, a former safety systems expert, analyzed the complete design and the alarm signals that were generating the malfunction and unwanted system startup. It did not take long for the designer to conclude that the cause of the alarms was related to a type of detector that, in theory, has high reliability: fusible plug detector.

This type of heat detector works with a low-melting-point metal at one end of a pipe pressurized with compressed air. When there is a fire, the high temperature melts the plug at about 70°C, releasing the compressed air and triggering the entire process of starting up the fire fighting water pumps. It was considered highly reliable by the designer because it did not depend on electronic sensors, the heat of the flames would suffice to cause the fusion and trigger the process. But the problem was that there was no flame and the plugs were not even melting.

Despite this, though, the pressure drop alarms in the compressed air lines were improperly triggering the system.

That was when the operations engineer explained that for each fusible plug a small-diameter pipeline is required, filled with compressed air. And this relatively fragile pipeline additionally had several connections along its length of several meters. The operator explained that on an oil platform, the space in some locations is very tight and despite that there is a need for the circulation of equipment and people. It only takes the impact caused by a ladder being moved or an equipment being loaded for these small-diameter lines to be misaligned thus creating small slacks and leaks. Over time, air slowly manages to leak from the pipeline, reducing the pressure to such a low level that the system interprets it as air release to the fusion of the plug, which in reality does not happen. As the compressed air lines in the system extends for tens of meters, it is nearly impossible not to have any air leakage that allows, after some time, to generate a spurious alarm. The operational problem was well understood. The small-diameter tubes that make up the fusible plug detection network were leaking. And the root cause of the leaks was practically inevitable impacts during the operational life of the platform.

Let us recall that for the design engineer in this case the fusible plug detection system is the most reliable of all. For the operations engineer, however, this system lacks the minimum reliability required, because any routine impact may trigger a spurious alarm and serious operational shortcomings. The design engineer is not wrong in his theoretical argument but if he had a minimum operational experience he would quickly understand that it is unreasonable to admit that throughout the operational life of an oil platform a long and fragile pipeline will not be exposed to damage caused by impact. Even if such impacts may happen due to negligence it is naive to expect that it will never happen.

The best solution would be to choose another means of detection, more up-to-date and more reliable. After all, immersed in his theoretical world, the design engineer can appreciate the simplicity of the working principle of the fusible plug detector which in theory seems indeed reliable. But if his vision horizon reached the day-to-day operational difficulties, he would realize that, in a broader context, the fusible plug is probably the least reliable detection system considering the real circumstances defined by the problem. The designer, without operational experience, insists on saying that operators should be more careful and never allow impacts and damage to the compressed air pipeline belonging to the fusible plug detectors.

Operators who need to keep the platform operating safely end up disabling the system and accepting the risk of such deactivation. Operators' rationale is that spurious alarms and false startups of the water-based fire protection system cause more disruption and danger to the platform.

Due to the designer's lack of operational knowledge, the platform operates under a risk condition always greater than that theoretically accepted in the design studies. After all, if the fusible plug detector system is working, the number of alarms triggered and spurious startups increase the risk of accidents and losses. Conversely, if the system is taken out of operation due to low reliability the platform will operate short of one detector resource which also increases risks.

2.5.2 The "best gas sensor in the world"

A risk management expert responsible for the approval of the safety design documents of the oil platform was taking part in a Hazard and Operability Analysis (Hazop) event (this analysis technique will be described in Chapter 7, Reducing Unpredictability). At this type of event, experts from each field of the project get together to assess the risks that can be generated during the operation of the platform that is being designed. Discussions can last more than a week and provoke some controversies. A designated expert is responsible for conducting this type of analysis. He or she leads the employment of the technique in search of a consensus on the recommendations to be followed by the design so that safety is preserved. Operators are necessarily included among the participants. Those are professionals who are used to operating oil platforms for years, some for decades. In this specific case, the design was about a general overhaul on a platform and so the participating operators were those who were already working on the platform that had been in operation for over 30 years.

Several experts from different design disciplines had already contributed to the assessments and the work proceeded normally, despite controversies that are always part of events of this nature. But then, towards the end of the analysis, one of the experienced operators noted that in the new gas detector location plan there was an area close to a pipelines assembly without any gas detector. That was then when he stopped the work to ask for the reason for the total lack of gas detectors for that area of the platform's production plant in the design documentation.

The designer and expert in the instrumentation discipline was responsible for the answer, since he was the one who approved instrumentation documents, including those that indicated the location of the gas detectors. The instrumentation expert opened the plan and explained that all detectors had been positioned based on a gas dispersion study (this type of study is described in Chapter 7, Reducing Unpredictability). The study had conducted simulations based on computational fluid dynamics, known by the acronym CFD. This type of CFD study, although it has been available since the 1970s, is still considered a high-tech resource and also represents a high cost. But this type of analysis requires significant operational experience to avoid distortions and evaluations of unrealistic scenarios.

In other words, the instrumentation expert justified the absence of the gas detector because a theoretical study, based on theoretical simulations, had shown that statistically the risk of a gas leakage in that region would be so low that it would not justify the cost of acquisition, installation, and maintenance of the detector. One of the objectives of this type of gas dispersion study is to optimize the number of detectors, reducing its application points, aiming at avoiding waste. But both the instrumentation expert and the computational fluid dynamics expert lacked operational experience and had only conducted the study and simulations using leakage frequencies from international databases. They did not know, in practice, what the demand for the equipment and pipelines would be in the area that was the source of the controversy and simply relied on the mathematical results of the simulations which from the academic standpoint were not wrong.

The experienced operator then made his remark and explained that the area in question where the design did not include gas detectors was the platform location with the highest incidence of leaks over the last 20 years of operation and, with the support from all the other experienced operators, considered absurd the lack of any gas detectors at the location with the highest incidence of leaks.

The atmosphere turned tense, because on one side the designers and computer simulations experts felt criticized regarding the quality of their study and, on the other, the experienced operators were feeling perplexed by the designers' resistance to accept the information from those who had effectively worked on the particular platform for decades (it was not a new project, the platform was already in operation and was being overhauled).

Realizing the situation, another expert, this one in risk management who was responsible for the design of safety systems was getting ready to express his opinion which at that point would add great weight to the discussion. During his preparation, the risk management expert, who had already worked in the operational area for years, did a review of his field experiences. He had never visited the platform that was being overhauled but had already embarked on many others of various types. He recalled the emergencies he experienced, the difficult times away from home in offshore confinement, problems to keep the platform in operation, safety concerns, the importance of the environment and the work climate to avoid accidents, including a specific recollection that came to his mind. He remembered one night when he slept on a platform after a day of safety concerns and recent accidents in other offshore operations companies. And that when entering his rest box, climbed the ladder up to the top of the bunk, and started to close the bunk curtain, but interrupted the sequence and asked himself a basic and obvious question: what if there is a fire here during the night? At that moment when insecurity hit the safety expert, he casually scanned the ceiling and, by chance, he found a typical cabin detector. He understood that that detector was there out of respect for risks and lives, including his own. He knew that behind that detector was a history of studies and analyses that justified the fact of being there. The expert understood that that detector meant a reason for minimal tranquility to close the bunk curtain again and sleep, to be prepared again for another hard day at work.

Awakening from that momentary journey through his memories of the operational experiences, the risk management expert posed the question about the number of years the operators have worked on the platform being overhauled. The answers were 8, 12, and 18 years. The second question was about what really led them to demand this detector so much. The answer was unanimous and referred to the smell of possible leakage and the presence of gas alleged by the operators. Then the risk management expert who had approved the documents related to the safety systems (including gas dispersion studies), recommended the inclusion of the detector in the area that was the source of controversy. He wrapped up by justifying that because it is a design being overhauled, already in operation for more than twenty years, with the platform operators' testimony about the concerns regarding leaks in that location the importance of the computer simulations becomes secondary since what they simulate in theory is what those platform operators had already experienced in real scenarios for almost 20 years. For this specific scenario in question, the best detector of gas in the world was the nose of each of those operations engineers who had worked for almost two decades on the platform under study.

Everyone, including the authors of the gas dispersion studies, reached a consensus. The studies continued to be utilized by the design to correct excesses and reduce the number of detectors, but the area that bothered the operators strongly started to have a gas detector with deep respect for risks and lives, including of those operations engineers who participated in that Hazop, the same way as it happened to the risk management expert during that worrisome night before sleeping in the cabin of an offshore platform.

2.6 Exercise

Indicate, for each topic, which component of the risk management strategic line refers to

1. Technical and Operational Knowledge
2. Hazard Reduction

3. Removal of Agents (People)
4. Emergency Control
5. Unpredictability Reduction?

( ) Risk analyses can be of a qualitative and quantitative type and they help designers and operators to make decisions about whether or not to accept risks.

( ) The escape and abandonment system is completely dedicated to protecting human lives

( ) In the event of a confirmed fire in a refinery, water systems for fire protection equipment must be available for operation.

( ) If there is an emergency alarm on a platform, even without smoke and fire signals, people must move to a previously defined meeting point (muster station).

( ) Working in operational activities is a very important training activity of a risk management expert.

( ) When there is a major emergency on a platform, one of the appropriate measures is the immediate reduction of a large part of the gas inventory of the process plant through flaring.

( ) Knowing the physical, chemical phenomena, and the practice of operational activities are basic requirements for performing risk management activities.

( ) Some risk analysis techniques allow that accidental scenarios be postulated and evaluated by specialists, anticipating situations in future emergencies.

( ) Safety systems should be designed to offer response conditions for the control of the "Design Basis Accident."

( ) One of the risk management strategies adopted on oil platforms is the segmentation of the hydrocarbon inventory during an emergency. With the segmentation of the inventory, the chances of uncontrolled escalation of fire are reduced, since portions of the inventory are isolated from each other.

( ) Operational experience allows more realistic risk management decisions.

( ) In order to reduce the number of victims, only the minimum number of operators and fire brigade members should control the emergency, while the rest must escape the scene as soon as possible and, if necessary, abandon it permanently.

( ) In the oil industry, hydrocarbon is the main source of hazard and the reduction or segmentation of inventories is one of the strategies adopted in emergencies.

( ) Computer simulations allow risk analyses to be performed that help designers to concentrate safety resources in the response to the most important scenarios.

( ) Safety systems help prevent the escalation of an emergency by promoting the cooling of pressure tanks and vessels, enabling the use of escape routes, directly fighting fires, among other response functions.

2.7 Answers

5 – 3 – 4 –3 – 1 – 2 – 1 – 5 – 4 2 – 1 –3 – 2 – 5 – 4.

2.8 Review questions

- What is the most important aspect in managing risk?
- What is an "absolutely necessary" risk?
- Explain the meaning of the term "unquantifiable risk."
- What is the concept of safety culture defined by the IAEA?
- What are the seven safety culture application principles?
- What are "human factors"?
- What does "human factors design" mean?
- What is the responsibility of the engineers for the reduction of human errors in the designs of technological enterprises?
- What are the seven application principles of human factors?
- What are the seven application principles for achieving risk management efficiency?
- Explain the term "region of undesirable risks."
- What are the five main components of the risk management strategic line? Explain them.
- Define Hazard.
- Define Risk.
- What is the difference between the concepts of "design-basis accident" and "beyond design-basis accident"?
- Which of the five components of the risk management strategic line is the most important?

CHAPTER 3

Technical and operational knowledge

It is not possible to acquire technical and operational knowledge only through books, courses, standards, and procedures. It is necessary to experience it in the field, which is the actual exercise of engineering. In this chapter, we will present the general information about the characteristics of some of the main types of facilities of the oil and gas industry. The objective is not to list all types of facilities or to detail the description of all operational activities. The content of this chapter aims at providing an information base that helps building a general "overall vision" of the oil and gas industry. The vision presented herein refers mainly to risk management and safety and serves as a preparation for an internship experience in the field.

The variety of facilities and equipment used in the oil and gas sector is enormous. Facilities in the same category have differences and particularities among them. For this reason, risk management needs to use the operational-specific history data about each unit as its main source of information. There are cases of oil rigs and other facilities that are built as if they were part of a family of "replicated" projects, that is, almost identical. Even so, that doesn't mean that the risk management solution that meets the demand of one of the "twin" units is applicable for achieving the same result in the other.

In practical training the professional should seek technical knowledge and also develop the perception of the impact of industrial activities on society and the environment. It is important to understand the degree of influence that a particular type of industry exerts over a neighborhood, state, country, and the planet. Balanced risk management requires a global vision of the social and environmental consequences caused by the relationship between the industry and the society where it operates.

In parallel to the culture that characterizes a particular industrial sector is also the internal culture of each organization. This internal culture must also be observed during the training period.

Risk Management in the Oil and Gas Industry.
DOI: https://doi.org/10.1016/B978-0-12-823533-1.00008-3

Some solutions, which are apparently suitable for managing risks in a given industrial sector, cannot be perfectly adapted to another type of industry. It is possible that some safety solution is perfect for one type of industry, whereas for another industry the same solution is not feasible. These subtleties can only be well perceived with operational experience, and this type of perception is essential to the production of good safety system designs.

3.1 Oil industry

We are contemporaries of the oil age. The basis of the world energy matrix remains highly dependent on the oil and gas industry. But it has not always been so. Only after the second half of the 19th century did the first entrepreneurs succeed in making their offerings of products refined from petroleum economically viable as replacement to fuel from plant or animal sources (coal or whale oil). Until then, these were the traditional massively consumed products for lighting purposes. The expansion of the use of kerosene obtained from crude oil refinement had to overcome a major risk management challenge for the oil industry to get to where it is today.

3.1.1 John Davison Rockefeller and risk management

The oil and gas industry has emerged after overcoming a major risk management problem that was a roadblock to its expansion. And one man was especially instrumental in the process: John Davison Rockefeller. He was a notable American entrepreneur, motivated by his Christian principles, and who had become the richest man in history as well as the greatest philanthropist of all times. Today, everyone wants to be an entrepreneur and examples are the ever-increasing inspirational stories of entrepreneurial success like those about California's Silicon Valley technology companies. When a company is formed, the entrepreneur may try to do the best economic analysis, evaluate the costs well, deeply analyze the market, and put in place the best administrative management. But every technological enterprise always needs some specific technical knowledge that only professionals who know how to make, assemble, and build can deliver it. Successful entrepreneurs need scientists. Refining oil in that decisive phase of the industry meant turn crude oil that comes out of the ground into kerosene.

When compared to the products of animal and vegetable source, it was considered cleaner and more practical for use for lighting purpose. But there were problems.

Kerosene emanates flammable vapors, and when refining doesn't use the best technique, it is not possible to obtain a pure, stable, and economically competitive product. Rockefeller needed to invest in the development of an oil refining technology, so that his product could offer better risk conditions during use and transportation.

For this reason, he partnered with chemist Samuel Andrews, with the objective of producing kerosene within a reliable quality standard. Based on a well-developed distillation technology, Rockefeller believed that he would position himself ahead of his competitors and gain control of the production process and the entire oil industry.

But to attract investors, Rockefeller would have to convince everyone that his refining technology would produce kerosene with standardized properties and, as a result, safer. Thus he would manage to repair its bad reputation regarding fires and explosions that kept investors away, which were indispensable for making his business viable. Kerosene's bad reputation at that time was caused by the perception that it represented an unacceptable fire hazard and risk. The front pages of American newspapers at the time often told stories of catastrophes with fires and explosions caused by kerosene lamps, which destroyed businesses and families. High demand for the product caused competitors to produce kerosene lacking quality standards, very volatile, and mixed with other by-products. The risks associated with the kerosene's low quality, as well as the high frequency of associated accidents, generated a bad reputation for the product, and that were the red flags that scared investors away.

The technical impasse would require a risk management solution that could change this scenario and thus allow the oil industry to be fully established. John Rockefeller viewed this problem as a great opportunity, realizing that it would be necessary to calm down the consumers by supplying a product based on a technology capable of ensuring quality standards. This also included filling process, storage, refinement, and transportation of kerosene. He then envisioned and created the first oil company that ensured uniform kerosene quality. Rockefeller called this company Standard Oil. Rockefeller did a brilliant job managing the fire and explosion risks that were preventing the emergence of the oil industry. He did it through his idea to always supply kerosene with the same properties and with more predictable risks. He started selling kerosene with a known quality standard. Although it would continue to be a hazardous product, Rockefeller started offering it to the market under acceptable risks, which removed the population's fears.

This immediately turned kerosene into the most desired product in the United States, attracting countless investors and definitively defining the road map for the oil and gas industry as we know it today. This was only possible through technical operational knowledge, which made it possible the development of the refining process, and a product of standardized quality had just dawned.

Operational technical knowledge is decisive in establishing the difference between success and failure of technological enterprises. John Rockefeller, to set up the oil and gas industry, had to overcome the difficulties associated with the rejection of its main product because of association with fires and catastrophes. He could have just followed most of the investors of his time, who gave up and switched to different business activities. But using technical and scientific knowledge, Rockefeller produced a standardized product, repaired the bad reputation of kerosene, and became in practice, out of necessity, the first risk management expert in the oil and gas industry (Fig. 3.1).

3.1.2 Components of the oil and gas productive chain

The activities related to oil and gas form a productive chain so complex that the industry has grouped them in components that are well established today. The term upstream refers to the sectors of the production chain that precede refining. It usually includes the sectors related to exploration and production. The downstream sector involves the refining process, distribution, and marketing of the refined products. There is also a less-used third term called midstream. Some oil-processing companies group as midstream the activities directly related to the transformation of raw materials (oil and gas) into products (refined, among others).

3.1.3 Onshore and offshore facilities

Two other terms that are widely used in the oil and gas industry, which also serve to group the oil and gas industry into different areas of activity, are onshore, which refers to activities on land, and offshore, related to maritime activities. With the technological evolution that took place for

Figure 3.1 John Davison Rockefeller, 1885, founder of Standard Oil and the biggest entrepreneur in the oil and gas industry. Rockefeller used the technology to repair the bad reputation of catastrophic fires caused by kerosene. From the standardization of production and brilliant risk management work, kerosene produced by Rockefeller has become the most desired product in the United States.

almost two centuries, oil started being explored both on land and at sea. Two other less-known terms have also been used: inshore—which means coastal or in sheltered waters such as bays, lagoons, and rivers—and at shore—which means exactly on the coast—as it happens with some marine terminals. As these terms are still not widely accepted, in this book the inshore and at shore facilities and equipment will be referred to within the context of offshore or onshore facilities, as best suited in each case, as installations of this type have both offshore and onshore characteristics.

Regarding risk management and safety, the major hazard for the oil industry is hydrocarbon, which occurs as liquid or gas. Both the offshore and onshore sectors are exposed to this hazard and are consequently subjected to its inherent risks. But the offshore sector can be considered more critical, because in addition to the hydrocarbon, hazard is another extreme scenario that involves survival at sea. The design of offshore rigs needs to consider all the difficulties associated with offshore activities, such as the influence of wind, current, and wave height. They also need to comply with regulatory requirements associated with maritime safety, buoyancy issues, and naval stability. In the offshore industry, designs need to consider the handling of hydrocarbons in processing plants on offshore platforms for exploration and production of oil and gas.

If we compare subject matters that are important to the safety of onshore plants and offshore rigs, mainly involving the main hazard, which is hydrocarbon, we will see that there is a good correspondence both in the scenarios and the safety systems, emergency control techniques, and risk management. But given that the offshore rigs aggregate the risks arising from hazards related to with the sea, we will prioritize them in this book, through examples, cases, and explanations using scenarios specific for offshore rigs, since in general, they are more complex from the point of view of risk management. When there is a particular aspect of onshore plants that needs to be dealt with in a specific way, it will be done accordingly. Our many years of experience in safety systems design in the oil and gas industry give us confidence to approach the matter of safety systems for both offshore and onshore facilities. However, particular situations of onshore facilities may require specific solutions. Although we can say that basic systems related to the hydrocarbon hazard are similar for both onshore and offshore facilities, the most critic ones are associated with the latter due to maritime activities. Our presentation of risk management in the oil and gas industry will accordingly focus on offshore applications, without losing sight of onshore facilities.

3.1.4 Accidents in the oil and gas industry

Data[1] collected from 1974 to 2013 by insurance and risk management experts in the United Kingdom allowed the production of a report with the 100 largest accidents in the oil and gas industry during this period (in terms of financial losses). In Table 3.1, we present a list of the 20 largest accidents in the oil and gas industry and the estimated financial losses for each event.

Analyzing all the 100 largest accident losses in the oil and gas industry raised in the same research, it is observed that 34% of these losses are a consequence of accidental events in the upstream sector (part of production chain that precedes refining). The breakdown of the

[1] Source: The 100 Largest Losses 1974–2013. Large property damage losses in the hydrocarbon industry. 23rd edition, Marsh & MacLennan Companies, United Kingdom, 2014.

Table 3.1 20 Largest financial losses from accidents in the oil and gas industry, based on the report by British insurance and risk management experts.

Date	Plant type	Event type	Location	Country	Property loss US$ (millions)
07/07/1988	Upstream	Explosion/fire	Piper Alpha, North Sea	United Kingdom	1810
10/23/1989	Petrochemical	Vapor cloud explosion	Pasadena, Texas	United States	1400
01/19/2004	Gas processing	Explosion/fire	Skikda	Algeria	940
06/04/2009	Upstream	Collision	Norwegian Sector	North Sea	840
03/19/1989	Upstream	Explosion/fire	Gulf of Mexico	United States	830
06/25/2000	Refinery	Explosion/fire	Mina Al-Ahmadi	Kuwait	820
05/15/2001	Upstream	Explosion/fire/sinking	Campos Basin	Brazil	790
09/25/1998	Gas processing	Explosion	Longford, Victoria	Australia	750
04/24/1988	Upstream	Blowout	Enchova, Campos Basin	Brazil	700
09/21/2001	Petrochemical	Explosion	Toulouse	France	680
05/04/1988	Petrochemical	Explosion	Henderson, Nevada	United States	640
05/05/1988	Refinery	Vapor cloud explosion	Norco, Louisiana	United States	610
03/11/2011	Refinery	Earthquake	Sendai	Japan	600
04/21/2010	Upstream	Blowdown/explosion/fire	Gulf of Mexico	United States	600
09/12/2008	Refinery	Hurricane	Texas	United States	550
06/13/2013	Petrochemical	Explosion/fire	Geismar, Louisiana	United States	510
04/02/2013	Refinery	Flooding/fire	La Plata, Ensenada	Argentina	500
12/25/1997	Gas processing	Explosion/fire	Bintulu, Sarawak	Malaysia	490
07/27/2005	Upstream	Collision/fire	Mumbai High North Field	India	480
11/14/1987	Petrochemical	Vapor cloud explosion	Pampa, Texas	United States	480

accidents that occurred related to activities in the downstream sector is 29% of accidents in the refining itself, 23% in the petrochemical segment, 9% in gas processing, and 5% in transfer terminals and distribution activities. The report also concludes that the accumulated total losses exceed 34 billion dollars, in figures adjusted to 2014. This shows the importance and economic and financial weight of risk management in oil and gas industry activities, not to mention the greatest of all that is the irreparable loss of human lives.

The accidents that occurred in the years 2012 and 2013 are shown in chronological order in Table 3.2. Most of these accidents occurred in the downstream segment. The two accidents recorded in the second semester of 2013 occurred due to upstream activities.

3.2 Getting to know upstream facilities

Upstream facilities are those whose applications precede activities refining within the oil and gas industry production chain. They consist of prospecting by geological and seismic methods, the onshore (terrestrial) or offshore (maritime) production, well drilling, operation of service vessels, subsea equipment activities, and natural and artificial pumping methods, among others.

There are various accidental scenarios that threaten upstream activities. The most well-studied scenarios involve fire and explosion related to hydrocarbon, which is the main hazard in the oil and gas industry. But for offshore activities, buoyancy and naval stability are also important. A classic accidental scenario of upstream activities is the blowout. Well activities require rigorous pressure control associated with its operation. One of the most serious events that can happen as a result of pressure imbalance is the establishment of an uncontrolled flow of hydrocarbons and other fluids from the oil well. During the well-drilling process, there must be a balance between the volumes of the fluid injected and the fluid returns. If the volume of fluid returns is greater than the volume of the fluid injected, it is an indicative of a blowout condition, more specifically, in operational language it is called a kick. An equipment of great importance for safety is the Blowout Preventer (BOP), which consists of a system of valves, spools, and equipment whose purpose is to quickly close the well and isolate it in case of blowout.

Table 3.2 Accidents in the oil and gas industry between 2012 and 2013.

Date	Plant type	Event type	Location	Country	Property loss US$ (millions)
05/05/2012	Petrochemical	Explosion/fire	Map Ta Phut	Thailand	140
07/04/2012	Refinery	Explosion/fire	Bangkok	Thailand	140
08/25/2012	Refinery	Explosion	Falcon State	Venezuela	330
04/02/2013	Refinery	Flooding/fire	La Plata, Ensenada	Argentina	500
06/13/2013	Petrochemical	Explosion/fire	Geismar, Louisiana	United States	510
07/01/2013	Upstream	Sinking	Atlantic Ocean, offshore	Angola	240
07/23/2013	Upstream	Blowout	Gulf of Mexico, offshore Louisiana	United States	140

The largest oil spill accident in the offshore sector occurred on April 21, 2010, in the Gulf of Mexico, United States. The oil drilling rig Deepwater Horizon experienced an operational transient that led to the release of well pressure. But the BOP equipment, considered one of the most reliable equipment in offshore oil rigs, was unable to close the well as intended, allowing 780,000 cm^3 of oil to leak in the Gulf of Mexico, causing material and human losses resulting from the accident. In addition to the blowout, many other accidental scenarios need to be studied as part of the design for upstream oil and gas facilities. These analyses are used in the assessment of risks and their reduction, as much as possible, through design recommendations and operational solutions so that the safety systems are able to offer a response compatible with the most severe postulated accidental scenarios.

The greater the number of professionals with operational experience who participate in the risk analysis and risk management process (from design to operation), the higher the quality achieved in the technical solutions of the problems. The pertinent operational experience needs to be acquired by the professional through the direct involvement in field activities. We will present some of the main equipment in the upstream sector that risk management professionals should be familiar with, so as to have at least a basic knowledge of their characteristics and their purposes in the oil and gas industry, always highlighting the main safety-related aspects.

3.2.1 Drilling rig and completion

Drilling rigs (Fig. 3.2) are terrestrial or maritime equipment for perforating the soil and rocks that remove fragments resulting from the operation through the flow of perforation fluid or mud. Rotary drilling rigs, as they dig deeper, also support steel cladding and cementing of the annular space of the well so that the operation can proceed safely. The main systems and associated equipment are as follows:

- Support system responsible for loads associated with the weight of the drilling column, consisting of a tower or mast, substructures, and pipe rack to organize the pipelines and equipment required for the drilling operation.
- Power generation and transmission system, consisting of an energy source (usually a diesel engine or gas turbine) and transmission system that varies according to the offshore drilling rig (mechanical or electromechanical).
- Cargo handling system to allow the assembly of columns, including winch and other additional equipment.

ALMIR JüNIOR/BANCO DE IMAGENS PETROBRAS

Figure 3.2 Onshore drilling rig.

- Rotation system used to rotate the drilling column, including rotary table, kelly (transmitting rotation of the table to the column), swivel (injection head equipment for transition between rotary and fixed elements), top drive (in the offshore drilling rig that don't use rotary table and kelly) and downhole motor, for high-slope or horizontal wells.
- Drilling fluid circulating system responsible for the injection, return, and treatment of the fluid that enables the operation.
- Well safety system—known as Wellhead Safety Equipment—for fast and safe closing in emergencies. The main equipment is the BOP, wellhead assembly, and annular preventer for closing the annular space. The BOP

is the main safety equipment and consists of a set of valves and the preventers that close when a kick occurs (uncontrolled flow of operational fluids). The BOP's main function is to close the well, preventing a well from producing through a totally uncontrolled flow, causing accidents with harm to people and damage to the environment.

Instrumentation and control system, responsible for operational data collection and data recording, deemed essential for the success of the operation.

The operation of drilling rigs requires the fulfillment of several safety requirements, and one of the most important aspects is kick control. The main causes of kicks are insufficient mud weight or insufficient sludge supply during the well operation. When the pressure of the confined fluid exceeds the mud pressure, we have a kick. The initial accident scenario of this event is the development of uncontrollable flow from the well to the surface. Special care during perforation must be taken to prevent kicks, such as avoiding swab (negative pressure during the removal of the drilling column), monitoring the presence of gas in the fluid or mud that is reducing its density, preventing interruptions in the fluid circulation, preventing cementing failures that cause hydrostatic pressure reduction, and preventing inadequate testing during the probe operation.

3.2.2 Primary processing equipment

The main purposes of the upstream sector are to identify reserves, the exploration, and production of oil and gas. But some basic processing activities, with respect to the context of the entire production chain, are also required in the upstream phase. Well production needs to be processed through a minimum set of operational facilities that allows the separation of water and the impurities contained in the oil and gas produced by the well. The high transportation costs of the production to facilities should not be increased further by including water and impurities. These could be separated, treated, and discarded through a primary oil and gas processing plant. Furthermore, the presence of impurities causes other problems such as the oversizing of pumps, pipelines, tanks, and transfer facilities, as well as increased energy consumption, corrosion, and pump scaling. Failure to separate water and impurities can also generate undesirable operational transients that can cause damages, accidents, and consequently material and human losses.

Primary processing plants have separators of two- (gas/liquid) or three-phase types (gas/oil/water). There are several different technologies associated with the separation process. Water separated from the oil needs to be treated before it can be discarded. Hydrocyclones and floats are normally used to reduce the residual oil to levels that allow safe disposal to the environment. Another option is the reinjection of water produced in wells when this appropriate and technically feasible.

The design and operation approaches to oil and gas primary processing plants, as well as their equipment, should be centered on process safety (see Section 3.4). Even though they are simpler in comparison to downstream processing plants, those that are part of the upstream activities should receive equal attention in terms of process safety.

3.2.3 Fixed offshore platforms

Conventional fixed-type platforms for offshore exploration and production are in general economically viable in water depths of up to 300 m. With the technological evolution of new designs, the depth can exceed 500 m. Fixed platforms contain a topside (main functional part of the rig, located above sea level) which may include a processing plant, various production, and utilities equipment, in addition to a superstructure with accommodation facilities (Fig. 3.3).

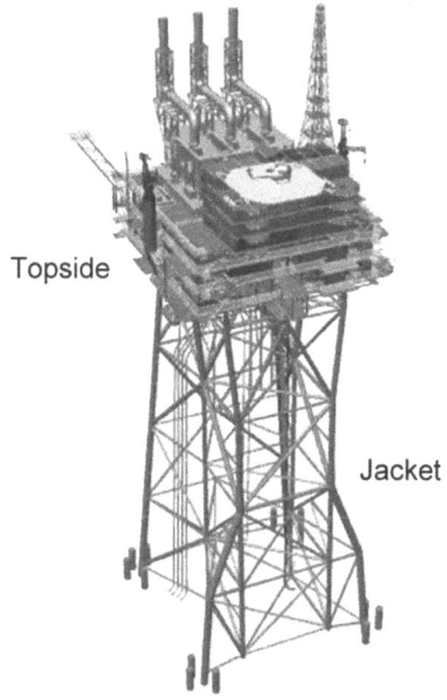

Figure 3.3 3D visualization of fixed rig design.

The complete assembly is supported by a jacket, which is a steel or another fixed concrete structure, and is connected to foundations on the seabed. The topside dead weight and other design loads supported by the jacket or concrete structure are transferred to the seabed. Obviously, fixed platforms don't float, and thus a series of maritime safety and naval design requirements involving stability and buoyancy matters are not applicable. But there is a major limitation, which is the water depth limit. Currently, designs are not viable for water depths much greater than 500 m. Even for shallower depths, for example, beyond 300 m, some fixed rig designs are economically unfeasible, where a different design concept can be considered a better option (Fig. 3.4).

Offshore platforms can be of two types, namely "normally inhabited" and "occasionally inhabited." Normally inhabited platforms always have a crew on duty, while occasionally inhabited platforms only receive a crew during the time required for the execution of specific services, when the presence of the operator is indispensable.

There are conceptual designs under development with the objective of reducing to a minimum the need for human presence on offshore platforms. According to these new concepts, the platforms would be built with a high level of automation and instrumentation and would be controlled remotely through a main control center that could even be on land.

Despite the positive aspect of such a concept, which is the reduction of the exposure of the human element to offshore risk scenarios, there may be some new problems. High levels of automation and instrumentation make equipment more complex and reliant on the flawless functioning of these technologies.

Another problem is the excessive distance between the plant and its remote control room. This can lead to a reduction of technical sensitivity due to more limited and, in some cases, even nonexistent operational field experience. For this reason the new remotely operated rig concepts should have their evolution backed by responsible and technically sound risk management so that not only the virtues but also the inconveniences are balanced in realistic risk analyses.

As fixed platforms don't float, safety design doesn't need to consider heeling scenarios (rig tilt). This can make a difference, for example, in the strategy of positioning lifeboats for deployment in situations that require abandonment, as in the case of floating platforms where it is necessary to consider that some of these lifeboats may be unavailable due to the excessive inclination caused by naval damage condition. On fixed platforms, it

BRUNO VEIGA/BANCO DE IMAGENS PETROBRAS

Figure 3.4 Fixed rig. Topside (modular assembly with equipment and operational facilities) supported by a jacket (metallic structure) founded on piles driven into the seabed.

is not necessary to take it into account due to its construction characteristics, which are incompatible with such naval accidents.

An important safety-related factor is that fixed platforms, in general, are located closer to shore. It makes it easier for the transportation of teams and possible support and rescue operations in emergencies.

It is also important to note that this type of rig usually keeps a smaller inventory of hydrocarbons, since they don't normally store oil and gas in large quantities as in other types of offshore rigs. But that doesn't mean that catastrophic accidents cannot happen on fixed platforms. Despite the smaller inventory, the units can store the production of oil and gas from several wells and export it to its destination on land. These connections to production and transfer are sufficient to justify a rigorous risk management design. Incidentally, the largest accident in the offshore industry to date in terms of human losses and materials took place on a fixed platform in the North Sea, United Kingdom (Piper Alpha Platform).

3.2.4 Semisubmersible offshore platforms

Semisubmersible platforms, also known as SS platforms, are floating rigs with a hull designed specifically to withstand the topside loads of an offshore rig. The hull can be built based on different concepts, but most hulls consist of vertical columns and pontoons that act as submerged floats. The positioning of the unit in the exact location for exploration and production is done using an anchoring system composed of moorings and anchors. Another higher cost option is the dynamic positioning system used in some types of floating platforms and based on GPS (Global Positioning System) satellite guidance. A sophisticated controlling system collects the reference position indicators obtained from the satellites and compares them with the actual rig location, making the permanent correction of its position through thrusters (Fig. 3.5).

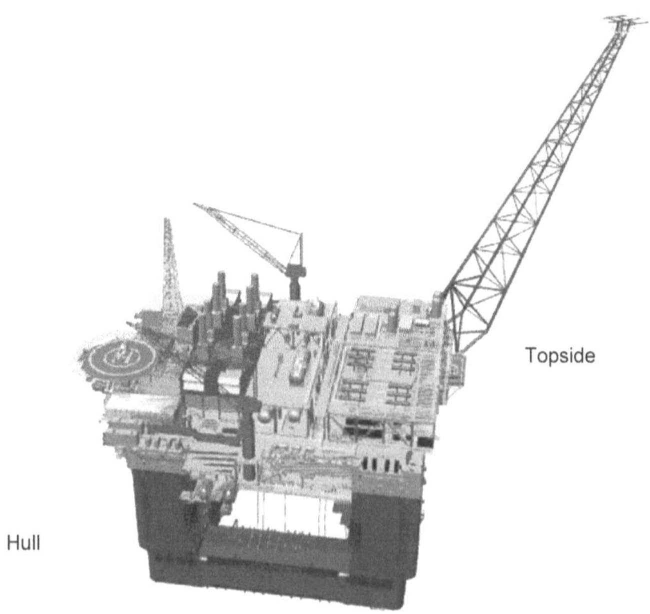

Figure 3.5 3D visualization of SS rig project. *SS*, Semisubmersible.

GERALDO FALCÃO/BANCO DE IMAGENS PETROBRAS

Figure 3.6 SS rig. *SS*, Semisubmersible.

From the risk management standpoint, due to its floating design, maritime safety requirements are applicable, including buoyancy and stability matters. Abandonment resources need to be designed to take into consideration the possibility of heeling caused by an accident or naval damage. This implies making resources available in quantity and location that can compensate for the possible loss of operation of some of the lifeboats. Such a situation may occur as a result of heeling that impairs the launching of lifeboats into the sea from mechanical launching systems located in the region affected by the slope. SS platforms, which are designed with the purpose of offshore exploration and production starting from the hull, in general have more space available for the arrangement of equipment and functionalities on the topside. This in turn allows for a more optimized layout of modules and equipment and most importantly more efficient ventilation, besides generating less congestion in the processing plant area. These aspects are essential for accident risk reduction related to fire and explosion (Fig. 3.6).

3.2.5 Floating production, storage, and offloading system platforms

FPSO (floating production, storage, and offloading system) platforms are floating units that have hulls with large tanks with high-volume capacity for production storage. The production is periodically exported (transferred) by flexible hoses to shuttle tankers that transport the product to its destination, as part of the typical sequence of stages in the production chain. Many designs involve refurbishing an oil tanker hull at the end of its useful service life. In this case the vessel is completely transformed so that it is adapted to receive a processing plant, the utility modules, and other equipment and operational facilities that characterize the arrangement of an offshore rig. Other projects start off from a completely new hull, not built to sail as in the case of oil tankers, but specifically to receive the topside. FPSO platform applications have been greatly successful, especially because many technical feasibility and economic studies indicate that this concept of offshore rig produces the best final result. But each case needs to be treated individually and only the specific evaluation of each project can determine the best option among the rig concepts available.

FPSO hulls hold large oil storage tanks and this restricts the area available for the topside in comparison with SS rig designs. When the hull of an oil tanker is reused, space limitations may be even greater. With respect to safety, FPSOs impose more difficulties in the optimization of the arrangement of the processing plant, utility modules, and other equipment in the available space. In some designs, special care needs to be taken to protect the areas underneath the main deck (in the hull interior). The movement of people is a little more restricted at these locations. In FPSOs, many important equipment are located underneath the main deck and need to be constantly serviced during operation and maintenance activities.

Similar to other floating rigs, in the FPSO the abandonment resources need to be distributed considering the possibility of a naval accident causing heeling. Beyond a certain slope threshold, there is a limitation or even a complete impediment to the launch of lifeboats on one side of the rig (portside or starboard). There are free-fall launching systems that allow more flexibility to overcome such a limitation. Free-fall lifeboats allow abandonment to be performed, for instance, through lifeboats located on the stern. This area is under less influence of the slopes, resulting from a possible heeling. Free-fall lifeboats don't require cables for launching and can also be used in other concepts of offshore rigs, according to each designer's criteria.

As in the case of SS platforms, FPSO offshore units can be positioned by means of an anchoring system or by dynamic positioning (for some special rigs that work together with the FPSOs as part of the equipment structure of the production chain). The production lines (riser pipelines) and injection from wells can be designed to reach the rig in different ways. This also influences the platform's anchoring concept. There are FPSOs designed with a tower that centralizes the entry of all risers (turret). The turret has a swivel whose function is to allow relative motion between itself and the rig, providing safe transition between fixed and moving parts. In this type of project the turret remains fixed, while the rest of the rig performs a rotational movement around the turret to accommodate the influences of ocean currents and wind. Another concept for the project of the entry of FPSO risers is called spread mooring. In this case the anchoring is performed by anchors and moorings, and one of the sides has a large empty area for the risers' entry. The FPSO that adopts the spread mooring concept always remains in the same position, similarly to an anchored SS rig (Fig. 3.7).

FPSOs concentrate immense amount of energy due to their large hydrocarbon storage capacity. Moreover, naval damage condition can result in loss of the containment of large tanks, causing accidents with significant environmental impact. As a result the maritime safety and process requirements for the FPSO equipment is very strict. The aspect ratio

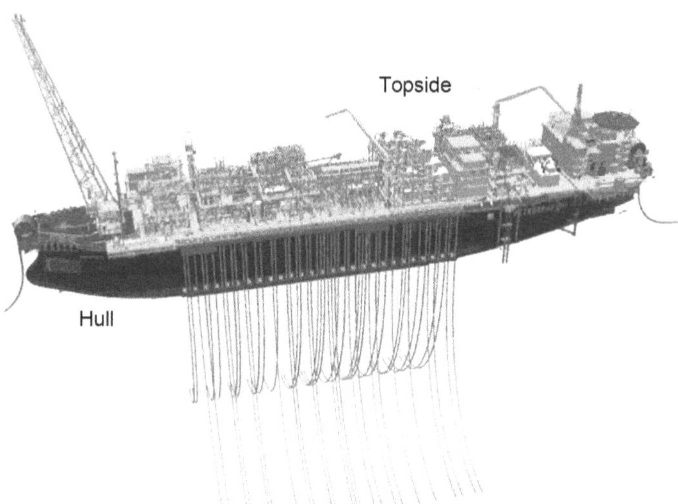

Figure 3.7 FPSO with anchoring and spread mooring risers entry. *FPSO*, Floating production, storage, and offloading rig system.

STEFERSON FARIA/BANCO DE IMAGENS PETROBRAS

Figure 3.8 FPSO rig. *FPSO*, Floating production, storage, and offloading rig system.

between the length of the main deck and its width raises the possibility, during a catastrophic accident, of isolation of people in the area considered less safe (bow and processing plant). This implies additional resources required for the protection of the main escape routes that are designed to allow communication between bow and stern. This feature also requires additional resources for a possible need for abandonment through the bow. Despite the amount of hydrocarbons stored and the critical operational activities that they perform, FPSO platforms are safe rigs as long as they are well designed and subjected to adequate risk management (Fig. 3.8).

3.2.6 Special offshore platforms

3.2.6.1 Submersible platforms

These are units whose use is restricted to shallow and calm waters, with depth limited to the height of the mobile hull structure. The topside is built on a floating hull that is towed to the desired location of the operation. Once positioned, the lower movable part of the hull is ballasted down to the bottom of the sea, lake, or river. The bed must be sufficiently uniform to allow the stabilization of the support structure. This type of rig can only be used in very shallow waters.

MICHEL REy/BANCO DE IMAGENS PETROBRAS

Figure 3.9 Self-elevating rig.

3.2.6.2 Self-elevating platforms

These are floating units towed to the desired location, equipped with steel structures, also called legs, capable of being submerged using a mechanical and hydraulic system down to the ocean's floor. Once the legs' supports are built and they are stabilized, the rig itself is elevated above the water's surface. The maximum water depth where this type of rig can be used is limited to the height of the legs, which in general don't exceed 200 m. From a safety standpoint the transportation of self-elevating platforms is complicated by stability issues, in some cases being required partial disassembly of the legs. An additional risk is posed by the influence of water motions on the stability and flotation during the self-elevation operation (Fig. 3.9).

3.2.6.3 Tension leg (TLP and SPAR platforms)

Tension leg platforms (TLPs) are floating platforms that combine the buoyancy forces (generated by displacement resulting from the draft—submerged part of the hull), with the tensile forces generated by tubular cables connected to the hull and anchored on the seabed. This combination of forces reduces the influence of water motions, causing the rig to remain in its location under the forces resulting from the thrust and tensile forces on the cables. As a result, such platforms can operate on water depths exceeding 2000 m. A variation of this type of concept is the single point anchor reservoir (SPAR) rig which, instead of being built with a hull similar to the semisubmersible platforms, is built from a single cylindrical hull that likewise keeps the rig under tensile forces.

3.2.6.4 Compliant tower platforms

The main feature of the compliant tower (CT) concept is that the topside is supported by a narrow tower built in a steel or concrete structure, anchored to the seabed, and with flexibility designed to withstand the forces associated with the sea and climatic conditions. On conventional fixed platforms the jackets are wider and, although they are also structures that share similar characteristics and functions, they don't have as much mechanical strength when subjected to bending forces as do the towers of the CT platforms. Although they resemble the fixed platforms, the CT units can operate on water depths from 450 up to just beyond 900 m. The tower's lateral flexibility provides structural strength to withstand forces from extreme sea conditions and strong winds like those that occur in hurricanes. At present time, due to their high costs beyond 1000 m, CT platforms are not economically viable. From the safety standpoint, this type of rig can be a technical solution for regions regularly subjected to severe weather conditions, where the water depth is within the limits of economic viability of the project.

3.3 Getting to know downstream facilities

We will consider that downstream facilities are all those that participate in the production chain processes of the oil and gas industry, starting with the refining processes. The specific refining activities may still be considered by some experts as a third segment, called midstream, but we will herein include refining as one of the stages of the downstream processes, as well as the activities of the marine terminals that have characteristics of both offshore and onshore facilities.

The oil and gas industry's productive chain is very vast and complex. We will not detail the procedures for each downstream activity or list all types of facilities in this segment. The objective is to describe some typical facilities to facilitate the development of the field experience, with a focus on issues related to risk and safety management.

3.3.1 Refining facilities and petrochemical plants

After the basic processing, the production of crude oil is transferred to the refinery, where the product is initially stored in large tanks. Crude oil can vary widely in its viscosity and contaminant characteristics, for example, by the levels of contaminants such as sulfur and hydrogen sulfide. Therefore samples of the product are analyzed in the laboratory to identify the characteristics of the oil to be processed. After that, it will be sent to the first processing stage called desalinization, with the purpose of removing water and dissolved salts in suspension form.

After desalinization, the product is pumped into a furnace to be preheated and sent to the first separation stage in the atmospheric distillation tower. The tower is basically composed of a main column with a stack of perforated plates arranged at several heights, according to the extraction requirements for each fraction. Gaseous hydrocarbons tend to rise due to the higher temperature at the bottom of the tower. At the top of the tower is a vapor collection system that directs the collected vapors to condensers through pipelines. The main fractions produced in this phase are gasoline and diesel, besides gas, naphtha, and kerosene. Heavier fractions are accumulated at the base of the tower, forming the atmospheric residue at the end of the first refining stage.

The residue from atmospheric distillation is pumped into another furnace to be heated again and transferred to a vacuum distillation tower, operating on the same principle as the atmospheric distillation tower, which also contains perforated plates to allow the extraction of the various fractions. In this stage, gas oils and a residue are obtained. The latter is used to produce asphalt or as fuel oil. The gasoil produced is sent as cargo to the cracking unit where LPG (liquefied petroleum gas) is produced, in addition to gasoline. In the catalytic cracking unit the gasoil gets into contact with a powder catalyst element, which is heated to high temperatures. The reaction causes the rupture, or crack of the larger molecular chains, resulting in smaller and lighter molecules that will be fractionated further to produce the noble oil products.

ANDRé MOTTA DE SOUzA/BANCO DE IMAGENS PETROBRAS

Figure 3.10 Overview of area occupied by an oil refinery.

The vacuum distillation process generates a residue that can be used as fuel oil, asphalt, or as a load for a delayed coking unit. The objective of the delayed coking process is to transform the vacuum distillation residue into higher added-value products with such as diesel, gasoline, and LPG. Petroleum coke can also be obtained by this process, which, among several applications, can be used in steel blast furnaces (Fig. 3.10).

From the refining activities, other products can be obtained from petroleum through processing in petrochemical plants. Some of these products are intended for direct consumption and others are for industrial applications. The primary petrochemical industry produces substances such as methanol, ethylene, toluene, and propylene. Intermediate petrochemical supplies are produced from the conversion of primary petrochemical products to more complex materials such as vinyl acetate, used for painting, vinyl chloride, PVC plastic, and styrene, to produce rubber and plastics. Petrochemical plants consist of different processing units. For example, an ethylene copolymer production plant requires a catalytic cracking unit that, through high pressures and temperatures, breaks down natural gas by repeated compression and distillation, whereas a petrochemical plant for the production of methanol requires a catalytic reforming process through high-temperature, medium-pressure steam.

Supply stock, final product types, production methods, and location are factors that influence the configurations of petrochemical plants. Nonetheless, there are some standard characteristics, for example, the need for large gas pipeline networks and the use of furnaces and rotatory equipment. Typically, the petrochemical industries are built near refineries to reduce operating costs associated with transportation and logistics.

Although the processes related to refineries and petrochemical industries have been succinctly presented in this section, such facilities are made up of complex equipment and operational routines. Thus refineries and petrochemical industries require vast areas of land for proper arrangement of the various processing units, in addition to areas for storage of both crude oil petroleum products and the final by-products.

All supporting activities that are essential for the successful process execution such as maintenance, transportation, and administrative activities need also to be included as part of refineries and petrochemical industries. Petrochemical facilities and refineries contain a large inventory of hydrocarbons distributed both in storage tanks and circulating through the pipelines and processing equipment.

There are also risks related to heat and ignition sources, and leakage of liquids and vapors. Accidental scenarios need to be studied and analyzed in advance, and a safety culture needs to be developed so that the right attention at the right time can be devoted to safety matters, both related to people as well as the environment and property. For all these reasons, risk management in refineries and petrochemical industries should be conducted with extreme responsibility and under the guidance of the technical and scientific process safety requirements.

3.3.2 Transportation and distribution

Just as important as producing oil and gas, it is to provide and operate means of transportation and distribution to transfer crude oil and gas to the refineries and to distribute the final products and by-products generated throughout the production chain. The transportation of oil and gas requires interconnections by rigid and flexible lines between subsea rigs and offshore platforms, in addition to rigid and flexible lines for export (e.g., oil and gas transport lines from the sea to the continent where it will be processed). Maritime transportation is also used by oil tankers, gas, and liquefied gas vessels, as well as on-land transportation by large intercontinental oil and gas pipelines. Other modes of transportation such as by rail and on-the-road transportation are also used. Petroleum and gas products produced along the production chain still need to be transported to petrochemical hubs and related industries, to domestic customers, and to the general population. Transportation and distribution activities are present from the oil well to the gas station, requiring enormous logistical capabilities, significant technology, and also involving considerable risks.

The operational difficulties that need to be overcome have a starting point in the subsea arrangement. In deepwaters (water depths beyond 300 m), only robotic equipment such as the remotely operated vehicle can work, because of the high underwater pressure. Offshore rigs require several types of product transportation lines, so that the demands of transport logistic and distribution are met. The construction of these lines uses ships specially designed for the launching of submarine pipelines, with a high-operating cost, strict schedules, and considerable risks. In deepwaters an enormous and complex network of transfer, transport, and export lines is built on the seabed, forming a submarine arrangement that is affected by interferences due to variations in seabed changes, including up and down paths, influences of the mooring of the rig anchoring systems, ocean currents, low temperatures in deepwater, and even the activities of the vessels and offshore equipment responsible for the oil and gas production themselves.

On land, transcontinental oil and gas pipelines also need to be built, operated, and maintained. These systems are utilized for the distribution of products and by-products, through the establishment of a complex network of pipelines and equipment. The variation of temperature along an oil or gas transmission pipeline can be very significant within a few-kilometer distance, ranging from high-temperature regions to below-freezing snow regions. The product must be delivered to its final destination within the required specifications, regardless of external influences in the pipeline transportation.

Petroleum products produced during refining such as diesel, gasoline, and LPG, in addition to natural gas, also need to be distributed to end customers such as industries, thermoelectric plants, and the general population. For this purpose, another logistical effort is required involving virtually all modes of transportation available, such as pipelines, tank trucks, tank wagons, barges, and ships for transport by sea and by river. All this operation involves the creation of intermediate bases and the transport of hydrocarbons in significant quantities through cities, rural areas, urban centers, residential areas, always involving a hazardous product and its risks.

Transport and distribution activities have an influence in the levels of risks and hazards under which society is exposed. Risks are also extended to workers, authorities, and the organization or company responsible for hydrocarbon transport. It is necessary to know the technical characteristics of the product and each by-product transported and distributed, and also the existing technology for the safe execution of all tasks associated with this major logistical effort. The hydrocarbon transportation business itself establishes a significant level of risk simply because of the hazardous characteristic of the product that needs to be transported. Each transport and distribution activity, from the oil well to the gas station, must necessarily be performed under responsible risk management, developed primarily based on technical and operational knowledge about the activities and products involved.

3.3.3 Marine terminals (inshore or at shore)

Marine terminals can be located on islands or platforms built in sheltered water areas such as bays. They can also locate on the continent in port areas. An uncommon nomenclature uses the term inshore for terminals on islands or platforms in sheltered water areas and the term at shore for marine terminals located on the coast (mainland). These facilities receive ships with crude oil, refined oil, gas, and liquefied gas. Arms specially designed for this purpose transfer the product to the terminal, where the product is stored in tanks or transferred directly to distribution lines such as gas or oil pipelines. All terminals for hydrocarbon transfer need risk management monitoring from the design phase to the end of its operational activity. Liquefied natural gas (LNG) terminals, in particular, have requirements involving cryogenics safety, in addition to the specific safety technology for the oil and gas industry. This is due to the fact that LNG is transferred at $-162°C$ temperature. LNG fires cannot be fought by conventional methods. The water used in fires at room temperature, when thrown directly on the product, heats it up and as a result increases LNG evaporation and the fire intensity will be increased. This requires special strategies for fire response systems, specific to the characteristics of the LNG product (Fig. 3.11).

Some terminals, most often located on the mainland or on islands, in addition to being designed to transfer product between ship and terminal, also store hydrocarbons and therefore meet additional protection requirements, applicable to storage facilities.

ANDRé MOTTA DE SOUZA/BANCO DE IMAGENS PETROBRAS

Figure 3.11 Terminal for transferring LNG. In addition to the safety requirements related to operation in the presence of hydrocarbons, specific safety requirements for cryogenics must be considered. *LNG*, Liquefied natural gas.

3.4 Knowing process safety

The processes of the oil and gas industry are diversified, which adds some complexity to the management of its risks when we analyze the production overall chain. The term process safety has been associated with technologies to protect human life, the environment, and property, not only in the oil and gas industry but also in the chemical industry in general. Process safety involves solutions based on knowledge of chemistry, physics, thermodynamics, biology, among other disciplines. Next, we establish an overview of the main topics related to process safety, as a theoretical basis for the implementation of technical solutions in the field of safety engineering and risk management.

3.4.1 Loss of containment (liquid and gas leaks)

The integrity of pipelines and equipment that transport and process oil and gas need to be maintained to ensure the containment of hydrocarbons within the means of processing and transportation. In the event of a containment failure, part of the hydrocarbons is released under the form of liquid or gas leaks, and depending on the accident conditions, an explosive atmosphere may form or cause a fire with possibility of human, environmental, and material losses.

There are different forms of loss of containment and, consequently, various types of thermodynamic transients associated with these losses, depending on the affected equipment and the area where they are located. A form of loss of containment is leakage of gas through an orifice. If this happens, for example, in an oil and gas separator, the pressure of about 1 bar will generate a compatible mass flow. If a hole of the same size occurs in a riser (pipelines connecting the oil well to the rig) of gas lift (gas that is injected to reduce the density of the oil in the processes of facilitating the raising of the fluid), the pressures involved may be about 100 bar and the mass flow generated will be much more critical. If the mixture is able to ignite during this type of leak, it will produce a jet fire, which is an accident that can lead to serious consequences. The study of gas leakage scenarios can be done using tools for simulation of dispersion of gases (e.g., computational fluid dynamics—CFD—or analytical codes), also taking into consideration process data, equipment characteristics affected, and environmental conditions. Other technical considerations are important as well, such as losses in the leakage flow as the accident develops. The flow reduction is associated with pressure drop that normally occurs during the leak itself. Other losses due to friction during the flow also need to be considered.

Liquid leaks are also a threat to the process safety. In the case of crude oil the liquid always contains a certain amount of gas (associated gas). The gas can be partially released into the atmosphere. For instance, if the loss of containment occurs through a hole at the bottom of a tank, the liquid leakage could form a puddle, thus generating the risk of a pool fire scenario. Computational tools, such as CFD, can also be used in this scenario to study the fire propagation as well as the dispersion of toxic smoke. In addition, there are leaks with two-phase mass, that is, simultaneous leakage of liquid and gas. Most gaseous hydrocarbons are denser than air. Methane is an important exception—the main component of natural gas—that is lighter than air. During dispersion, the gas cloud is dissolved, for the most part, to a concentration that doesn't represent risk of fire ignition. But there may be regions in the gas clouds with different gas concentrations, being of fundamental importance for safety to investigate regions where concentrations may be between the lower flammability limit and the upper flammability limit. These are the regions at risk of fire and explosion. An investigation can also be conducted through studies of dispersion of gases for specific scenarios using CFD.

Losses of containment (leaks), for gas, liquid, or biphasic gas, are always the permanent focus of attention for risk management activities, from the design to the operation. It is important to understand that a loss of containment also triggers a thermodynamic transient inside the leaking tank or pipeline and it causes cascading effects that are propagated to other interconnected equipment and to the environment where the fluid is released. The consequences of such transients need to be investigated because the concurrency of events can aggravate the initial scenario conditions in a catastrophic manner. From the conceptual design to the operational phase, preventive resources must be provided to reduce the risk of loss of equipment containment (leaks). Pressure, volume, mass, temperature, levels, and other operational parameters will change under the influence of the loss of containment, what makes it necessary to study the most likely scenarios in detail, aiming to protect facilities from catastrophic accident escalation, thus reducing the human losses, environmental and material costs to the absolute minimum.

3.4.2 Stable or explosive burning combustion

A chemical reaction can result in combustion when it releases, in significant proportion, high amounts of energy per unit of mass in the reaction. However, the type of combustion depends on many other factors such as reactivity of the oxygen/fuel mixture, capacity of heat transfer to the environment, degree of confinement, ventilation, and amount of energy of the ignition source. Variations of influence of these parameters in a scenario can make a difference in the outcome: either steady or explosive burning and either deflagration or a detonation type of reaction.

The heat generated by combustion depends on the fuel type, but in general, oil products generate twice as much heat than, for example, dry wood (considering the heat released per kg of fuel). The flame temperature associated with the burning depends on the heat transfer from the flame to the environment. The maximum flame temperature possible will be the one at the limit where the burning becomes adiabatic, with no heat transfer to the exterior, that is, all combustion heat is kept in the form of sensible heat in the reaction products.

3.4.2.1 Flash point

The safety of activities related to transport, storage, and processing of hydrocarbons is evaluated based on the flash-point parameter. Hydrocarbons can be simple compounds or a blend (a mixture of several compounds) as in the case of crude oil, for example. Vapors released by compounds or blends form a flammable mixture when they come in contact with an ignition source. Flash point is the lowest temperature at which a fuel, simple compound or a blend, releases sufficient vapors to produce a flammable mixture.

3.4.2.2 JET fire

It is a combustion scenario where the loss of integrity (small rupture, hole, or flanged joint failure) also establishes the loss of hydrocarbon containment, generating a gaseous or biphasic flow under pressure. In this type of scenario a literal, intense fire jet is formed after ignition that can have a far-reaching effect and lead to serious accident cascading effects, depending on the pressure and the size of the causing failure.

3.4.2.3 Pool fire

It is a combustion scenario in which the loss of integrity of a tank, pipeline, or other equipment allows the leakage of liquid hydrocarbon forming a puddle. In this type of fire the flame front contains a high proportion of hydrocarbon vapors and high temperatures, a situation that is aggravated by the increased evaporation of the puddle under the effect of the heat that is generated by the fire itself.

3.4.2.4 Fireball

The term fireball refers to the rapid combustion of a significant mass of hydrocarbons resulting from a catastrophic (large-scale) loss of containment. In this type of scenario, leakage doesn't occur through a small hole or a crack, but there is a sudden release of a large hydrocarbon mass that is burned in a few seconds.

3.4.2.5 Boiling liquid expanding vapor explosion

This type of combustion can be considered a special case of fireball, in which the hydrocarbon related to the accident, despite assuming gaseous state under ambient conditions, in the process in question it is stored in liquid state. BLEVE (boiling liquid expanding vapor explosion) can be triggered by external heating of a tank containing hydrocarbon. As the temperature rises, the hydrocarbon generates internal vapors, which considerably increases the internal pressure beyond the strength limit of some components and thus creating loss of containment (leakage). As the internal pressure is very high and the hydrocarbon in this case, based on its characteristics, vaporizes under ambient conditions, the tank inventory quickly expands after being released due to the loss of integrity, creating a large cloud that goes into catastrophic combustion after coming in contact with an ignition source. LNG is a cryogenic product and, although it occasionally burns like a fireball, it doesn't technically generate a BLEVE. It is not expected that BLEVE events occur in offshore rigs due to the nature of the products, processes, and offshore safety systems, but fireballs can occur in the event of sudden release of gas or in biphasic inventory.

3.4.2.6 Vapor cloud explosions and flash fire

Vapor cloud explosion (VCE) is the explosive burning of a vapor cloud generating a shock wave, unlike flash fire, which is also the burning of a vapor cloud, but that doesn't generate a shock wave. One or more sources of loss of containment can contribute to the formation of a hydrocarbon cloud in a proportion that can create an explosive atmosphere. One of the main factors that influences the cloud combustion type, either VCE or flash fire, is the degree of confinement and obstacles that intensify turbulence, which increases the likelihood of occurrence of VCE. A single leak can generate both combustion phenomena. One region of the cloud may enter into combustion as a flash fire, while another in the same cloud may enter in combustion as VCE. In both cases, the consequences are serious and pose a threat to human life, either by a shock wave (VCE) or by burns caused by a flash fire. The propagation of combustion of gaseous clouds can lead to a deflagration condition, when the speed of the reaction is subsonic, or to a detonation condition, when the speed of the reaction is supersonic. During detonation, a shock wave with considerable destructive power is generated.

Explosives exhibit propagation by detonation in general, while accidents with formation of hydrocarbon clouds in the industry most often generate deflagrations.

3.4.3 Safety in physical and chemical operations with hydrocarbons

The transportation and storage of hydrocarbons in tanks require special attention regarding the vapor pressure generated within the tanks due to temperature fluctuations. Designs include devices for controlled vapor relief to avoid a more severe scenario, which is the rupture of the tank due to catastrophic failure. Certain types of hydrocarbons need to be stored at above atmospheric pressure and this is also is taken into consideration for the assurance of operational safety. Static electricity is another relevant matter due to the risk of producing a spark and also the effects of solar radiation that also need to be taken into consideration.

Refining activities involve the circulation of hydrocarbons across processing units under various sources of hazards and risks. A wide range of specific technical knowledge of process engineering is required for the proper functioning of heat exchangers, cooling systems, cracking units, distillation towers, among other equipment. Risks need to be kept within acceptable levels with respect to hazards related to intoxication, asphyxiation, fires, explosions, and threats to the environment.

Process safety depends on the design of pressure vessels, pipelines, equipment supports, and other components. The operating conditions of vessels and equipment, and the hydrocarbon flow in pipelines need to be studied taking into consideration the basic parameters such as density, viscosity, and enthalpy. Not only during the design phase but also to maintain the operational routine, many other parameters need to be controlled and continuously monitored to ensure process safety.

Automation and instrumentation systems are developed for data acquisition and for control of important process-related parameters. Flow, pressure, temperature, and level are examples of such parameters.

Each product and by-product in the oil and gas production chain has specific safety requirements, according to their individual characteristics, in addition to the general requirements applicable to the safety of processes that involve hydrocarbons. Accidents related to dissolved gas contained in crude oil usually involve pool fires.

Natural gas, basically methane, is less dense than air and has low reactivity. This causes slow speed of flame propagation, which reduces or nearly eliminates VCE combustion (with shock wave generation). In general, accidents with natural gas generate flash fire combustion (without shock wave formation). On the other hand, LNG is a cryogenic product that associates other risks to traditional processes due to its characteristics.

LNG can be transported by ships, which requires to be transferred between ships and terminals, each of which with its own requirements and safety measures. LPG, basically propane, has energy content greater than natural gas and may cause an accident with BLEVE in certain scenarios. There are also a variety of products for which risks are associated with intoxication hazard. This is the case with chlorine, ammonia, hydrogen fluoride, and other hydrocarbon-derived or associated products such as benzene, toluene, xylenes, vinyl chloride, acrylonitrile, sulfide hydrogen, and carbon tetrachloride. There are also unconventional methods for obtaining hydrocarbons, such as shale oil and shale gas. These processes still divide experts' opinions due to the suspicion of very high risks related to the harmful consequences to the environment associated with this type of exploration. Each product in the productive chain of the oil and gas industry should have all the information relevant for safety documented in an Information Sheet of Chemical Product Safety. In some countries this document is known as Material Safety Data Sheet—MSDS. The following is an example of the main information registered in the GLP MSDS.

3.4.3.1 Identification of hazards

- Classification: Flammable gas—category 1 and liquefied gas
- Classification system used: Standard ABNT-NBR 14725–2:2009 (Brazilian standard) (corrected version 2: 2010)
- Globally Harmonized System for the Classification and Labeling of Chemicals, UN.
- Other hazards that do not result in classification:
- It causes asphyxiation by reducing the concentration of oxygen in the air. Contact with liquefied gas can cause frostbite. The product contributes to the formation of photochemical smog.
  Signal word: DANGER
- Hazard statements: Extremely flammable gas. Contains gas under pressure: it can explode under the action of heat.
- Precautionary Statements: Keep away from heat, sparks, open flames, hot surfaces. Do not smoke. Flame gas leak: do not extinguish unless the leak can be safely contained. Keep out of sunlight. Store in a well-ventilated place.

3.4.3.2 Composition/information on ingredients

- Common chemical name or technical name: Liquefied Petroleum Gas—LPG

Petroleum substance group: substances in this category contain mainly low molecular weight hydrocarbon molecules, which are the dominant hazard in petroleum hydrocarbon gases. Their physical and chemical characteristics require that they be maintained within strictly closed systems. Unlike some other gases that are present in the processes of a refinery, petroleum hydrocarbon gases do not contain inorganic compounds (e.g., hydrogen sulfide, ammonia, and carbon).

Synonym: Petroleum gas, liquefied Registration # CAS: 68476-85-7

Impurities that contribute to hazard: This product does not contain impurities that contribute to hazard.

3.4.3.3 First-aid measures

Inhalation: Remove victim to fresh air and keep at rest. Monitor respiratory function. If the victim is breathing hard, provide oxygen. If necessary, apply artificial respiration. Seek medical attention. Take the MSDS.
Skin contact: Remove contaminated clothing and shoes. Wash exposed skin with large amounts of water for at least 15 minutes. Seek medical attention. Take the MSDS.
Eye contact: Wash with running water for at least 15 minutes, keeping the eyelids open. If using contact lenses, remove them if it is easy. Contact a TOXICOLOGY INFORMATION CENTER or a doctor. Take this MSDS.
Ingestion: Not applicable (gas).
Most important symptoms and effects, acute or delayed: Hypoxia caused by asphyxia can cause fatigue, visual impairment and motor coordination disorder, impaired judgment, cyanosis, loss of consciousness, and, in severe cases, death. Contact with liquefied gas can cause frostbite, making the skin white or yellow, with a waxy appearance.
Notes for physicians: Symptomatic treatment should include, above all, supportive measures such as correction of hydroelectrolytic and metabolic disorders, in addition to respiratory assistance. In case of contact with the skin, don't rub the affected area.

3.4.3.4 Firefighting measures

Suitable extinguishing media: Compatible with chemical powder, alcohol-resistant foam, carbon dioxide (CO_2), and water fog.
Unsuitable extinguishing Media Water jets. Don't pour water directly at the spill location, as freezing may occur.
Specific hazards arising from the mixture or substance: Extremely flammable gas. Risk of explosion if ignition occurs in a closed area. Spontaneously explosive in sunlight with chlorine. Explosive mixture is formed with air and oxidizing agents. Combustion can generate anesthetic fumes.
Protective measures for the firefighting team: Self-contained breathing apparatus with positive pressure and complete protective clothing. Stay as far as possible or monitor the nozzles. If possible, firefighting to be done from upwind. Don't extinguish fire before the leak is contained. For large fires, use hose holders or monitor nozzles if this is impossible leave the area. Cooldown containers with large amounts of water until the fire has been put out. Remove containers from the fire area, if possible, without taking additional risks.

MSDSs are valuable tools for designers and operators. They provide much more information than those illustrated above and may even be composed of a set of 15–20 pages with relevant information for designers to create safe and compatible environments for each chemical product. MSDSs are also useful during the operation phase. Immediate access to MSDS allows the identification of the best approach to help victims or to handle emergencies in cases of accidents.

The responsibility for the safety of processes in the oil and gas industry lies not only with risk management professionals but also with all engineers, designers, and professionals from all areas of the industry's production chain. No one is better prepared than an expert from each field, equipment, and processing unit to know what requirements must be met for the operation to be successful in terms of results and safety. Risk management professionals work as part of an interdisciplinary effort, taking a broad and multidisciplinary approach to problems related to safety. In this sense, some of the process safety subject matters mentioned in this section will be revisited throughout this book, but always with a practical approach based on scientific knowledge and records of standards and technical guidelines. But, above all, we always consider the most important source of knowledge: operational and previous project experience.

3.5 Knowing operational practice (field experience)

The only way to get to know operational practice and obtain field experience is to actually work in the field. This can be accomplished through a visit plan, an internship, a specific training during a predefined period of time, or accumulated during years of routine work in the operational areas. The main field activities (directly associated with the operation of the facility) are maintenance, planning and control, technical operational support, inspection and testing, quality control, safety, and the operation itself (control room and in the field). Operational knowledge and field experience are essential for risk and safety management activities. Over years of practical work, it is possible to acquire the perception of how far the designed environment is from the actual built environment. With additional practical experience, it is also possible to notice the differences among the original, built, and start-up-ready delivered environments, and also the ever-changing environment, as the result of years of services and operational activities.

From the theory postulated in the conceptual projects, through the difficulties encountered during construction and assembly, in all stages, adaptations are required. Some systems and equipment don't meet the minimum requirements specified. In some cases, adjustments can be made to restore the conditions specified in the project. In others, this may be simply unfeasible. Some improvements developed during the commissioning phase may even mischaracterize the original project, and thus generating significant differences between the original design intent and what was actually commissioned (tested, approved, and delivered for operation). In the years along the operation timeline, new problems can only be identified through routine operation. In the operational dynamics, these problems are reformulated and in turn generate proposals for improvements, some of which might get implemented. Formal and informal operating practices that are deemed possible and acceptable get established based on their unique characteristics and become part of their own operational culture of a specific technological enterprise. This culture becomes indispensable to ensure operation continuity and the availability of the facility.

This is the reality of engineering activities. Despite all the calculation exactness, accuracy of the safety margins, and tolerances, only operational practice will determine what really needs to be done to operate a specific facility. And this is a continuous process, which spans from the conceptual project to the final decommissioning of the technological enterprise. A complex process and far less logical and accurate than many laypeople may suppose. This reality reinforces the importance of operational knowledge and field experience, especially for risk management professionals.

The human element is at the center of the complex web of relationships among branches of knowledge, companies, organizations, societies, countries, and other interests. With the participation of the human element the human error problem is introduced, with its consequences for the systems and equipment that need to be operated. It would be impossible the complete control of all the variables involved, but in terms of risk management, a strategy is needed to deal with the complex reality of the factors that may cause failures of both equipment and people. There are substantial scientific work and many researchers dedicated to the study of these problems. One of the most widespread and widely accepted models is the "Swiss cheese" model, in which each stage of a process (and by analogy, each discipline of a project) is represented by a slice of Swiss cheese. A set of aligned slices forms the complete process. The holes that are normally found in a slice of Swiss cheese represent the occasional errors and failures of each stage or segment of the production process. Each slice represents a step in a process, with its own errors and failures. But when the holes in the slices are aligned, then the errors and failures add up and thus increasing the chances of a catastrophic accident happening.

The Swiss cheese model is a representation of how failures and errors are present in different stages of a technological enterprise and how they can add up to the detriment of safety. But a more specific approach to engineering methods has led us to develop another model, which better represents operational and project activities, and more strongly based on operational knowledge and field experience. We call this model a "membrane diagram" (Fig. 3.12). In this diagram, we replace the slices of Swiss cheese with permeable membranes that represent technical barriers that

Figure 3.12 "Membrane diagram." Each discipline has a technical barrier to avoid the spread of errors. Each barrier is represented in the diagram as a permeable membrane. We use the term "membrane" because some types of failures are "invisible," difficult to detect. Certain types of errors and failures that are difficult to detect are able to permeate these barriers causing the accidental event. Additional barriers are represented as part of safety systems that are activated during emergencies. If these barriers also fail, the accident reaches a condition of extreme degradation in a "beyond design-basis accident" scenario.

attempt to prevent the propagation of errors in each typical project discipline and each operational activity. The rationale for this substitution was that with the evolution of the technological enterprises, failures and errors have become better observed preventively, and constant focus of attention by those involved in all stages of the processes. In this specific organizational context of the engineering processes the holes in a Swiss cheese would be a coarse representation of the threats to the safety of processes. In complex technological enterprises the failures that can lead catastrophic escalation of accidents can be much more subtle and more difficult to be detected. They may be related to the safety culture and human behavior. For this reason, we herein present the proposal for a new model, where the Swiss cheese slices are replaced with membranes permeable to human errors and errors resulting from voids in the safety culture.

Currently, in organizations and engineering companies, there are administration and risk management systems, also called safety systems or HSE (health, safety, and environment) or QHSE (also including quality) systems. Organizations, companies, and professionals involved in technological enterprises are continuously monitoring and identifying possible failures. For each little "Swiss cheese hole" that may pop up, a hand will appear immediately afterward trying to cover it, or at least a few pairs of eyeballs to report them.

Being more realistic with respect to the current state of development of technological enterprises, the errors and failures that represent important threats are not as easily visible as the "holes in Swiss cheese"; on the contrary, these are failures that are difficult be detected and identified. These failures are often related to the safety culture and error-inducing human factors. We consider the permeable membrane model currently to be the best representation to be adopted, through which failures and errors that are difficult to be detected in some cases, unfortunately, manage to permeate through it. When this happens in each stage of the processes of technological enterprises, then we have the scenario for a catastrophic event.

The membrane diagram (Fig. 3.12) presents the barriers against failures in each technical discipline of the technological enterprise, as a specific membrane through which errors and failures that are difficult to detect can permeate and reach other disciplines and their respective processes, and thus compromise the safety of the overall enterprise. When errors and failures permeate through all the membranes, the conditions for the catastrophic accident are established. In the sequence, following the catastrophic accident, the evolution of event comes in contact with other membranes related to barriers against specific errors in the safety systems designed to respond to emergencies. These membranes (from the safety systems) may also fail and, in such a case, the event will escalate beyond the level of degradation postulated by the project. We call this scenario "Beyond Design-Basis Accident" (Chapter 6: Emergency Control), where only some minimal and general guidelines remain available as the final attempt to provide a response to the catastrophic event.

Knowing the operational and field activities is to understand how the dynamics represented in the membrane diagram occurs in practice and on a daily basis, from the small local adjustment and the monitoring of tasks up to the decisions made in the main control rooms of the technological enterprises. The field experience is acquired by tackling the small and large failures that, unfortunately, are part of the operational life, and understanding how they develop and can be overcome. This knowledge is essential for professionals working in the risk and safety management field. Only with solid operational knowledge, it is possible to come up with realistic solutions regarding risk and safety management.

3.5.1 Safety barrier

It is a *physical* barrier capable of interrupting the flow of energy that can generate an accident (loss) in a dangerous scenario. If the barrier is intact, it by itself will not allow the flow of energy capable of causing the accident.

Note: The act of requiring a safety barrier does not constitute a "barrier" because the "safety barrier" must be a *physical* barrier, capable of *physically* interrupting the energy flow that causes the accident.

3.5.2 Professional work in operational activities and in the field

Industrial facilities need to be kept in operation, with the greatest possible availability, without compromising safety. The professionals who work in the operations are totally focused on meeting the following objective: to keep the facility running safely. This makes operational and field activities highly centered on practical actions, with less room for assessments and theoretical analyses, which may require time unavailable for such activities. Fast decision-making is required at all levels of operational activities.

Therefore professionals who work in the field must have previously acquired the minimum knowledge and skill set necessary to operate safely. Operational situations lack sufficient time for the study and continued analysis of problems. At most, it is possible to delay a decision for a short period of time for a technical consultation that necessarily has to be accommodated within such acceptable delay period. Decisions need to be made quickly to ensure operational continuity and safety. Some operational situations may require immediate interventions that depend on the operators' decision-making ability, even if based on incomplete information available on a given problem. If operators hesitate to make an operational decision despite the immediate need for it, such a lack of decision becomes, in a way, a kind of "consequence-based decision." Probably this is the worst decision of all: not to act. It can have the same effects as deliberate action against safety that cripples the business continuity.

The context of the operational activities turns professionals in the field into demanding and objective professionals regarding the promptness of the approach to technical problems, which may generate some conflicts with professionals from other engineering fields such as design, research, and development. For these, on the contrary, their work foundation is basically studies and in-depth analyses, and more time is spent on the solution of each individual problem. For operations professionals, an adequate solution is one that immediately maintains operational availability and safety, other more elaborated and optimized solutions that require too much time are deemed of secondary importance, putting two objectives at risk, namely, availability operational and safety.

Even with these components influencing the operational safety culture, there are situations where operations engineers need to give in and accept the need for more detailed analysis on certain problems. This happens when a high-risk decision needs to be made, which may impact both safety and operational availability, should a decision be made hastily. In such cases, risk management systems provide resources to streamline the analysis process, through techniques that allow for minimal verifications before the decision for process intervention is made. One of such techniques is the adoption of Permit to Work (PW) for tasks involving risks and/or outside of the operational routine.

The PW is obtained by filling out a form with the most important information about the technical task that needs to be performed. The information provided allows the evaluation of the type of service and which safety requirements need to be met. Through a PW system, it is possible to keep control over the system's operational situation or equipment that is involved in the task to be performed. Information about the isolation of equipment, pipelines, and valves, as well as information about the state of completion of the service can be monitored by the PW progress. Many accidents can be prevented through an efficient PW system. Depending on the complexity of the task, the preparation of the PW may require prior meeting with several experts. In this type of meetings, it is possible to identify the need for further in-depth analyses of the task. All this care prevents the execution of tasks without proper preparation and overlapping with incompatible tasks. It is also possible to control the isolation of systems that have interfaces with the task to be performed, as well as providing the safe return of these systems to operate.

3.6 Knowing the project routine

The routine of project activities is a counterpoint to the operational routine. Professionals involved in project activities work under pressure imposed by document delivery deadlines. The consistent completion of each project step allows safe progress of the construction, assembly, commissioning, and operation phases of the facility, which ultimately is the main objective of all efforts dedicated to a technological enterprise. The difference between project routines and operational routines is significant. Nonetheless, the membrane diagram shown in Fig. 3.12 applies to any technological activity, whether project, operation, or even research & development. In project activities, each discipline prepares its own set of documents according to the respective safety requirements. Thus each discipline creates its own barriers to hinder the propagation of errors. Projects are also subject to errors that are difficult to be detected such as those caused by error-inducing human factors and errors resulting from voids in the safety culture of the organization and society.

The project activities have their own routines and professionals with different profiles when compared with their counterparts from the operational activities. An advantage is that, while the technological enterprise is in the design phase, the consequences of any errors can be corrected with less impact in terms of costs and deadlines than during construction or operation. Risk management professionals need a vision of the project routines and the professional profiles of those working in this field. This justifies, as part of the training of risk management experts, a period working in a design office for the acquisition of practical experience; likewise, a period of operational activities is required. Below, we will present some preliminary information for guidance regarding the main aspects of the project routine, which are important to obtain practical experience in this field.

3.6.1 Project routines

The nomenclature used in a project breakdown may vary depending on the organization or the company responsible for the project, but in general, the projects follow a similar typical routine. A technological enterprise usually starts with a document from the client requesting the project. This document contains the problem statement, the basic requirements, and the objectives of the project. It may also include hypotheses, definitions related to concepts being adopted, and basic parameters that are fundamental as guidance for designers. This is how the "conceptual project" solicitation is made, which is generated at the customer's request. In a conceptual project the teams of the disciplines responsible for each part of the project prepare a descriptive document in which the technical content of the work plan for each discipline is presented. The parameters and equipment of greater importance that can influence the project viability need to be identified and described in the document. On the basis of conceptual project, specialists and administrators are able to assess costs, technical and economic viability and corroborate the feasibility of the enterprise continuity to subsequent phases.

If the conceptual project is considered feasible and approved, a second phase, usually called "basic design," begins. In this phase, more complete project teams are assembled, one for each discipline, under a general project coordination. These teams are composed of technicians and engineers who will work on the preparation of a set of more detailed documents. The set of documents include drawings and plans, technical specifications, descriptive documents, bills of materials and lists of documents, technical reports, and complementary studies. In the basic design phase, there is intense interaction among the disciplines, and collaboration among the teams, due to integration and compatibility requirements of the various systems. Also during the basic design, the most important decisions are made about the technical concepts that will shape up the technological enterprise throughout its life cycle. With more information available than during the conceptual project phase, basic design teams can research the different technical options for the solution of each problem and request complementary in-depth assessments (external if necessary). This happens mainly in technical assessments of the safety and risk management discipline.

Once the basic design is completed and approved, it becomes the input to the next phase, called "detailed design." In the project execution phase the enterprise concepts are well established and shouldn't undergo major changes. New teams are then formed and the designers' tasks are to work on the final design and incorporate detailed information into the documentation produced by the basic project so that, resulting in a new and more complete "executive project" at the end of this phase, which will allow the physical construction and assembly of the technological enterprise.

During all project phases, perfect coordination is required for the definition of schedules and deadlines, as well as the exchange of information across disciplines. There is an interdependence of the teams in each discipline. Part of the documentation related to one discipline can only be prepared based on documents from other disciplines. Risk management plays an important role in the dynamics of projects of oil and gas facilities. In general, the end result of documents and safety-related discipline activities is the integration across other disciplines. Safety documents and activities provide an assessment of the risks of the complete project, considering the relationships among all disciplines and their influences on safety.

3.6.2 Professional work in project activities

There are several engineering disciplines involved in the preparation of documents that make up an engineering project. This discipline-based organization, created in the project phases, is also repeated in other phases of technological enterprises, starting with the operational phase. Maintenance service can be subdivided into electrical, mechanical, instrumentation, etc. Operation is subdivided into several systems, and each system is associated with specific disciplines.

But it is in the project phases that the diversity and distinction among engineering disciplines are more evident. Each project team has its specialists with different backgrounds, and from this wealth of knowledge, a complete and efficient technological enterprise is born. We will present some aspects related to the contribution of each specialization, with emphasis on those related to risk and safety management. It should be remembered that the work of these specialists is not limited only to the project environment. The organization of teams based on disciplines establishes a structure with specialized technical sources. This structure will be repeated in the administration of the technological enterprise, from the design phase going through the operation, until the decommissioning and permanent facility closure.

All engineering disciplines have important contributions to risk management, within the subject matter of their respective specializations. Planning and execution of tasks related to the safety of their systems are part of each discipline. Engineers work mostly in a preventive manner with respect to safety and thus, they need to identify hazards during the project execution phase for their effective reduction or elimination. Under the human factor aspect, engineers need to create environments, through their designs, that reduce system-induced human errors. Also regarding human factors, the designs need to include means of mitigating the effects of unsafe behavior and human error, by controlling and reducing its consequences for the enterprise safety.

3.6.2.1 Civil engineering and architecture

Civil engineering is responsible for the structural design, construction, and integrity of buildings, bridges, and buildings of industrial facilities. It also encompasses the design of means of safe disposal of waste, liquids, and gases, preserving the quality of water, air, and the environment. In addition, it addresses safety issues related to rail, road, sea, and air transport modes. Architects design buildings, office, and dormitory accommodations, operating environments such as control rooms, service, and support areas. Civil engineers establish safety requirements for anchoring offshore rigs, mainly fixed platforms, and also perform structural design of related structural components and facilities. They design layouts considering technical safety requirements such as ventilation and clearance between hydrocarbon inventories, to avoid escalation of accidents.

3.6.2.2 Industrial engineering

It is a discipline whose main design considerations are related to safety requirements for industrial processes and operations, aiming at adapting the work to people and making the methods and environments for the safe execution of tasks. Industrial engineers usually complement their training with specialization in safety and occupational health, safety systems engineering, ergonomics, and human factors.

3.6.2.3 Mechanical engineering

It is a discipline that is most widely involved in the establishment of safety requirements for machines in general, boilers and pressure vessels, transport machines, and all types of mechanized equipment in the industry. Mechanical engineers have a long history of standardization of safety rules and specifications, and for some of these systems and equipment, they date back to before 1900. They are also responsible for structural analysis and design of mechanical components, pipeline design, gas and oil pipelines, construction and assembly activities, welds, inspection, quality control, and other areas related to safety. Air conditioning, heating, and ventilation systems are designed by mechanical engineers and include specific safety requirements to reduce the consequences of accidental events such as the spread of fire, smoke, and dispersion of gas clouds.

3.6.2.4 Electrical engineering

This discipline is related to the design of electrical safety devices, electrical interlockings and safety protections, electrical grounding, power transmission grids, besides other safety-related equipment and systems.

Currently, electrical engineers also use electronics and computing knowledge to create sophisticated protection systems for power generation systems.

3.6.2.5 Electronic engineering

This discipline is related to the design of data acquisition systems, process monitoring, automation, and control systems. It also includes the fields of instrumentation, fire detection, and alarms. Electronic engineers participate in the design of the interfaces (human–system interaction) such as screens and consoles located in the control rooms.

3.6.2.6 Chemical engineering

Its contribution to safety is associated with the hazards reduction in the projects of industrial plants. It also includes employment of techniques for the creation of process safety systems and development of processes for waste treatment for the protection of the environment. It establishes the operational parameters that limit the operation within acceptable risks.

3.6.2.7 Risk and safety management engineering

This discipline is devoted to the application of scientific knowledge and the engineering principles, as well as the creation of methodologies for eliminating and controlling hazards. Risk management experts need to have knowledge of and be very familiar with different engineering disciplines, which make them multidisciplinary professionals. The training and development of the technical background for this type of occupation require years of previous experience in their original area of the basic engineering field, complemented by years of specialization in the recognition and control of hazards. In addition, risk management experts need the ability to work with professionals from other engineering disciplines and also have knowledge and work with professionals from different other fields, such as psychology, people management, technological management, administration, biology, and sociology. Risk management experts also work toward the minimization of all types of losses so that they remain within acceptable levels, or at least as low as possible in terms of economic viability. They often participate in accident prevention programs, work in risk analysis, accident investigation, consequences analyses, and other safety studies.

Manageable risks of a technological enterprise are partially transferred to third parties through insurance contracts. This also establishes and association between the risk management expert and insurance programs. Also related to risk management activities, many professionals also specialize in loss control management in the development of programs to prevent or minimize major business losses and to reduce the economic consequences for organizations operating in speculative markets. Controlled losses include damage to people's health, damage to property, fires, explosions, thefts, vandalism, terrorism, industrial espionage, environmental pollution, occupational injuries, and product defects.

3.6.2.8 Human factor engineering and ergonomics

Even though there are similarities between human factor activities and ergonomics, the latter is more closely associated with the physical characteristics of the human body, while human factors refer to the error-inducing environment created by each project. The experts in these fields apply the information obtained through biology and behavioral sciences to the design of systems and equipment that aim to reduce the risk of accidents and workplace injury. Its main objective is to improve the performance, safety, and satisfaction of the people involved with technological enterprise. These experts try to improve people's adaptations to equipment, environments, systems, workstations, and communication media. They also try to improve human performance and safety through the reduction of errors in the performance of tasks and stress reduction related to physical and psychological activities. Ergonomics is strongly focused on biomechanics and psychology, whereas human factor engineering places emphasis on the environment and the cognitive aspects related to the performance of tasks.

3.6.2.9 Fire prevention engineering

This is focused on system designs whose objective is to safeguard life and property against losses resulting from fires, explosions, and related hazards. Professionals in this field are specialized in prevention, protection, detection, alarm, fire control, and fire extinguishment. Such systems are applied to structures, buildings, equipment, processes, and other systems. These experts design escape and abandonment systems, which are the most important life-saving fire safety systems.

3.6.3 Safety systems design documents

Safety-related design documents can be distributed across disciplines or concentrated on a specific discipline, usually called "safety." There is wide variation in the organizational format and presentation of design documents. We present below a list of basic documents for an offshore oil and gas rig project (Table 3.3).

3.7 Lessons learned

3.7.1 Avatar for "experts" without operational experience

Professionals planning to work in the risk management field need to invest time and dedication toward the acquisition of genuine operational activities experience. Just as important is a practical experience period related to project activities, but nothing is as important as the operational field environment to provide genuine technical engineering experience. Some engineers manage to avoid field activities throughout their careers and prefer the comfort of offices. Despite that, they manage to climb the professional ladder within fragile organizational structures, often exploiting other people's experience. One of the easiest ways to accomplish this is to become a simple good rule-follower. The rules, standards, and guidelines are developed based on works by operators and technicians who learned

Table 3.3 Basic documents for an offshore oil and gas rig project.

Document title	Application/usual nomenclature
Area Classification-General	Maps of zones for classification of explosion risk
UFD – Fire Water Supply System	Fire water supply fluxogram
P&ID – Fire Water Pump Set (Diesel Hydraulic Unit)	Fire water pump fluxogram
P&ID – Fire Water Supply System	Fire water supply system pipe & instrumentation diagram
P&ID – Fire Water Distribution-Main Deck and Production Plant	Fire water supply system instrumentation detailed diagram
P&ID – Fire Water Distribution-Accommodation, Engine Room, Pump Room and Helideck	Fire water supply system instrumentation detailed diagram
UFD – Foam Supply System	Foam system fluxogram
P&ID – Foam Supply System	Foam system instrumentation diagram
P&ID – Foam Distribution	Foam distribution system instrumentation diagram
Rescue Boat and Davit	Used to rescue a person at sea
Escape Route	Subdivided into main and secondary routes
Inflatable Liferaft	Inflatable liferaft used when lifeboats are unavailable
Totally Enclosed Lifeboat And Davit	Lifeboat after the helicopter, the most important means of abandoning an oil platform
Safety Signalling	Safety signaling to meet international maritime safety rules
Fire Fighting Equipment	Firefighting equipment set and accessories available for firefighting brigades
Life-Saving Equipment	Life-saving equipment set and accessories available to assist anyone on board to survive at sea
Safety Philosophy	Safety philosophy standards, rules and regulations that represent the project's strategy to respond to emergencies and mitigate risks
Safety Applicable on Machinery Area	Safety requirements on machinery area
Fire Protection for Machinery Hoods	Protection for confined machinery
Passive Fire Protection System	Passive protection
Water/Foam Fire-Fighting Systems	

(*Continued*)

Table 3.3 (Continued)

Document title	**Application/usual nomenclature**
	Water/foam firefighting systems is the description of the entire firefighting water system and its action strategies
CO_2 Fire-Fighting Systems	Description of CO_2 systems
Safety Data Sheets	Safety requirements specific to each installation location
Diesel Hydraulic Fire Water Pumping Unit	Diesel hydraulic fire water pumping unit data documentation
Jockey Pumps	Pressurization pump
Area Classification Data List	List of equipment that classifies the area in which they are located
Preliminary Hazard Analysis (PHA) Report	PRA report—preliminary risk analysis
Evacuation, Escape and Rescue Analysis Report	Escape and abandonment risk analysis report
Gas Dispersion Analysis Report	Gas dispersion risk analysis report
Fire Analysis Report	Fire risk analysis report
Explosion Analysis Report	Explosion risk analysis report
HAZID – Report	Hazards identification risk report
Smoke Dispersion Analysis Report	Smoke dispersion report
Hazop Report	Hazard operability risk report

by trial-and-error that certain things don't work, in other words those who actually try to perform an operational engineering task and document what works and what doesn't in the normative technical documentation. We can say that *the rules, standards, and guidelines are the records made by those who know how to engage in engineering activities, so that those who still don't will not repeat the mistakes already made in the past.* Some engineers manage to present themselves as risk management experts without having ever worked on the operation of a technological enterprise. In general, these engineers specialize only in rules and compensate their lack of operational experience with the use of the experience of other engineers (manifested in the standards), and thus they become totally dependent on the standards and insecure in their scientific arguments, despite all their attempts to cover up their deficiencies. It is possible to build an apparently successful career with this wrong posture. But it is impossible that throughout the career of this type of professional, an enormous embarrassment caused by such limitation will not happen. The worst embarrassments are the mistakes made against people's safety, the environment, society, and property. When facing desperate situations, these professionals might seek help from other professionals with real operational experience not as a gesture of humility but to use them as shields. In this type of situation the engineer who lacks operational experience, albeit being considered a senior risk management expert, needs an "avatar."

At an event where large companies from the oil and gas industry gathered to share technical knowledge related to the risk management field, one of the participating companies (the host) hired an "avatar" to present their contributions. The hiring idea came from the engineer who was in charge of organizing the event. The operating companies' representatives had previously exchanged topics of interest for discussions at the meeting. It was an international event where several countries would be represented. As the topics of interest were getting submitted, the event organizer realized that the interest by the external experts was in discussions related to specific operational aspects, for which standards lack definitive guidance. That is to say that the experts were interested in matters where recent operational experience could make it possible to include proposals for changes to existing standards.

Due to little operational experience by the event organizer, the engineer started to feel insecure to lead the meeting, especially for being an employee of the organizing company and host of the event. On the verge of despair, the engineer called a meeting with the technicians and risk management professionals to plan a strategy to make feasible the company's leading role in the meeting. But the company culture, built over years, was based on the predominantly legal use of the technical standards. The company meetings resembled homeowners association meetings, where residents discussed the rules, especially internal rules, seeking their fulfillment or to exploit the loopholes.

This had been the company's tradition for decades dealing with safety-related matters. Operational professionals mainly avoided risk management administrators because of their legalist stance, whereas risk management professionals avoided contact with field engineers as much as possible. In an overwhelming majority, risk management professionals were people with no field experience whose work always consisted of exploring their ability to use standards, even without practical knowledge regarding the purpose of those standards. It was not unusual when an overly meticulous, very detail-oriented engineer regarding rules, almost popularly considered "annoying" to be identified as someone with the ideal profile to work in the field of risk and safety management. An unfortunate error! A professional working in this field needs to be aware of the importance of overcoming risks and take it as a challenge. One needs to know that only multidisciplinary technical knowledge about the task to be performed will allow the identification, reduction, and, in some cases, even elimination of risks.

The possibility of having to discuss technical matters, without being able to use the enormous amount of internal rules as references, generated a real panic in all the team members preparing for the participation in the event. After all, if an expert from one of the other companies asked a question or presented a technical argument for discussion, the expert team from the organizing company wouldn't know how to carry on with the discussions barring their bureaucratic rules, based solely on the rule-interpretive methods. In general, technicians discussed their opinions internally with statements such as "According to the internal standard xyz we must do this," or "it cannot be done because there is a standard kwz that doesn't allow it." Fearful, the technicians then decided to hire an external professional (an "Avatar"), well known in the business, and with a proven track record of operational experience prior to his specialization in risks and safety management.

Finally, the day of the event had arrived. The "avatar" had already learned about the subject matters for several weeks in advance. Everyone was gathered in the auditorium and the opening presentations had already wrapped up. It was then when the technical work session started, where each company made a general presentation of their topics of interest for discussion, and the most important part would happen afterward: direct questions among the companies' experts. The host team didn't submit any questions. After all, according to the company's prevailing culture, the source of the answers were their own rules, and experts from other companies couldn't help much in this regard. At the same time, the experts from the other operating companies were eager to ask technical questions so as to understand the rationale for certain strategies adopted by the host company.

When the main part of the session related to direct questions got underway, the experts team from the largest guest operating company asked the first question, specifically to the host company's team, which was actually unprepared for this type of event, despite not openly acknowledge it. The host company had previously designated the "avatar" hired for the event as the one who would answer any questions on its behalf. But the question asked by the main operating company complicated their whole strategy. Offshore platforms have a very important equipment as part of the abandonment plan in an emergency. This equipment is called lifeboat and serves to remove people from the rig when the accidental scenario escalates beyond a critical point. There are some worldwide-accepted concepts of lifeboats, but the host operating company had excluded a specific type of lifeboat (free fall) and did it systematically through its internal standards, which was polemical and generated many internal discussions. Oil tankers that were operating companies themselves had already been using the same type of free-fall lifeboat regularly because the risk management teams were different and the vessels were not subjected to offshore rig standards. But internal standards related to offshore platforms were prepared by the small group that was present at the event, and since they had no field experience and were not receptive to the opinion of field engineers, the group had considered, based on their own experience, that this specific type of lifeboat was dangerous.

Finally, the leader of the largest operating company participating in the event asked the following question: "Why does your company prohibit the use of this type of lifeboat on its platforms if it has never being used, so the company has no historical data that can undermine its safety, given that this type of lifeboat is widely accepted by the other companies?" At that moment, the "avatar" couldn't provide an answer and requested that the team that hired him answered the question. It was a big embarrassment, because the expert hired for the event, being there in the professional capacity to answer questions from the other operating companies, was unable to provide a technical justification for the decision contained in the company's internal standards, which lacked any technical basis. And that was exactly what the question had exposed. Amid the embarrassing atmosphere, the engineer had no way of escaping from the attendee's gaze and the microphone that was put at her disposal. Very insecure, her answer was limited to: "those were internal standards," and thus exposing all the technical fragility of the risk management team established by a legalistic safety culture and poor technical operational knowledge. At times when technical arguments become vulnerable, risk management teams dominated by a legalistic culture often use the naive argument that the subject is "confidential" for security reasons. But for the more seasoned interlocutors, this type of response actually means: "we cannot provide an answer due to our lack of technical and operational knowledge."

Therefore all those who plan to specialize in risk and safety management need to consider the valuable investment in their career development that is the work related to the operational field. Only experience in field activities and complementary experience in project activities can provide a solid foundation for the development of true experts. The field experience provides the minimum practical information for helping in the decision-making process regarding whether or not to accept risks. The success of technological enterprises relies on these decisions, as well as the safety of people, the environment, society, and property.

3.8 Exercises

1. How can the activities that are part of the oil and gas production chain be subdivided?
2. What are the main types of upstream facilities?
3. What are the main types of downstream facilities?
4. What are the typical accidental scenarios that involve combustion in the oil and gas industry?
5. What is the main difference between James Reason's "Swiss cheese" model and the membrane diagram?
6. In a succinct form, what are the main phases of technological enterprise projects?
7. What are the main engineering disciplines utilized in the subdivision of teams into project activities and operational activities?

3.9 Answers

1. The term upstream refers to the part of the production chain that precedes refining. The entire exploration and production activities are routinely called upstream. The remaining parts that succeed the refining are identified as downstream activities. Refining, transportation, and marketing processes are part of downstream. There is also a third, less-used term called midstream. Two other terms widely used in the oil and gas industry, and which also serve to subdivide the oil and gas industry into sectors of different activities, are onshore, which refers to on-land activities, and offshore, which refers to maritime activities.

2. Drilling and completion rig, primary processing equipment, fixed offshore platforms, semisubmersible offshore platforms, FPSO offshore platforms, submersible offshore platforms, self-elevating offshore platforms, tension leg offshore platforms (TLP), SPAR offshore platforms, compliant tower platforms (CT).
3. Refining facilities, petrochemical plants, oil and gas pipelines, transportation vehicles such as tanker trucks, vessels and tank cars, maritime terminals, land terminals.
4. Jet fire, pool fire, fireball, BLEVE, VCE, and flash fire.
5. In James Reason's "Swiss cheese" model, the holes that represent the failures during a process are easily identifiable. In the membrane diagram, failures are considered difficult to detect and are related to the error-inducing environment caused by human factors and voids in the safety culture. In such a diagram, the failures permeate the protection barriers of each discipline until eventually causing the accident. Even on the heels of the accident, the safety systems that mitigate the event also have protection barriers against failures. If these are also permeated, the accidental event escalates to the condition of "beyond design-basis accident," with harm to people, damage to the environment, society, and property.
6. Conceptual project, basic design, detailed design, executive project. Such steps are followed by construction and assembly, commissioning and operation.
7. Civil and architecture (the latter is also called layout), mechanical, electrical, instrumentation and automation, process, risk management (also called safety systems), naval, structural.

3.10 Review questions

- What was the major problem of risk and safety management in the establishment of the modern oil and gas industry, how was this problem solved, and who proposed its solution?
- What do upstream and downstream terms mean?
- Define onshore, offshore, inshore, and at shore facilities
- Which segment of the oil and gas production chain was responsible for most of the accidents listed among the 100 biggest accidents recorded in the last 40 years?

- In which segment of the oil and gas production chain did most accidents happen between 2012 and 2013?
- What was the biggest accident in terms of volume of oil spilled in the offshore industry?
- Describe the main systems and equipment associated with the operation of drilling and completion probes.
- Explain the function and the meaning of the acronym BOP used in well safety equipment.
- What are the objective and the main equipment of a primary processing plant?
- Describe the basic construction characteristics of the main concepts of offshore platforms.
- What does the acronym FPSO mean and what is the equipment function?
- What is an SS offshore rig?
- What are the main limitations of self-elevating platforms and the related safety aspects?
- Describe the stages of the refining process and identify their main equipment.
- What are the main modes of transportation used for transport and distribution in downstream activities?
- What type of downstream facility adds the presence of cryogenic material as a hazard? Explain.
- Explain the meaning of the term loss of containment.
- Explain the difference between combustion with stable burning and explosive burning.
- Define flash point.
- Explain the difference between jet fire and pool fire, showing the characteristics that generate each of the scenarios.
- Explain the difference between fireball, BLEVE, VCE, and flash fire.
- What are the main safety aspects relevant for physical and chemical operations with hydrocarbons?
- Explain the membrane diagram.
- What is a permit to work and how is it used?
- What are the main phases of technological enterprise projects?
- What are the main differences between the tasks performed by engineers working in the operational and design fields?
- Describe the main disciplines in design and operational activities identifying the most important contributions of each discipline for risk and safety management.

CHAPTER 4

Hazards reduction

As defined in Chapter 2, Fundamentals of Risk Management, hazard is a threat to the technological enterprise that can cause losses. In Fig. 2.6, it was shown that hazard can be compared to the scenario of an operating theater with its characters. Depending on the sequence of events that occur within this scenario, there may be property, social and environmental losses, in addition to harms to people. The hazard is the scenario itself, while the risk can be associated with a numerical value, the probability of occurrence of the scenario caused by a given sequence of events. In the oil and gas industry, the main player in the hazard scenarios is hydrocarbon, the raw material that is the driving force in the industry and the main input in the production chain. The establishment of the oil and gas industry activities is akin to establishing hazard scenarios where the main element is hydrocarbon. Risk management associated with the hazards arising from hydrocarbon inventories is essential for the viability of this industry.

Although large hydrocarbon inventories are required in oil and gas facilities and processing plants, when other elements that make up the hazard scenarios are detected (fire, containment failures, leaks, and operational failures) it is possible to reduce, and in some cases to eliminate the presence of hydrocarbons in the accidental scenario through operational maneuvers and equipment layout strategies previously put in place.

Hazards reduction is one of the top most important components of the Risk Management Strategic Line (Fig. 4.1), second only to the "technical and operational knowledge" component, and higher in importance than the "removal of agents (people)" component. Technical and operational knowledge is considered the most important component because it is responsible for a safe operation, capable of preventing an accident. But, at the same time, hazards reduction is higher in importance than the removal of agents, because its actions are preferably performed by automated systems, as soon as the accidental event is confirmed, which may increase the chances of a successful escape and abandonment operation. The final decision for the abandonment of the facility is dependent on the

Risk Management in the Oil and Gas Industry.
DOI: https://doi.org/10.1016/B978-0-12-823533-1.00011-3

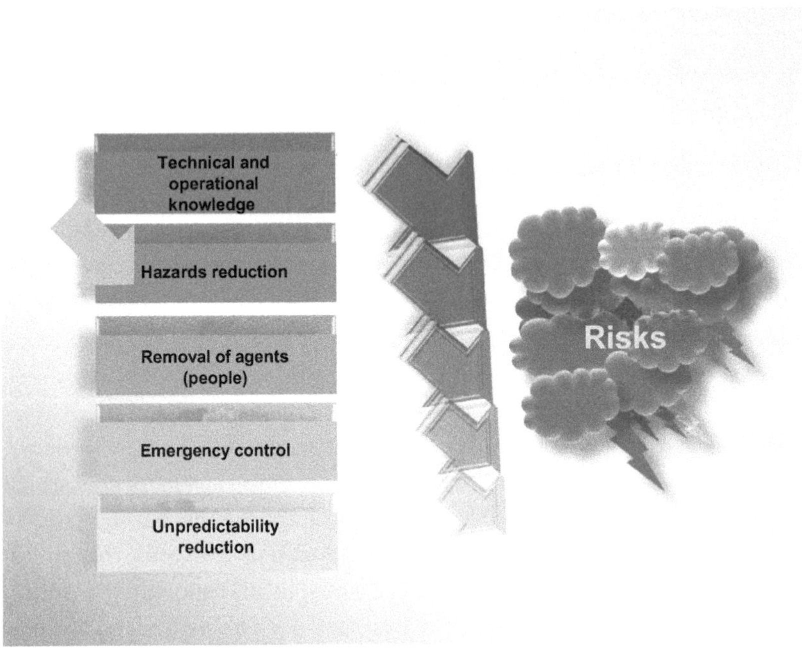

Figure 4.1 Hazards reduction is the second most important component in the risk management strategic line. While the diagnosis of the accidental scenario is performed to assess the need for escape and abandonment, actions to reduce the hazard must be taken, to increase the chances of success should the escape and abandon operation be required.

completion of specific diagnostics of the accident scenario. The hazard needs to be reduced within such time window.

An intrinsic characteristic of the oil and gas industry is its coexistence with the hazard of hydrocarbon inventories. These risks need to be managed, and one of the strategic components of such a management is the reduction of hydrocarbon inventories, to be started as soon as an accidental scenario is established. In this situation, the operational continuity becomes secondary and the hazard sources can be drastically reduced at the expense of the operation halt. It may additionally be interrupted in order to reduce the hazard through the management of the hydrocarbon inventory. We will present next, conceptually, how the hazards reduction can be achieved in the oil and gas facilities and processing plants.

4.1 Segmentation of the hydrocarbon inventory

One of the most important techniques for reducing the hazard of hydrocarbon inventories is the segmentation of these inventories as much as possible. Onshore facilities store large amounts of liquid and gaseous hydrocarbons in areas composed of large tanks and containment basins to accommodate hydrocarbons in the event of leaks. The positioning of the tanks in the area preserves safety distances, keeping the segmentation of the inventory, and hindering the spread of fires.

Pipelines may also contain significant inventory of hydrocarbons and so they are also considered as part of inventory segmentation strategies. Shut-off valves are located along the pipelines so that they can be shut off during an emergency, thus eliminating the communication between inventories and limiting the spread of fires. Segmentation difficulties are much more pronounced in offshore rigs. A FPSO (floating production, storage, and offloading system) platform stores large volumes of oil. Its tanks are located in the hull area, under the main deck, where there is usually a primary processing plant. In the event of oil spills into the sea, environmental damage is a major threat, and if a large-scale fire breaks out, people will have limited options to protect themselves within the unit, since as the result of its isolation, the oil cannot be disposed of, even during a fire. Accidental scenarios, mainly due to fire in offshore rigs, may require the abandonment of the platform and exposure of people to another risk scenario: survival at sea.

4.1.1 Layout techniques

The objective of the oil and gas facility layout design is the distribution of equipment, modules, stocks, operating units, machine rooms and other rooms in order to facilitate the operation, preserving the best possible safety conditions. Some techniques are widely employed in projects and enforced throughout the operational life of facilities in the oil and gas industry. We will present some general layout techniques of particular importance to safety. We emphasize that each type of hydrocarbon facility needs to be analyzed on a case-by-case basis, being insufficient just the use of general layout techniques without the consideration of the specific characteristics of each project. It is necessary to assess the need for development of specific solutions and techniques for each facility.

A general layout technique is the separation between "safe area" and "unsafe area." Alternatively, they can also be called "hazardous area" and "nonhazardous area." We prefer to use the latter nomenclature because the expression "unsafe area" can create interpretation problems, suggesting that such an area lacks safety features, which is not true. Thus we call a hazardous area where significant presence of hydrocarbons is required by the facility operational processes while a nonhazardous area is the area reserved for utility and support processes where hydrocarbon inventories are lower or nonexistent. In general, the hazardous area is the area where the hydrocarbon processing and storage plants are located. The nonhazardous area is one that maintains a safety distance or a physical protective barrier to the hazardous area. In the nonhazardous area are located the places where people spend the most time, such as offices, cafeterias, cabins, accommodation, leisure areas, and areas for remote operational activities (control rooms, radio and communication rooms, etc.). In hazardous areas, people should stay as briefly as possible, just long enough to perform essential tasks. Marine salvage resources infrastructure needs to be available in both areas, but considering the higher concentration of people in the nonhazardous area, it is the area to be given highest priority for the main allocation of marine salvage resources and establishment of meeting points (muster stations). In the facilities layout design, each equipment, module and operational unit is positioned either in the hazardous area or in the nonhazardous area, according to its characteristics and, mainly, according to its associated hydrocarbon inventory.

Another important layout technique regarding safety is to take advantage of natural ventilation wherever possible. Adequate natural ventilation significantly reduces the chances of the buildup of an explosive atmosphere. Therefore process equipment should be installed preferably in natural ventilation areas.

It is also necessary to take into consideration that overcrowding of equipment and supports can affect natural ventilation, resulting in insufficiently ventilated areas including those that would otherwise be considered adequate. The final layout is reached after an analysis is conducted to identify possible regions of poor ventilation. As these regions are identified, the designer does the necessary relocation of equipment until the possibility of regions with poor ventilation is finally eliminated. In summary, the layout design aims for the separation of equipment and facilities with significant hydrocarbon inventories, placing them in open and well-ventilated areas that are considered to be hazardous areas. Other equipment and facilities with lower hydrocarbon inventories are grouped in another area considered to be at lower risk when compared to the process areas, where it is also occupied by people during the operation, most of the time.

Special care should be taken to reduce connections in pipelines and equipment as much as possible. Each connection is a potential pathway for leaks that could cause containment failure and consequently represents higher probability of occurrence of an accident, such as explosion, puddle fire or jet fire. Equipment that operates with open flame, such as some types of furnaces, should not be installed in areas subjected to gas leaks, but instead should be isolated from the process plant and hazardous areas so that they do not become sources of ignition related to possible formation of an explosive atmosphere. The insulation is done through bulkheads whose fire resistance rating required is determined based on safety studies that assess possible accidental scenarios.

4.1.2 Blocking segmentation technique

A very important technique for the reduction of hazards is the segmentation of inventories through special shut-off valves. These valves are designed to block the interface between pipeline segments and equipment, thus reducing the possibility of accident cascading effect resulting from the connection between the hydrocarbon inventories of each segment. They are robust valves, protected to remain operational even when exposed to fire. They ensure leak tightness between inventories in each segment. Some operating companies use the shutdown valve (SDV) nomenclature to identify them (Fig. 4.2).

Another type of valve [Blowdown Valve (BDV)] is installed in each pipeline segment and equipment delimited by the blocking valves as part of the segmentation technique. This valve's function is to reroute the hydrocarbon inventory to a safe destination during an emergency. They are called relief and depressurization valves or, alternatively, BDV, see Fig. 4.3. The strategy of the blocking segmentation technique is to split lines, pipelines and equipment through blocking valves that will shut off, usually automatically, as soon as an accidental scenario is identified. After closing the SDV, the inventories for each segment will have become separated. But it will still be necessary to reroute the inventories of each segment to a safe destination through the relief and depressurization valves. The displacement of hydrocarbon inventories must occur in a programmed fashion, avoiding transients and undesirable overflow.

Figure 4.2 SDV during maintenance. *SDV*, Shutdown valve.

Figure 4.3 Relief and depressurization valve or BDV awaiting maintenance procedure. *BDV*, Blowdown valve. *Courtesy J.C. Melchior's personal archives.*

The location of shut-off valves and relief and depressurization valves should be studied in advance to facilitate operator's access in case of emergencies that require local manual operation. For this purpose, ergonomics requirements should be considered when choosing the installation location for each valve. The definition of the segments affects the number of valves to be installed. An aspect to be taken into account during segmentation is the layout of the equipment to ensure the separation of inventories during an emergency, both in terms of physical distance and hydrocarbon volume. For this reason, the application of the segmentation technique is a joint effort between risk management and layout experts.

During the application of the blocking segmentation technique, it is necessary to pay special attention to the location of tanks or vessels that are interconnected to the processing of oil and gas, among other particularities. The rationale for it is to prevent the equipment from being installed in the interior of structural components solely based on convenience. Even if under normal operation is not expected that the hydrocarbon inventory be significant in these containers, simply due to the interconnection with the process plant, a very high risk condition may develop, should any lack of operational control cause the improper transfer of significant volume of hydrocarbons to such equipment. There are records of serious accidents in which tanks and vessels interconnected with the process plant exploded within structural components like semisubmersible platform columns and pontoons. For this reason the equipment in question needs to be considered as possible repositories for hydrocarbon inventories, and should not be installed within structural components.

4.2 Disposal of the hydrocarbon inventory during an emergency

Through to the employment of the segmentation technique, the process plant and the entire hazardous area of the facility has its hydrocarbon inventory subdivided into smaller quantities, which are separated by segment blocking valves. This layout is achieved by a joint effort of risk management experts and experts in oil and gas facility arrangements. In the event of an emergency, the shut-off valves close automatically, ensuring the physical separation of the inventories of each segment. But hydrocarbon will still remain contained in the pipelines and equipment until the relief and depressurization valves for each segment are opened, allowing the preprogrammed sequential removal of inventories, as designed. In summary, the process plant is divided into sections through SDV. Additionally, each section has a relief and depressurization valve (BDV) to allow the removal of its local hydrocarbon inventory. But where will end up this massive volume of hydrocarbon to be discarded? Coming up next.

4.2.1 Pressure relief and depressurization

A safe destination for the hydrocarbon inventory is required. Liquid hydrocarbons are temporarily stored in limited quantities in tanks specially designed for this purpose. But the gas needs to be discarded immediately to avoid expansion and explosion, a scenario that may be aggravated by the fast heating of pipelines and equipment caused by a fire. Liquid hydrocarbons contained in the pipelines and equipment, when heated externally by the flames, will also generate significant volume of vapors and increased pressure. Consequently, gas relief is urgently needed. Gaseous hydrocarbon relief is the most effective method for depressurization and fast reduction of its inventory. This required special equipment for the safe disposal of large volumes of gas.

4.2.2 Controlled burning and dispersion

There are two main types of equipment responsible for the safe disposal of hydrocarbon inventory during an emergency. An option is the atmospheric vent (Fig. 4.4), which discards the gas inventory when the volumes are compatible with the location and natural ventilation of the environment. Atmospheric vent is designed to make controlled dispersion of the gas when it can no longer be contained in the pipes and equipment.

Atmospheric vent is applicable for the disposal of small volumes of hydrocarbon inventory. The objective is to transfer the inventory from areas where they may cause potential hazards to others with low possibility of accidents with losses. Hydrocarbon processing plants are usually built with relief systems for prevention of operational transients that may lead to uncontrolled increase in gas pressure, especially in situations where there is accidental overheating of the pipelines and equipment, as well as in vessels and pipelines supply overloads. In addition to the main atmospheric vent, storage and transport tanks and vessels also have their own atmospheric vents, designed for the relief volumes required by the equipment to be protected. In these cases, when the operating condition

Figure 4.4 Atmospheric vent (indicated by the *arrow*) in a process plant: there is no burning. The gas to be discarded is dispersed to the open area exploring the location of the disposal point (free of ignition sources) and exploring the conditions of natural ventilation.

reaches the nominal relief pressure, these atmospheric vents operate by releasing the gas to preserve the equipment safety. The gas can be released directly to the atmosphere without the need for a relief line, provided that the gas release site has adequate natural ventilation, compatible with the volumes postulated in the design. The vents need to be designed for fault-free operation when the release pressure is reached.

Because of the intrinsic safety-related aspects, atmospheric vents need to be designed with high operational reliability.

Atmospheric vents are usable only in cases where the estimated volumes of hydrocarbon release can be quickly diluted down to a concentration below its lower flammability limit. This implies that designers should adopt criteria based on comparison between the density of the hydrocarbon to be released by the atmospheric vent and the air density. During the design phase it is necessary to study the gas released through the atmospheric vent, considering possible formation of areas of different hydrocarbon densities in its surroundings (Fig. 4.5).

Another option for safe gas disposal is the gas flare (flare stack), equipment designed to burn the excess gas to be discarded in normal operation and emergency activities. The flare (Fig. 4.6) is a combustion device designed for controlled and safe burning of large gas inventories, either as part of normal processing activities or as an emergency response action. Many hydrocarbon processing facilities have a gas flare with continuous burning, for example, in onshore or offshore oil production facilities where the gas associated with the production process cannot be used economically due to transportation limitations, low volume available, or contamination issues. In other chemical processing facilities, flare burning may be intermittent. In such a case, the flare can be used to burn toxious gases or those an undesirable odor (off-gas), both of which are rejected in the production process.

There are several concepts and designs for different types of gas flares, some quite complex, based on location difficulty, volume to be burned, the type of gas, and interference with other areas of the facilities. Some offshore projects, due to the limited space available on the platforms, may require closer proximity between the flare stack and the helideck. That is an additional complicating factor, considering that the wind conditions, combined with the gases from the flare combustion, may create regions with different air densities, which in turn may result in difficult conditions for helicopter flights. It may even prevent approaching aircrafts from landing on the platform. There are engineers who specialize in complex flare designs who use computational tools to study the effects of burning and temperature generated in the process.

The basic operation of a flare system consists of receiving the inventory of gas to be disposed of, removing remaining oil and water and burning the gas safely. Fig. 4.7 presents a simplified didactic flowchart about the basic operation of a flare.

Initially, when the emergency shutdown (ESD) system, detects an emergency, it automatically sends signals to the process plant to close the shut-off valves (SDV) and, subsequently for the opening of the relief and depressurization valves (BDV). The pipelines located downstream of the BDV valves redirect the hydrocarbon inventory to be disposed of to the Knock Out Drum (KOD; final disposal tank) in which the liquid mixture (oil and water) still remaining is captured and later released through the drain systems (oily or clean) or reused in the process plant, if technical resources are available for this purpose.

Figure 4.5 Detail of the support structure of an atmospheric vent. The hydrocarbon inventory is safely discarded to the open area in a controlled manner, using natural ventilation and away from sources of ignition.

The KOD gas inventory can also be partially reused by the process plant, if a recovery system is available. The gas inventory to be disposed of goes to a seal tank that prevents backflashing (through a water seal).

Figure 4.6 Flare (flare stack), designed to make controlled and safe burning of large gas inventories.

It prevents the event of flare system malfunction. The flame backflashing needs to be stopped before it reaches the KOD, where significant hydrocarbon inventory is expected. The seal tank serves as a second barrier, in case of failure of the other adjacent fire prevention systems.

From the seal tank, the gas stream to be disposed of is redirected to the flare stack and rises by pressure and differential density until it crosses the backflashing prevention section. Finally, the gas is burned under the conditions postulated in the design, ignited by the flare pilot flame. The pilot flame burners are fed by a gas supply system with a built-in ignition source and air supply for mixture control and burning. The design of the assembly of devices responsible for the pilot flame needs to ensure high reliability. In the event of the pilot flame failure, several accidental scenarios may occur. For example, a large gas cloud can be formed, and regions with mixture concentration within the flammability limit range are susceptible to ignition by a heat source of sufficient energy.

The most important function of the safety-related flare system is for burning of the hydrocarbon inventory to be disposed of during an emergency. The flare generates a nonpremixed turbulent flame, which releases intense heat with associated hazards to people and facilities. When a safety

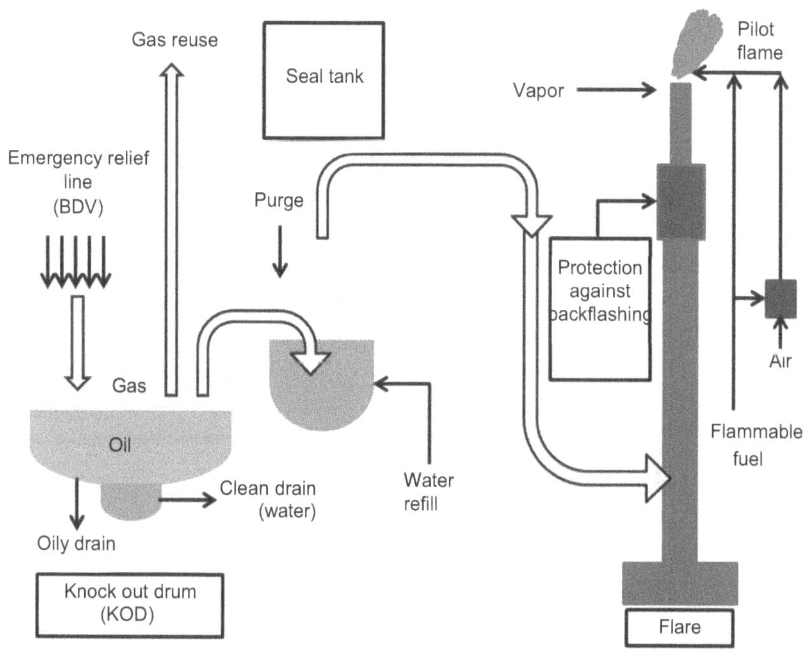

Figure 4.7 Didactic flowchart of the basic operation of a flare system. The system receives the gas inventory to be disposed of, removes remaining oil and water and burns the gas inventory safely. The pilot flame must always be kept lit so that the volume of gas disposed is immediately consumed by the burning.

study identifies risks to structural integrity and to people's safety caused by the heat generated by the flare transient burning, heat dissipation systems are installed for the protection of areas potentially at risk. Heat shielding systems can be composed of heat sink panels that absorb and dissipate thermal radiation safely (Fig. 4.8).

Operational flare failures may lead to serious consequences. One type of malfunctioning that may cause flare explosion is the presence of air in the pipelines and valves that redirect the gas for burning. The malfunction causes air to mix with the hydrocarbon flow, and this can generate undesirable levels of gas concentration in the flare, within the flammability limit range. The heated surfaces and the flame itself can be a sufficient source of ignition of the mixture, thus causing the flare explosion. Another type of malfunctioning occurs when there are problems in the pilot flame fuel supply. When the pilot flame goes out without the immediate and complete purge an explosive atmosphere may be generated inside the flare. Another critical problem for the safety of the flare is flame

Figure 4.8 Heat dissipation systems in the flare region. Also known as heat shielding, they are made up of panels similar to screens that protect structures and people from excessive heat radiation.

backflash. The air burners must be fed with discarded hydrocarbons within a supply rate limit compatible with flare heat delivery capacity to avoid backflash, which poses extremely high safety risks. This can be circumvented by the addition of a supplementary gas supply line, with the objective of complementing the gas flow if necessary. Thus the adequacy of the total thermal output required to maintain the proper operation of the flare burners is ensured, including when the gas flow rate is lower than the design value. Another possibility of flare malfunctioning is the blocking of the flow stream of gas due to freezing. This could happen when the flare purging subsystem uses water vapor as an agent or by freezing of organic compounds such as benzene and cyclohexane, which have higher freezing point than water.

But considering the overall system functioning, before the gas reaches the vent or the flare, the maneuvering of SDV block valves and relief and BDV depressurization valves need to be synchronized, as part of a broader a safety strategy, with the objective of providing pressure relief and depressurization of the entire facility. This strategy is known as ESD.

4.3 Automatic emergency shutdown

The objective of ESD is to provide safe shutdown of the process plant, segmenting and reducing the hydrocarbon inventory in the areas affected by the event in order to avoid emergency progression or minimize its consequences. ESD should happen automatically in facilities with significant oil and gas inventory, based on data collected by the field instrumentation. Detectors are able to locate leak gas or confirm the presence of fire. The instrumentation monitors operational parameters such as pressure and temperature, which may reach values beyond the safety limits. The automation system uses the available data to start the ESD process. It may also be started manually via a shutdown button, from the main control room and from strategic locations planned by designers. Designers must create safety interlockings capable of generating the automatic ESD signal directly from instrumentation and process data. Actions that can be performed automatically during shutdown are those that are not contingent on an operator's diagnosis. In general, automatic actions are the result of logical interlockings based on data that confirm the action immediacy. Additionally, whenever time is available for assessment, the most complex actions, which require more detailed analysis of the developing scenario, they need to be corroborated by the operators before they can be performed.

A means to avoid production interruption unnecessarily long due to important safety-related events that do not require complete shutdown is the strategy of subdivision of the shutdown into levels, creating the so-called shutdown levels, each of which compatible with the type of emergency being detected. The number of shutdown levels may vary according to each designer's strategy and according to the complexity and type of facility. Before a complete ESD takes place, shutdown levels may be sufficient for controlling and reversing the emergency scenario. An example of subdivision by shutdown levels is:

ESD:

Level 1: Equipment shutdown

Level 2: Process shutdown

Level 3: ESD (complete shutdown)

Level 4: Preparation for abandonment.

4.3.1 ESD level 1

Shutdown (automatic or manual) of a specific equipment, with partial impact on utility systems (main process support systems) or part of the process plant. Only the affected equipment is shutdown, but its removal from operation may generate some limitations, malfunction, equipment overload or operational demand in other equipment or systems.

Example of Trigger Condition: process variables reach values outside the set point range (design reference).

Example of Response Action: shutdown of the equipment that presents parameters outside the set point range.

4.3.2 ESD level 2

Complete shutdown (automatic or manual) of the process plant without affecting the utility systems (main power generation etc.), due to variation of any process variable (pressure, temperature, level, etc.) that has exceeded the reference set points. ESD Level 2 includes the shutdown of the shut-off valves (SDV), which triggers the start of the inventory segmentation process.

Example of Trigger Condition: high level in all oil storage tanks on a FPSO offshore platform, preventing the continued storage of the oil being produced (Note: High level in only one tank, usually, does not generate a ESD 2 signal, but only results in the interruption of the oil flow that fills the respective tank, while the oil produced continues to be stored in the remaining others tanks available).

Example of Response Actions: automatic shutdown of the shut-off valves (SDV). Permission for localized depressurization (in the area affected by the problem) by the relief and depressurization valves (BDV). This means that in addition to the segmentation of the inventory in the affected area (corresponding SDV valves being shut off), the automation systems enable the depressurization, that is, based on operational criteria the opening of the BDV valves to send the gas to the flare.

4.3.3 ESD level 3

Shutdown (automatic or manual) generated when there is a signal from the fire and gas detection system (FIRE CONFIRMED SIGNAL), resulting in the shutdown of the main power generation system and the process plant. Only those considered to be emergency systems remain operational (Note: Some designers subdivide ESD 3 in total and partial shutdown creating in fact a shutdown sublevel. In this type of design, ESD Level 3 is subdivided into partial ESD Level 3, when power generation is preserved, and total ESD Level 3, when it is turned off during the emergency. The ESD Level 3 subdivision is not recommended, because it can generate an error-inducing environment during the emergency, obfuscating the actual event and causing difficulties in scenario diagnosis).

Example of Trigger Condition: leakages with confirmed release of methane or H_2S gas. Fire confirmed in the risk area.

Example of Response Actions: shutdown of nonessential utility systems, shut-off SDV valves, opening of the BDV valves for depressurization (in the event of a confirmed fire), shutdown of the main power generation system start-up of the emergency power generation system.

4.3.4 ESD level 4

Activated manually, it begins the preparation for the abandonment of the facility. Applicable if the emergency has reached a safety degradation level that requires occupants to abandon the facility immediately.

Example of Trigger Condition: manual activation, following emergency diagnosis and risk assessment for people.

Example of Response Actions: in the case of an offshore rig all subsea valves are shut off along with complete depressurization of the facility. As a last resort, ESD Level 4 is generally started manually in most designs. Complete depressurization means a long down period until production can return to normal, causing significant economic impact. ESD Level 4 also needs to generate a general audible alarm for the entire facility, as guidance for people on board in "preparation for abandonment."

4.3.5 Example of an emergency shutdown sequence

Table 4.1 shows an example of a sequence of events that triggers an ESD. Shutdown can reach different levels, depending on the shutdown trigger actions. The response actions differ based on the shutdown level. We note that any level of ESD above 1 also includes the response actions for the lower levels. For example, if there is an ESD Level 3, not only the level 3 response actions will be generated, but also all the response actions associated with the levels 2 and 1. Each designer following the standards and applicable international references for each type of facility, will develop its own sequence of actions for safe shutdown of oil and gas facilities. The following example is shown for illustrative purpose only.

Table 4.1 Example of a sequence of trigger conditions for emergency shutdown (ESD), and associated response actions for each level of shutdown.

Possible trigger conditions	Possible response actions
ESD Level 1	
ESD Level 2 causes ESD Level 1. Equipment malfunctions. Process parameters and variables outside the operational safety limit range. Automatic stand-alone equipment shutdowns. Manual equipment shutdowns.	Interruption of the operation of the affected equipment. Secondary interruption of associated equipment caused by the failure of the first equipment. Limited functioning of the processes and systems associated with the failed equipment.
ESD Level 2	
ESD Level 3 causes ESD Level 2. Total loss of the main power supply without fire and gas leakage events. Failures in gas purge systems. Unacceptable oil level in Flare KOD. Instrumentation air underpressure. Hydraulic fluid underpressure in subsea manifolds (offshore rigs) and topside manifolds (affecting SDV valves). Failures in the inerting systems. Pressure values in oil export lines outside acceptable limit ranges. Manual action by the operator.	Shut off of SDV valves on the topside (offshore rigs). Shut off of dry Christmas tree valves. Authorization for partial depressurization (of specific equipment and pipelines) by the logical interlocking (automation).
ESD Level 3	
ESD Level 4 causes ESD Level 3. Combustible or toxic gas confirmed. Fire confirmed. Manual action by the operator.	Transfer of the main power supply to the emergency power supply. Start-up of the fire-fighting water pumps. Fire dampers shut in the HVAC ventilation and air conditioning system. Shutdown of nonessential power consumers. Shut off of valves of the subsea manifold (offshore rigs). Shutdown of gas pipelines. General alarm on the entire unit.
ESD Level 4	
Manual action by the operator.	Shut off all safety valves and isolation systems. Fully automatic depressurization of all process equipment and facilities (sequential opening of BDV valves).

Note: Each designer develops their own shutdown sequence, with as many levels as necessary to maximize the plant availability without safety risks.

Table 4.2 Consequences of failure of the main utility systems in oil and gas facilities.

Failures in utility systems and equipment affected	
System failure	**Equipment affected**
Power generation	Cooling water pumps, heat exchangers, ventilation and air conditioning system fans, cooling towers, combustion air. Instrumentation air compressors, process steam, cooling system compressors, vacuum systems, electrical instruments, valves with electric actuators.
Cooling	Process or utility condensers, process fluid coolers, lubricating and sealing oil coolers, coolant fluid supply in rotating machine jackets.
Instrumentation air	Transmitters and controllers, process regulating valves, alarms and emergency shutdown systems.
Steam feeder	Turbines that drive pumps, compressors, insufflators, combustion air feeders, power generator actuators, equipment that require steam injections, eductors, reheaters, heat exchangers.
Fuel supply (oil, gas, etc.)	Heaters, reheaters, power generator and pump motors, compressors, turbines.
Inert gas supply	Seals, catalytic reactors, purge systems.

4.3.6 Shutdown requires caution

The design of safety shutdown systems needs to consider in its strategy the importance of keeping the plant available as much as possible. But safety priority should not be threatened at any time. When a safety threat is confirmed, shutdown should become the highest priority. And when there is no confirmation of a safety threat, the shutdown scope should be such that the operational needs are met and the impact on the plant availability is kept to a minimum. Even when the failure occurs in systems not directly associated with oil and gas processing, the consequences may impact safety significantly. Table 4.2 shows an example of the possible impact of failures shutdowns on utility systems.

4.4 Lessons learned

4.4.1 Piper alpha hazards reduction failure

One of the most influential accidents regarding offshore safety projects was the fire, explosion and collapse of the Piper Alpha platform in the

North Sea, in 1988. After decades it still remains the offshore industry's accident with the highest number of fatalities. The cascading effect of failures involving safety systems and risk management made the catastrophe possible and led experts to deep reflection on the concepts adopted in the risk management and offshore safety projects. Some concepts developed after Piper Alpha have also started to be adopted in the projects of terrestrial facilities.

One of the most important factors responsible for the accident aggravation was the inability to interrupt the gas flow to the platform, even after the confirmation and communication of a major fire at the unit. One of the main objectives of the hazards reduction component of the risk management strategic line is to segment and, if possible, reduce the hydrocarbon inventory as soon as the accidental scenario is confirmed. The lessons learned from the Piper Alpha provide numerous relevant technical aspects be considered in projects of in oil and gas facilities. Technical aspects related to abandonment and escape routes, ESD, safety culture, permits to work and risk management, among others, got more attention from authorities and designers in the wake of the Piper Alpha accident.

There are many sources about the event and documentaries with footage from the accident. Survivors have reported the dramatic moments experienced on that offshore platform. There is a wealth of information about the Piper Alpha accident, nonetheless we will present herein an eight-step summary of the main facts, highlighting the technical aspects associated with the Hazards Reduction item from the risk management strategic line. The hydrocarbon inventory, not reduced at the start of the emergency but rather increased during the event, worsened the consequences of the unfortunate accident.

Step 1: The main losses

Piper Alpha was a fixed platform located in the North Sea, approximately 110 miles from Aberdeen, Scotland, with 226 people on board at the time of the accident, 165 of whom died (in addition to 2 rescuers). The platform was completely destroyed.

Step 2: The beginning of the catastrophe

The disaster started with a routine maintenance procedure. On the morning of July 6, 1988, a propane condensate backup pump in the process area required its pressure valve to be inspected. The work did not get completed by 6:00 p.m., and the workers requested and were granted a permit to work (PW) to be able to leave the outstanding service for the next day.

Step 3: Work shift transfer failure

Later in the evening, during the following shift, the main condensate pump failed. No one had noticed that an essential component of the line had been removed (safety valve) and they decided to start up the backup pump. The gas started to leak through the flanges that had been left open when the safety valve was removed.

Step 4: Design, maintenance and operation failures

Due to the high pressure, the gas leak, which was audible, ignited, and caused explosion, damaging the fire walls. The fire spread through the damaged fire walls, destroying a few oil lines and quickly causing a large volume of the stored oil to burn out of control. The automatic deluge system, designed to mitigate the accident, was not activated because it had been taken out of operation previously by the administrators.

Step 5: Other platforms supplied gas to the fire

About 20 minutes after the first explosion, at 10:20 p.m., the fire had spread and reached such a high temperature that the gas risers, coming from the other platforms interconnected with the Piper Alpha, started to lose their structural strength and failed, and also causing explosion of the gas sent by the other platforms. These steel lines were between 24 and 36 in. in diameter, and carried gas and flammable products at pressure of 2000 psi. When these lines collapsed, the result of the JET FIRE action increased the severity of the accident dramatically.

Step 6: Disorientation in escape and abandonment

The accommodation facilities were not smoke-proof, and poor training caused people to open and close doors repeatedly, making the situation worse. The conditions became so grave in the accommodation area that some people understood that their only chance of survival would be to go immediately to the abandonment points where the lifeboats were located.

Step 7: Failures in official procedures

However, they found out that all escape routes directed to the lifeboats were blocked by smoke and flames and, for lack of additional instructions, decided to jump into the sea in the hope of being rescued. Such a procedure was not included in the Piper Alpha safety standards, but it resulted in 62 lives being saved. The majority of the other 167 people died of carbon monoxide poisoning or suffocated by smoke inside the accommodation facilities.

Step 8: Complete destruction of the facility

The power generation and utilities module, which also included the fire-resistant living quarters, slipped into the sea. Subsequently, most of the platform followed suit. The accident lasted 22 minutes, from the beginning to the last instant described.

4.4.2 Lessons learned from piper alpha

The lessons learned from the Piper Alpha catastrophe are presented below, which are summarized in thirteen lessons. Decades after the accident occurred, designers and operators are still prone to repeat these unfortunate failures. Only through the development of a solid safety culture can the risks associated with the problems identified as part of lessons learned be continuously reduced.

Lesson 1: Regulatory control of offshore rigs

This accident was instrumental in the elaboration of the safety regulation for offshore rigs and was the basis for the creation of a Safety Case.

A Safety Case is a written document in which the company needs to demonstrate that a risk and safety management system (such as Environment, Health and Safety) is effective in the operation of the offshore rig. Safety Case was implemented in 1991. However, this practice has not yet been adopted by all operating companies.

Lesson 2: Adherence to the PW system

This is a system of forms designed to promote adequate communication among the parties involved and affected by any maintenance activity performed on the platform. At Piper Alpha, workers had become complacent about the system, relying too much on informal communication, and allowing gaps during the shift transfer. Had the system been used correctly, the initial leak would never have occurred.

Lesson 3: The quality of risk and safety management was critical

The Piper Alpha Cullen Report[1] was emphatic about the operating company's risk and safety management deficiencies. Managers had minimal technical qualifications, and established poor safety practices, as well as inadequate audits.

Lesson 4: Unavailability of post explosion protective equipment

The fire walls installed at Piper Alpha should have prevented the fire from spreading. However, they had not been designed to withstand shock waves from explosions. The first explosion alone was able to destroy the fire walls, allowing for rapid fire propagation.

[1] *The Public Inquiry into the Piper Alpha Disaster*, Cullen, The Honourable Lord, HM Stationery Office, 1990.

Lesson 5: Safety training needs

Platform workers were not adequately trained in emergency procedures and crisis management. There was an overall technical deficiency that prevented the lack of training from being overcome, and as a result the workers were unable to make a realistic assessment of the scenario during the crisis.

Lesson 6: Importance of audits

The company conducted regular audits, but without the required technical quality. Only a few problems, if any, had been addressed. But even though serious issues related to the pipeline corrosion in the deluge fire protection system and other serious problems were identified, in practice they were simply "ignored," even after being identified.

Lesson 7: Isolation in maintenance operations

The disaster could have been prevented if the pump whose safety valve was being serviced had been effectively isolated. Isolation is not achieved by simply shutting off valves, but it requires additional actions such as the insertion of a blind flange or removal of sections of the line.

Lesson 8: Fire and explosion protection

Fire prevention, automatic fire extinguishing systems, as well as direct fire-fighting systems are all of special importance in facilities such as Piper Alpha, since external help is not immediately available on an offshore rig.

Lesson 9: Safe temporary refuge (STR)

STR of each offshore rig needs to be supported by a system that ensures a breathable atmosphere by blocking smoke and provide fire protection. Escape and abandonment routes and effective abandonment locations should be defined through the Safety Case studies. The smoke barrier system from the STR should be provided with smoke and gas detectors with signal generation capability for automatic closure of fire dampers.

Lesson 10: Wind tunnel tests and computer simulations of explosions to be included in the design phase

Wind tunnels are useful for the assessment of the effectiveness of ventilation and the gas detection system. Computer simulations of explosions help to investigate the possible effects of different layouts in the response to shock waves resulting from explosions. This makes it possible to assess the efficacy of the explosion-safe bulkhead design.

Lesson 11: Escape and abandonment

It is required that more than one route be available for access to the helideck and to the lifeboats in order to ensure the escape and abandonment of the platform during emergency and crisis scenarios. Aiming to facilitate escape and abandonment in dangerous situations, escape routes need to have self-luminous or electroluminescent signs and provide heat protection, allowing the availability of access through smoke and flames. Secondary escapes such as ropes, ladders and nets should also be available as an alternative option in the lack of more sophisticated capabilities.

Lesson 12: Limitation of inventory in facilities and pipelines

The immense hydrocarbon inventory of the pipelines connected to the platform helps to increase or sustain a fire. Regardless of any technical problems, the platform should have been designed for the reduction of the amount of hydrocarbon (HAZARDS REDUCTION) at the start of the emergency.

Lesson 13: ESD valves

A suitable location for the emergency SDV and their redundancies are essential for interrupting the fuel supply in the event of a fire and to prevent it from spreading. These SDV valves need to be designed to withstand and be operational under fire, and be tested regularly, in addition to being accessible for their continuous monitoring.

4.5 Exercises

1. Why is the "Hazards Reduction" component at a higher level than "Removal of Agents (people)" in the "Risk Management Strategic Line?"
2. What is the difference between "hazardous area" and "nonhazardous area?"
3. What is the importance of natural ventilation for Hazards Reduction?
4. How do the SDV and BDV valves operate in the event of a confirmed fire or gas in the hazardous area?
5. Explain the difference between atmospheric vent and flare and in which situations each equipment is recommended.
6. Describe the basic operation of the backflow seal tank in a flare system.
7. Based on Table 4.1, is the main power generation system automatically turned off in a level 2 Emergency Shutdown (ESD Level 2)?
8. Does a level 3 Emergency Shutdown include shut off the SDV valves of the topside of an offshore rig as part of the response actions?

9. If in the design of an oil and gas facility all redundancies of the firefighting water pumps are powered by the main power generating system only, considering the logic of emergency power shutdown presented in Table 4.1, what would be the response of the emergency shutdown system in the case of confirmed fire in the process plant?
10. What is the main reason for designers to establish the subdivision by levels of emergency shutdown in oil and gas facilities?

4.6 Answers

1. In general, the "Hazards Reduction" actions are generated by the automated system and take place during a shorter period of time than is required to execute the "Removal of Agents." Thus the actuation of the SDV valves to block the gas, and by the BDV valves to direct the hydrocarbon inventories happen quickly and reduce the hazard in the plant, in preparation for the escape and abandonment operation. Due to the fast automatic actions to block and then depressurize pipes and equipment, the focus becomes on the "Hazards Reduction" actions immediately after the confirmation of the accidental scenario, even before the escape and abandonment, in case this operation becomes necessary.
2. In the "hazardous area" a significant inventory of hydrocarbons is expected and so are possible releases due to normal operation or containment losses (leaks). In the "safe area," on the other hand, no large hydrocarbon inventory is expected.
3. In the case of gas releases, the natural ventilation quickly dissipates emissions, avoiding the formation of areas with a concentration within the flammability limit range. Gas concentrations in this range can cause explosions and fires, should a spark with sufficient energy come into contact with the mixture. For this reason, designers, mainly those responsible for the layout documents, explore natural ventilation as part of the definition of the location of equipment, pipelines and modules of the process plant.
4. As soon as the instrumentation signals confirm the presence of gas or fire, the SDV shut-off valves close. Then, according to the sequential

plan defined in the design, the BDV relief and depressurization valves accept an opening signal and redirect the inventories to the safe destination.

5. The atmospheric vent does not burn the discarded gas inventory, but rather releases it in a safe place, away from possible sources of ignition. The flare burns the discarded gas through burners on its release to the atmosphere. The atmospheric vent is recommended when the volumes of gas to be disposed of are not too high, making it feasible for the inventory to be safely dispersed in the environment. The atmospheric vent can only be recommended when it can be installed in area with efficient natural ventilation and sufficiently distant from sources of ignition. The flare is recommended for large volumes of gas to be disposed of, including in places where the distance to the process plant is not ideal. Although the flare is less dependent on natural ventilation, it is still important for the definition of its location in the design, considering the formation of combustion gases and the heat generated by the flare burners flames.
6. One of the main types of accidental scenarios in flare systems is the flame backflow produced from the burners and through the interior of the flare equipment, until it reaches the KOD vessel. In this vessel there may be a significant volume of hydrocarbons susceptible to ignition. Near the burners is a flame backflow protection system, but in case of failure allowing the flame to backflow to the seal tank, the level of the sealing water will prevent the flame to reach the gas line from the KOD vessel, thus interrupting the burning chain reaction and avoiding a more serious scenario associated with internal combustion and explosion in the KOD.
7. Not necessarily. Although the total loss of the main power supply may cause an ESD Level 2, the automatic shutdown of the main power supply is not an automatic response action of an ESD Level 2. There are trigger conditions, such as failures in the gas purge system, which cause ESD Level 2, without necessarily disconnecting the main power supply.
8. Yes. All response actions at the lower shutdown levels (ESD Level 2 and ESD Level 1) are performed and the closure of the topside SDV is expected as a response action to an ESD Level 2.
9. Based on the shutdown logic presented in Table 4.1, we can confirm that the design is incorrect for supplying power to all fire-fighting water pumps from the main power generating source only. In the case

of a confirmed fire, an automatic signal associated with ESD Level 3 would be generated. The expected response action includes simultaneously turning off the main power supply and the start-up of the fire-fighting water pumps. As all pumps were incorrectly designed to receive power exclusively from the main power source, there would be no means to start up the pumps and fight the fire. The design would need to be revised to make a sufficient number of pumps available to fight the fire, even with the loss of main power. This could be done by specifying diesel pumps and/or pumps powered by the emergency power source, whenever it is technically viable.

10. The main rationale for the use of the strategy of subdivision in levels of emergency shutdown of oil and gas facilities is to maximize the availability of the facility under conditions of acceptable risk to the safety of people, the environment and property. Depending on the degree of threat posed by each emergency, intermediate levels of emergency shutdown may be sufficient to regain control and restore operational safety without the need for prolonged periods of downtime, usually required for complete emergency shutdown.

4.7 Review questions

- What is the main component of the hazard scenarios in the oil and gas industry?
- What hazards reduction technique preserves distances and includes shut-off valves to prevent the escalation of accidental events?
- What are the objectives of the layout designs related to the safety of oil and gas facilities?
- What is a "hazardous area" and what is a "nonhazardous area?"
- What types of equipment, rooms and modules are located in hazardous and nonhazardous areas?
- Explain the technique of exploration of natural ventilation.
- What precautions need to be taken regarding the definition of the positioning of pipe connections and connections between pipes and equipment?
- How can the positioning of the equipment interfere with the natural ventilation of the process plant?

- What is an SDV valve and what is its function?
- What is a BDV valve and what is its function?
- Where is the destination of the significant amount of hydrocarbons to be disposed of during a depressurization related to an emergency shutdown?
- What is the difference between the dispersion process and the controlled gas burning process in emergency stops (ESD)?
- Explain how an atmospheric vent works.
- Explain how a flare works.
- What are the main limitations of the atmospheric vent?
- What care should be taken regarding the location of the flare and the helideck?
- What is heat shielding?
- Describe the main operational failures of the flare system.
- What is the meaning of the term emergency shutdown?
- What is the purpose of the subdivision into levels as part of the emergency shutdown strategy?
- Give an example of an emergency shutdown system subdivided into four levels.
- Why does ESD Level 4 need to be manually deployed by the operator?
- What action is expected from the operator, when a threat to safety is confirmed?
- Give examples of the main equipment affected by failure of the following systems: power generation, cooling, air for instrumentation, steam supply, fuel supply and inert gas supply.

CHAPTER 5

Agents (people) evacuation

The effective evacuation of agents is the most important operation to reduce the number of victims in accidents. It is the third level of importance in the risk management strategic line (Fig. 5.1). Technical and operational knowledge is the basis for accurate emergency diagnosis and, in tandem with hazard reduction, allows for proper preparation for the successful evacuation of agents.

This chapter will present the hazard escape and scenario abandonment system, its strategies, main equipment and the importance of this system for the reduction of the number of victims caused by accident of any nature, including those unrelated to the oil and gas industry.

The adopted name "Hazard Escape and Scenario Abandonment System" is based on the nomenclature defined by International Organization for Standardization (ISO), no 13702 (Petroleum and Natural Gas Industries—Control and Mitigation of Fires and Explosions on Offshore Production Installations—Requirements and Guidelines). The term evacuation is often used by risk management professionals, standards and procedures. But, according to ISO 13702, the term evacuation refers to the planned method of leaving the rig during an emergency, whereas the act of effectively leaving the facility during an emergency is defined by ISO 13702 as abandonment. The meaning of the term escape also in ISO 13702 is the act of personnel moving away from a hazardous event to a place where its effects are reduced or removed. To escape does not necessarily mean to leave the accident scenario, while abandoning means exactly that, even if the abandonment of the accident scenario is to become part of another risk scenario, provided that it is not under the influence of the first.

Escape and abandonment systems can be designed using previous designs, as reference, and based on standards and procedures. But currently, given the importance of escape and abandonment systems for the reduction of the number of victims, new tools have been developed that offer higher quality design of escape and abandonment systems. Some of these tools are based on computer simulations. More sophisticated software products support consideration of some behavioral factors and

Risk Management in the Oil and Gas Industry.
DOI: https://doi.org/10.1016/B978-0-12-823533-1.00001-0

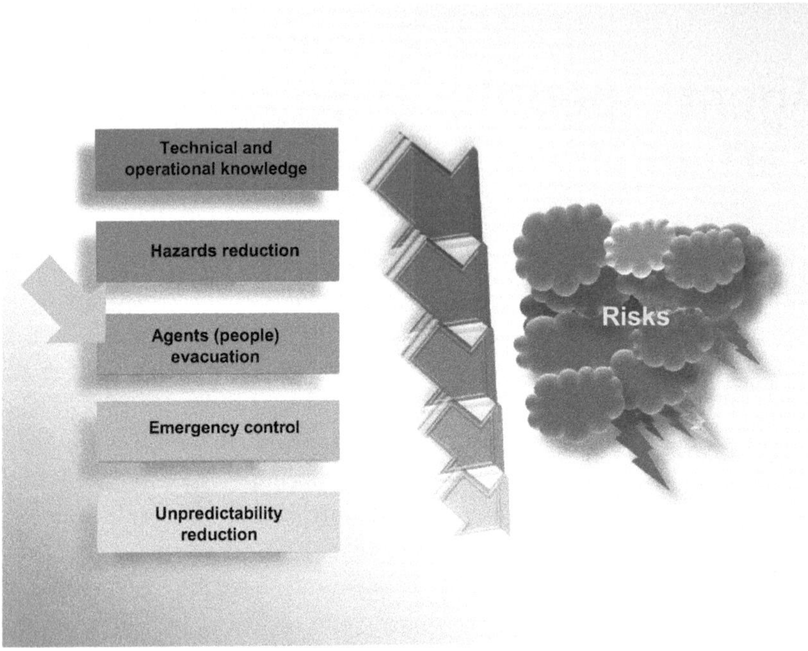

Figure 5.1 The agents evacuation is the third most important component defined by the risk management strategic line. Technical and operational knowledge allow the correct diagnosis and attitude in response to the accidental scenario. Hazard reduction reduces the risks arising from large hydrocarbon inventories in the shortest possible amount of time, providing better time and safety conditions for the effective evacuation of agents. It is the most influential component for reducing the number of accident victims.

cultural components associated with the projects to create more realistic simulations where it is possible to identify failures and opportunities for critical improvements to reduce the number of victims in the accident scenarios being studied.

Some types of computer simulations allow the study of people's response during the accident and, based on these results, it is possible to reassess the theoretical methods and plans. We recommend the adoption of the terms escape and abandonment as they are more accurate with regards to the nomenclature in ISO 13702 standard. The simulations study ESCAPE, which primarily refers to the act of people moving away from immediate hazard, that is, moving away from the place where the event has immediate consequences to people. Simulations also study

ABANDONMENT, which occurs after people escape from immediate hazard through escape routes (or "escape route from hazard," as adopted in some nomenclatures). If required by scenario conditions, people will effectively abandon the facility and, consequently, will be considered outside of the original accident scenario, even if included in other risk scenarios, but no longer under the direct influence of the original accident scenario.

Therefore we herein adopt the denomination of Hazard Escape and Abandonment of Scenarios System or, in simplified form, system of escape and abandonment, for its accuracy and coherence with the ISO 13702 standard.

5.1 Importance of the systems of escape and abandonment

Considering principle 1 of human factors (centralization of objectives in people), the safety system with the highest potential to save lives in an accident is the escape and abandonment system. When facing a hazardous situation, moving away from it is the natural attitude, and this applies to any technological enterprise, including those outside of the oil and gas industry.

Quite often, when the subject matter is safety, there is a natural tendency to focus attention to fire-fighting water systems, flame detection, and heat and gas, because they are directly associated with the mitigation of the most basic accident postulated in technological enterprises: fire. In the case of offshore rigs, fire is one of the most important scenarios considered in the design of safety systems, and perhaps the greatest threat to these facilities. But the approach based on principle 1 of human factors (Section 2.2) allows us to realize that, if the accident happens, the fire-fighting and detection systems protect the facility itself much more efficiently, leaving the protection of people more closely associated with the escape and abandonment system. Obviously, fire-fighting systems also have the role of protecting people, but indirectly. Conversely, the escape and abandonment system's explicit objective is to protect people, thus being directly in line with principle 1 of human factors: centering objectives on people.

5.2 Accidents in facilities with hydrocarbon inventories and survival

Escape and abandonment systems are essential for the safety of people in any type of facility. They are particularly critical for offshore rigs in the oil and gas industry.

New projects have positioned offshore platforms at progressively greater distances from the continent. The fact that deep-sea offshore rigs operate in isolation is a complicating factor in the event of an emergency that requires the facility to be abandoned. In offshore rigs, abandonment occurs from an accidental scenario (e.g., a fire on the platform) to a scenario that is also very critical: survival at sea.

Escape and abandonment on offshore platforms is more complex and critical, for this reason we will focus next on offshore application in the following examples and explanations.

As presented in Section 4.4, the accident in an offshore rig with the highest number of fatalities was the explosion, followed by fire and ultimately, destruction, with total loss of the fixed platform Piper Alpha, on July 6, 1988, in the North Sea. There were 167 deaths and 62 survivors. Most of the fatalities occurred in the living quarters awaiting a rescue that never arrived. This behavior was predicted in standards and the operators had been trained to act accordingly. At the same time, the 62 survivors did not follow the procedures because they were able to identify factors, as the accident developed, that would lead to a likely failure, if they did so. Most of the survivors jumped directly into the sea, in a clear violation of the training and standards of the time. Based on the survivors' account, the attitude of the first operator to jump into the sea played an important role, who by taking such a risk led others to the same right attitude at the right time, and by this extreme action managed to abandon the facility in a scenario of fire and extreme degradation, moving to another high-risk scenario of survival at sea, but whose acceptance by those operators made the difference between life and death.

While awaiting rescue in the living quarters following established procedures, the fire protection systems were functional, albeit with limitations due to failures. But these systems could not cope with such a large fire, nonetheless. The fire-fighting systems of an offshore unit are technically designed to mitigate "initial" fire, being efficient in the first seconds, or for a few minutes, with the objective of avoiding the immediate escalation from the initial fire category to a full-blown large fire scenario. Due to the volumes of hydrocarbon inventories in offshore rigs for oil and gas exploration and production, there are technical limitations for fighting a major fire. There is a big difference between the energy that the hydrocarbon tank may provide to add fuel to the fire and the energy that can be removed by cooling down and smothering through fire-fighting systems, whether using water, foam, gases, or any source available for offshore application.

In offshore rigs the escalation to a full-blown blaze can happen very quickly. Under these circumstances, the system that offers the greatest direct potential for saving lives is the escape and abandonment system. The remaining systems are very important to give people more time to leave the facility before the situation worsens, thus increasing the efficiency of the escape and abandonment system.

This is also applicable to other accident scenarios, not necessarily involving a fire, as in the case of structural failure and compromised structural stability that can lead the facility to undergo serious structural and naval damages and, consequently, sink. In these cases as well the system most directly associated with lifesaving is the escape and abandon system, while the others also likewise responsible for providing extra time, if the escalation is inevitable, from a simple naval damage to a critical and irreversible scenario with loss of stability and buoyancy.

5.3 Human–system interaction during escape and abandonment

Considering the principle 3 of the human factors design approach (Section 2.3), the human–system interaction should be controlled so as to limit the consequences of human errors to prevent that they trigger a catastrophic accident scenario. In case of an emergency in an offshore rig where people are able to escape and leave the facility in a timely manner without physical damage, despite the accident, the evacuation of unharmed people can be considered a win for technical efficiency of the overall safety of the unit, in addition to minimizing the negative impact of the accident on the company's image, which represents costs.

The escape and abandon operation maximizes human–system interaction at all levels. As soon as the process is triggered, the automated systems intensify the communication with the operators of the control room through their safety interlocks, and they start to work in tandem in the attempt to obtain the most complete identification of the accident scenario so that it can be correctly evaluated and produce the right operational response at the right time, according to the principles of safety culture (Section 2.1).

Besides the interaction that takes place in the control room (with greater cognitive load), people will also interact with the machine (platform) throughout the unit. This happens during the perception of the emergency associated with alarms and audible signals, and in the activities of immediate preparation for escape and abandonment. Such an interaction includes physical and ergonomic aspects related to difficulties in identifying and moving through the best escape and abandonment route available during the accident. Also included in this type of emergency scenario is the people's interaction. Although everyone may have the same objectives, individual attitudes and decisions are not always compatible.

The escape and abandonment operation justifies the use of research tools capable of considering the complexity of hundreds of simultaneous human—system interactions, motivated by the quest for survival in a critical scenario and taking into account the particularities of each design. An analysis of the escape and abandonment system based solely on the calculation of travel time from the farthest location from the abandonment point, considering only the operator's expected average travel speed, does not capture the complex technical and behavioral reality of what occurs in the escape and abandonment of an oil and gas facility.

A realistic and efficient design of the escape and abandonment systems requires large amount of data. It needs to be taken into account the possibility that at the time of an emergency people can be located in all accessible places of the unit.

Individual response time varies from person to person. Depending on the location and technical role of each worker, there may be tasks to be performed prior the beginning of the escape and abandonment itself.

The probable location of each person may also vary according to the time of the accident, whether during the day or night. We need to recognize that the response time to alarms and to initiate the action of escape and abandonment is also influenced differently between daytime and nighttime, and according to each individual.

Speed also varies from people to people. Whether due to the level of technical knowledge about the unit, operational experience or even by age, gender, anthropometric characteristics or simply by psychological behavior. There may also be speed variations for the same person, and it could be reduced by stairs, doors, and in accidents with naval damage the unit may suffer heeling, tilting, and likewise for possible flooding.

Smoke and high temperature can have effect on people during escape and abandonment and, unfortunately, lead to fatalities, especially if such elements are present in congested areas. Also, previously established operational procedures and the training of people may require, for example, the displacement to cabins in search of life vests, causing further congestion and delay.

Computational analysis tools allow the simultaneous management of these and many other factors associated with an escape operation and actual abandonment. Through a simulation it is possible to include the characteristics of each person, in addition to the distribution of people in different configurations in a 3D model of the facility, considering all the parameters mentioned, in addition to others, such as the effects of fire propagation and the ocean water motions, when required.

Escape and abandonment simulation software allows repeated simulations performed in batch mode, covering as many variations in the distribution of people as possible. The results, after undergoing statistical analyses, may offer important answers for the safety prioritization in the project. In adherence to the principles of safety culture, systems of escape from hazards and abandonment of scenarios should provide the right attention at the right time to the anticipated difficulties related to the displacement of people during emergencies. This improves the chances of survival in accident scenarios in oil and gas facilities. In Chapter 7, Reducing Unpredictability, the study of escape and abandonment utilizing computer simulations will be approached in a specific way, as a tool to reduce unpredictability.

5.4 Escape and abandonment operation

There are differences between the procedures and strategies adopted in the design of escape and abandonment systems of facilities in the oil and gas industry, especially when comparing offshore facilities with onland installations. But there is a basic strategy that works for any escape and abandonment operation, as shown below.

Differences due to the particular characteristics of each type of project and facility need to be considered in the studies for each specific project.

An important point is to separate the operation into two distinct phases, namely: ESCAPE and ABANDONMENT. As presented at the beginning of this chapter, ESCAPE is associated with people's movement away from the immediate hazard, that is, from the place where the event causes consequences that affect people immediately. ABANDONMENT occurs after people escape the immediate hazard and, in theory, it depends on an explicit decision by a facility authority.

When an emergency happens, it may generate a risky condition that may or may not be perceived by people. For example, if a fire has erupted, the presence of smoke or an explosion would easily lead people to realize the hazardous condition and would naturally begin the procedures to escape from immediate hazard. But if an event such as an odorless, colorless gas leak were detected by an instrument, it would be more difficult for people to perceive the risky condition they were facing. A precondition for the justification of a planned escape and abandonment operation is the emergency diagnosis. It can be prepared directly by people or produced from the signal processing of the detection and automated system. The more efficient the emergency diagnosis is, so will the escape and abandonment operation be. A wrong diagnosis falsely identifying an emergency, would disrupt operational activity, reduce risk management systems reliability, and hinder future reactions to alarms. If the diagnostic error, on the contrary, were the failure to identify an actual emergency situation, then precious time could be lost with consequences and losses for the technological enterprise. The ability to accurately identify an emergency, as it happens, is the first objective of designers in the development of an efficient escape and abandonment system.

When an emergency is properly identified, it means that there is a condition beyond-acceptable risk that needs immediate response action. If the hazard is caused by an easily identifiable source such as fire or explosion, then people who are directly exposed to it will perceive the risk and will naturally begin the escape procedure. For this there is no dependency on an alarm and, in some cases, the scenario itself and the people involved will trigger the alarm for the entire facility.

Even before that happens, though, it is necessary to start the escape procedure considering those who are present and are part of the accidental scenario itself. Therefore escape routes are required for all facility locations. These routes need to serve both the locations with the highest concentration of agents and those working areas with as little as a single operator. For areas where fewer people are expected, secondary routes are designed to provide an immediate access to primary escape routes. The primary, or main, routes are designed to accommodate a greater number of agents, including those coming from the various secondary escape routes. The primary escape routes are very important for the success of the escape and abandonment operation and need to be protected of fire-fighting and fire-resistant means, thermal and smoke protection at locations where it can become critical.

Primary escape routes need to have at least one redundancy, that is, at least two independent primary routes should be available as protection, in case one of them is out of service due to the accident itself. Temporary refuges should be created along primary escape routes to serve as a shelter, in the event that a severe emergency prevents the displacement of people, the rescue would require a fire brigade, and sufficient time for the brigade to carry out the rescue operation.

The decision on the abandonment order should be the responsibility of the facility authority. Ordering abandonment does not always ensure greater safety. In offshore rigs, rushed decisions by the facility authority may result in people's displacement to a scenario of considerable risk unnecessarily, which is survival at sea. Conversely, hesitation and a lack of sense of the real severity of the scenario can also cause irrecoverable delay in the abandonment operation to be ordered and consequently compromise the efficiency of its execution. For this reason the facility authority needs to be trained in crisis management and have solid technical and operational knowledge required timely scenario diagnosis and with the greatest accuracy. There is a time window within which the escape and abandonment operation can be successfully performed. Both premature and delayed decision on the abandonment order prevent exploring such a window, and as a result in threatening consequences for people.

In terms of crisis management, missing the appropriate time window for the abandonment order can also cause additional problems. The lack of a firm decision by the facility authority for just a few minutes may be sufficient to leave room for independent initiatives taken by the agents. The likelihood of such initiatives will depend to some extent on the level of operational and technical knowledge of the people who make up the scenario as well as the safety culture in which the accidental scenario is defined. As these independent initiatives become adopted by agents, the facility authority naturally loses control of the emergency, which characterize a crisis management failure. In reality, during emergency situations, formal leaders have a few minutes to confirm their leadership through accurate diagnostics and technical commands. If this does not happen, natural leaders will fill the void and people will follow the orders and technical references that they feel most convinced about the final outcome of the operation. Emergency scenarios may dramatically alter human behavior and make the hierarchical systems extremely fragile, especially when these hierarchical systems are established without considering technical authority as a prerequisite for exercising leadership. In situations of severe crisis, people will not follow the orders of formal bosses solely for hierarchical reasons, but rather what will really determine the actions will be the technical authority coming from whoever can convincingly demonstrate to have the technical solution that will save them.

The escape and abandonment operation does not end with the arrival of all people at the meeting points (muster stations). The facility authority's emergency diagnostics should be as accurate as possible leading to the decision whether or not to effectively abandon the facility. People are displaced to a muster station to receive clear guidance from the facility's authority regarding the decision on possible abandonment. In the case of escape, the design should plan primary routes from the muster stations and meeting points to the physical boundaries of the facility, which we call "abandonment points." So, the escape consists of moving away from the exposure to the immediate hazard of the accidental scenario through the secondary and primary escape routes up to a meeting point (muster stations). Once the decision for abandonment is confirmed, people will have to move to the abandonment points through other escape routes. These routes need to be designed based on the same concepts adopted for the design of the main escape routes, but larger number of people and the arrangement of groups of people should be considered. Unlike the escape operation, where people depart from different locations of the facility toward meeting points, in the abandonment operation people are gathered together and depart in large groups, creating much more difficulty and obstruction along the abandonment route.

In offshore rigs, its physical boundary is the sea. Therefore resources and vessels need to be provided for the abandonment effectively take place, and this will be presented in greater technical detail in subsequent sections. The project of some onshore facilities as well as at shore/in shore facilities may also require the inclusion of additional means of displacement beyond its physical boundary. That is the case of some marine terminals located in sheltered waters isolated from the main land, which do not have their own rescue boats, and depend on external resources to provide for the effective abandonment of the facility. Additional examples are terrestrial facilities located in areas of difficult access such as forests, which likewise need to incorporate in their projects of escape and abandonment systems external means to facilitate the escape and abandonment of the facility.

5.5 Technical recommendations for escape and abandonment system

There is a wide variety of strategies adopted by designers to create efficient escape systems. There are also many international and corporate standards and rules and even cultural issues that will to some extent influence the design of escape and abandonment systems. Risk management experts need to consider all of these factors and most importantly the technical particularities of each facility, to design efficient escape and abandonment systems. The best way to investigate alternative solutions to achieve the required efficiency is through escape and abandonment computer simulations. But a precondition for such an analysis is a minimum level of maturity reached by the project. And when that has not yet happened, how can the design of escape and abandonment systems get started? Next, we will present some concepts and useful criteria to guide designers during the conceptual phase of the design of escape and abandonment systems. With these concepts and criteria, it is possible to make progress in the design and create secondary and primary escape route systems, meeting points, abandonment points and strategies for moving people in emergency scenarios. In a later phase, these projects must be submitted to computer simulation analyses, with the ability of identifying possible failures and provide recommendations for improvements aiming at the optimization and efficiency of the system.

5.5.1 Possible operational sequences

There are different technical opinions regarding the best strategy for the sequence of steps in an escape and abandon operation. These strategies vary mainly if each agent decides to go to his cabin or not before going to the meeting station. There are also the least likely cases in which the accident makes access to the booths or muster stations unavailable. In this case, each agent needs to go directly to the dropout point (e.g., region close to lifeboats). Following are the most widely adopted sequences:

- *Option 1 (with intermediate task)*: Escape from immediate hazard, Secondary Escape Route, Primary Escape Route, Control Room or Cabin (offshore) to fulfill task, Meeting Point (Muster Station), Waiting for Abandonment Order, Secondary Escape Route, Primary Escape Route, Abandonment Point, Exit through the Border or through External Transport Resources (e.g., rescue boats in offshore rigs).
- *Option 2 (without intermediate task)*: Escape from Immediate Hazard, Secondary Escape Route, Primary Escape Route, Meeting Point (Muster Station), Waiting for Abandonment Order, Secondary Escape Route, Primary Escape Route, Abandonment Point, Exit through the Border or through External Transport Resources (e.g., rescue boat in offshore rigs).
- *Option 3 (direct)*: Escape from Immediate Hazard, Secondary Escape Route, Primary Escape Route, Abandon Point, Exit through Border or through External Transport Resources (e.g., rescue boat in offshore rigs).

5.5.2 Basic dimensions and recommendations for escape routes

There are variations in the dimensions of the escape routes (Fig. 5.2), according to the standards applicable to each project. We herein present the recommended values for the escape routes; however, through analysis using computer simulations, the dimensions of the escape routes, doors, and stairs included in them need to be corrected at the locations identified as critical with respect to congestion. The adjusted values need to be tested in the simulations in an iterative process until congestion that compromises the escape and abandon operation is eliminated.

- Primary escape routes: 1.20 m wide × 2.10 m tall.
- Secondary escape routes: 1.00 m wide × 2.10 m tall.
- The dimensions above need to be adjusted in situations where a scenario with large number of people (more than 50 people) needs to be supported, based on escape and abandonment computational studies.
- The same dimensions need to be used for stairs, doors and throughout the route that is part of escape routes.

Figure 5.2 Main and secondary escape routes in an offshore rig.

- Escape routes should contain exit floor marking bands 100 mm wide, and arrows indicating travel direction, painted in white and with a nonslip finish.
- There should always be two main route options between the hazardous and nonhazardous areas.
- Opening doors and other routines for movement of loads and people cannot create obstacles within escape routes.
- People at any location in the facility should always have at least two escape options for immediate hazard that provide access to the escape route, except in the administrative and accommodation areas and in special service rooms such as freezer rooms. Rooms with less than 10 m^2 may not require two escape options when technically acceptable.
- The maximum travel distance to reach an exit should not exceed 7 m. This also applies to corridors, which should not have sections longer than 7 m without an exit.

- Escape routes should contain emergency lighting and signaling markings (powered by the emergency power supply) and should be available even in the event of loss of the main power supply.
- In the case of semisubmersible (SS) offshore rigs, structural legs need to be supplied with ladders with direct access to the sea as an alternative to lifeboats should them be unavailable. These stairs should be designed to allow access to the sea (jump) from a height of 2 m, where there should be a landing designed for two to four people.
- Escape routes need to be unobstructed regardless of equipment such as elevators, cranes, baskets, ropes, etc. When it is imperative that an escape route depend on auxiliary equipment, they should be powered by the emergency power supply and their availability should be ensured under the conditions considered in the accidental scenarios postulated by the project.
- All escape routes and its components such as doors, stairs and passages need to consider the feasibility of moving injured people on stretchers.
- Guardrails should be considered in technical assessments of escape routes. They should be designed to avoid falls from heights greater than 0.80 m, that is, whenever the difference in height between floors is equal to or greater than 0.80 m there should be guardrails to prevent falls.
- Based on escape and abandonment studies, temporary refuges should be provided at strategic locations along the escape routes, where people can stay protected for up to an hour, making rescue crew access viable.

5.5.3 Evacuation, escape, and rescue analysis

International standards (ISO 13702) require evacuation, escape, and rescue analysis (EERA). The EERA studies do not produce the same level of results based on escape and abandonment computational simulations. We will cover this type of analysis specifically in Section 7.2.4. EERA studies need to consider that the means for escape and abandonment should be available in all accidental scenarios postulated in the project.

5.5.4 Spaces with limited access and machine rooms

Some rooms and locations may have very limited access due to the facility conditions. This is very common in offshore rigs, where rooms that house machinery and equipment may be located within a local area of the submerged hull. These locations need to be evaluated take into consideration the level of access limitations and difficulties, so that people are able to leave these areas. Some recommendations are part of the standards applicable to projects, mainly offshore projects. Following are the main points to be considered.

- Spaces and rooms can be classified into categories, according to the degree of safety requirement demands. In offshore projects, rooms with equipment of rated power greater than 375 kW are classified as "machine spaces of category A." For example, this is the case of boilers and internal combustion engines. For these spaces, in addition to a main staircase from the lowest level to the top level, with access to the open space, an independent staircase needs also to be provided to allow access to all floors. This second staircase should be protected by a stairwell from the lowest level to the outside area, accessible by door and protected by bulkheads, thus allowing its use during an emergency in the own machinery space. Another alternative is a set of two main stairs, one of which is protected from the bottom to the top area with to access the open space.
- All spaces in the facility need to be studied during the design phase and their access limitations be identified. Additional means of escape and abandonment should be designed, besides those required by standards, to ensure the efficiency of the escape and abandonment system regardless of specific access limitations of each space created by the designers.

5.5.5 Applicable materials in escape and abandonment systems

The reference material to be used on escape routes should have strength equivalent to concrete or steel. Grid floors and nonmetallic materials can be used, as long as their strength can be technically demonstrated for the intended application and there are no conflicts with the standards adopted in the project.

5.5.6 Meeting points (muster stations) and abandonment points

- Meeting points should be strategically located and provide protection for the personnel during the time allocated for the decision-making process related to the abandonment.

- Rescue resources should be kept at the meeting points, based on the characteristics of each oil and gas facility. For marine facilities, life jackets should be available to all the people on board (POB) for use during abandonment.
- Beyond each meeting point, there should be at least two independent route options.
- Facilities equipped with helideck, train station, bus station, etc., should provide specific routes to these strategic locations.
- Offices, accommodation areas, cafeterias, cabins (offshore) should be provided with two independent escape routes, bound for the meeting point.
- Areas of large concentration of people such as offices, accommodation, cafeterias, superstructures (offshore) should be surrounded by an escape route that facilitates the evacuation of the area in case of emergency.
- If there is a need to gather people temporarily at the ABANDONMENT point in preparation for abandonment, floor markings need to be added to facilitate the formation of lines and organize people. An example is the boarding area on a land-based bus station or the abandonment point in offshore rigs, where people need to be organized in line to get on buses or rescue boats before launching into the sea.

5.6 Sea survival equipment

Sea survival equipment is used in marine facilities, ships, and offshore platforms. The main equipment are lifeboats, life rafts, and rescue boats, in addition to rescue equipment such as life jackets and buoys. This equipment is vital for the safety of offshore rigs because they allow the effective abandonment and rescue in accident scenarios of people jumping into the sea.

The requirements and technical specifications associated with this equipment differ according to the requisites of the maritime authority of each country, the requisites of classification societies, practices and internal standards adopted by the operating companies and offshore design companies. Following are some general recommendations for this equipment that are accepted internationally by the regulatory organizations.

Each designer, however, needs to consider the requirements of specific reference standards of each project and facility. Through safety studies, the limitations and technical characteristics imposed by the accidental scenarios postulated in each offshore project need to be analyzed, with the objective of ensuring the compatibility of the survival equipment at sea with these scenarios. The compliance with the regulatory requirements is not sufficient, as these are generic. It is also necessary to analyze the conditions, hazards, and risks of each offshore project. Thus specific risk situations intrinsic to each project can be considered and effectively safeguarded, regardless of whether or not being part of official standards.

5.6.1 Lifeboats

An alternative term used for lifeboats is the acronym TEMPSC, totally enclosed motor propelled survival craft, as this type of equipment is completely closed, fireproof, equipped with external sprinklers to allow temperature control inside the vessel even under intense flames. They also include oxygen supply (to maintain breathable air conditions under intense smoke), drinking water and minimal food supply, and are also equipped with a diesel propulsion engine. Lifeboats are built using special fire-resistant materials, and design to withstand fire and smoke both during launching and when sailing in areas of oil puddle fire at sea. The naval design of lifeboats ensures stability and buoyancy conditions even when they are completely filled with water.

Helicopter is the primary means of abandonment of an offshore rig provided that the scenario conditions allow the aircraft's approach. When this is not possible, lifeboat is the main equipment for the evacuation of people from the offshore rig to the sea. Every offshore facility should be equipped with enough lifeboats to allow POB to abandon the unit during an emergency.

The number of lifeboats is dependent on its size (number of seats), lifeboat type (releasing mechanism for launching into the sea) and type of offshore rig [fixed rig, SS, FPSO (floating production, storage, and offloading system), etc.]. The positioning of lifeboats in the unit needs also to take into account escape and abandonment strategies and the different accidental scenarios that were studied in the risk analysis conducted during the rig design phase. Factors such as main current and wind, availability of access and ease of boarding should also be considered to determine the boarding and location of boarding stations and launching of the lifeboats.

They use winch-type equipment called davits as a lifeboat launch system. Their function include also to lift lifeboats from the sea and positioning the lifeboat on the deck, upper-deck, or on the side of offshore rigs.

Lifeboats need to undergo maintenance for testing after launching, which for some types may be an annual requirement. The davits also serve to launch conventional lifeboats based on older technology that still relies on launch cables.

There are two main types of lifeboats, distinguished by the releasing mechanism for launching into the sea: conventional lifeboat (Fig. 5.3) and free-fall lifeboat (Fig. 5.4). Conventional lifeboats are based on older technology in which they rely on cables, motors, support structures, and a set of hooks for connecting cables to the lifeboat. Without this equipment being in fully operational condition, it is not possible to launch a conventional lifeboat, because its concept is based on the descent into the sea supported by two steel cables connected by the hooks. The cables must withstand the lifeboat dead weight plus the live load corresponding to the people being saved. The loads are transferred from the cables to the davit's hooks, motors, and the structural components. The descent movement is

Figure 5.3 Conventional lifeboat (rescue vessel). The concept of launching of the lifeboat into the sea includes support by steel cables during the descent operation. The hull is not wedge-shaped and does not withstand the impacts of a free fall. *Photograph courtesy of the supplier norsafe: www.norsafe.com.*

Figure 5.4 *Free-fall* lifeboat type. The concept of launching of the free-fall lifeboat into the sea uses the force of gravity as an ally and dispenses control through cables and descent mechanisms, which increases the reliability of the operation. The hull is wedge-shape, which allows the transfer of the launch energy at the time of contact with the sea. *Photograph courtesy of the supplier norsafe: www.norsafe.com.*

a delicate balance that is affected by wind conditions, sea conditions, distribution of loads between cables, presence of flame and smoke along the route, heeling (slope) resulting from the damage to the offshore rig, among other factors. The conventional lifeboat concept considers gravity a force to be counteracted for the control of the descent, and therefore the cables and davits are designed to withstand the descent, as the conventional lifeboat's structure is not designed to withstand the free fall into the sea. Under this concept, steel cables control the effect of the gravitational force so that the lifeboat does not reach the water with a speed higher than that supported by the hull at the instant of contact. During the descent, possible movements of the ship or rig may produce dangerous pendular oscillations, with the risk of impacts between the lifeboat and the side or structure of the offshore rig.

Another type of lifeboat, called free fall, is based on the concept of aligning the force of gravity with the direction of motion of the lifeboat launch. Thus there is no need for cables and davits for launching the vessel, which reduces the number of moving parts, equipment subject to failures and destabilizations during launch, besides unbalanced loads between cables. The free-fall lifeboats are designed with the location of its center of gravity such that during free fall the lifeboat angle assumes the ideal value at the point of impact at water entry. The launch time is significantly reduced, because there is no need to control the distribution of forces by cables. The free-fall lifeboat structure is reinforced, with a wedge-shaped hull, capable of absorbing the energy from the impact with water without deformations that could harm its occupants. Free-fall lifeboats are considered more technologically advanced and safer, although in some designs conventional lifeboat can also be specified without compromising safety.

The concept of free-fall launch is more in line with principle 2 of Risk Management Efficiency—Respecting Natural Laws (Section 2.4). General accident databases contain records of fatalities and serious accidents involving conventional lifeboats, but free-fall lifeboats are also not exempt from accidental records, albeit with lower frequency. The entire offshore abandonment operation is considered high risk and requires excellence in training, crisis management, and equipment maintenance. The choice of the type of lifeboat depends on many factors and the characteristics of each project. In the following sections, we will describe each type of lifeboat in more detail, which likely helps in identifying the best technical option for each project.

5.6.1.1 General safety requirements

In general, the seating capacity of lifeboats varies from 10 to 120 people, but in offshore rigs, most of the time, lifeboats with capacity between 50 and 90 people are used. The choice of capacity and as a result the size of lifeboats depends mainly on the maximum number of POB.

Fixed (nonfloating) offshore rigs should have a sufficient number of lifeboats to accommodate the entire POB and also have an extra spare lifeboat and with the capacity to replace any unavailable lifeboat. The location of the lifeboats is chosen taking into consideration that in all accident scenarios postulated in the safety studies, sufficient lifeboats are always available to meet 100% of the POB. The projects of offshore rigs of type SS also need to meet the requirement regarding the positioning of lifeboats, however in installations of type SS the total amount of seats provided by the sum of lifeboats must be at least 150% of POB. FPSO and similar rigs need to provide seating accommodations for 100% of the

POB on each side of the facility (port and starboard), totaling 200% of the POB, for the case of conventional lifeboat specification. This is required because, should a list occur due to a naval accident, the angle of list may impede the launching on one side, requiring that there are enough seating spaces to accommodate 100% of the POB using the lifeboats located on the opposite side.

The davits of conventional lifeboats can be of two types: fixed or foldable. In the fixed davits, part of the rigid structure of the equipment protrudes out of the rig and maintains the lifeboat always with the bottom of the hull projected over the sea, in the position in which the lifeboat is to be launched (Fig. 5.5). In folding davits the conventional lifeboat is kept stowed on the deck, and in preparation for the launching operation, a movable part of the davit's structure performs a rotational movement toward the lifeboat launching position over the sea. Both types of davits and lifeboats should be positioned on decks as close to sea level as possible to shorten the launch path and consequently reduce the risks. For this, designers should take into account aspects such as access and gathering of people, strategies for the movement of people in the escape routes and the operational procedures adopted by each offshore company during the abandonment operation.

The free-fall lifeboat davits are used as initial guiding rails directing the lifeboat first movements. As the free-fall lifeboats do not use cables for launching, a hydraulic piston is activated by the operator from within the lifeboat with the purpose of releasing it from the davit and leaving it subjected solely to the action of the gravitational force. If the free-fall lifeboat needs to be recovered after launch for continued use on the rig, the davits are equipped with a winch-type mechanism capable of lifting the lifeboat from the sea to the stowed position. However, contrary to many professionals' thoughts, in special situations it is also possible to launch free-fall lifeboats through the cables and motors of its dedicated davit, in a reversed cable operation in comparison with the recovery of free-fall lifeboat. In this situation, it is necessary for at least one person to remain on board the rig to operate the davit, and another means will be required to abandon the offshore rig after the launching. This is not the ideal procedure for launching free-fall lifeboats and should only be reserved for extreme situations, when free-fall launch is hampered by the presence of debris or floating objects resulting from explosions and other accidents. Another abandonment scenario that requires the use of cables is in frigid zones where icebergs can prevent free-fall launch.

Figure 5.5 Construction and assembly phases of the FPSO rig. Bottom view of a set of two conventional lifeboats supported by the cables of a fixed davit in the launching position. The lifeboat keels are always positioned directly over the sea, with no need for the rotational movement of the davit during launch. The detail view shows the view from the deck of another fixed davit. *FPSO*, Floating production, storage, and offloading system.

It is very important during decision making regarding the launching point to be taken into consideration factors such as currents, winds, waves and risks of the collision of the lifeboat with the rig itself after launch. The ideal objective is for the launch point to allow immediate and safe distance from the lifeboat in relation to the platform. With this purpose, exclusion areas and launching envelope areas should be preserved throughout the complete rig lifecycle, during which the availability of lifeboats for emergency use is required. The launch envelope area ensures the ideal angle at which the lifeboat enter the water for perfect contact with the sea.

5.6.2 Life rafts

Life rafts (Fig. 5.6) are types of inflatable equipment, much simpler and more limited than lifeboats. They are used as redundancy for situations where it is not possible to reach the lifeboat abandonment points or when such equipment is unavailable due to the accident itself. Life rafts are not a substitute for lifeboats. They are not equipped with motors or are resistant to fire and smoke, but their advantage is that they are able to be launched to the sea directly, by simple and easly operation equipment. Generally the life rafts are stowed within cylinders and when launched from the rig they are automatically inflated by the action of an interlink cable connected to the rig structure. Once loaded with people, the life raft is detached from the cable and should be moved away from the offshore rig as far as possible while awaiting rescue by a vessel rated at a higher level of safety.

In fixed offshore facilities the number of life rafts should be sufficient to accommodate at least 50% of the POB. For SS platforms and FPSO type or similar, the required percentage is 100%. Life rafts should be

Figure 5.6 Life rafts positioned inside the storage cylinders, ready for launching from the main deck of the offshore rig. The detail view illustrates the appearance of the inflated life raft after being launched into the sea.

positioned adjacent to lifeboat launch points, as a redundancy in case lifeboats cannot be launched due to problems. There are cases in facilities of type SS, FPSO and others where the set of lifeboats may be located more than 100 m from some work zones. People (e.g., located in the bow) may have difficulty to access lifeboat launching points during emergencies such as fire with large flames and smoke. For this reason, in such cases additional life rafts are included with capacity of at least six occupants in the areas located the furthest away from the lifeboats launching points. Due to the characteristics of manual life raft launching and access difficulties, they should be positioned very close to sea level.

5.6.3 Rescue boat

The rescue boat (Fig. 5.7) is a type of speedboat, used to rescue people who are at sea due to falls or other accidents. They are usually powerful speedboats, suitable for the sea conditions where the offshore rig operates. They should not have exposed propulsion propellers that could cause injuries during rescue maneuvers. Rescue boats are fast and have superb maneuverability when compared to lifeboats, which allow them to

Figure 5.7 Rescue boat used to rescue people waiting for help at sea. *Photograph courtesy of the supplier: www.norsafe.com.*

quickly reach the victim even in adverse sea conditions, with rough waves and currents. Rescue boats are launched into the sea by dedicated davits. At launch, the rescue boat is supported by a single cable. Rescue boat certification tests include verification of stability and buoyancy requirements such as the ability to turn over and restore its stable position, to withstand a fall in the sea of up to 5 m and not be submerged even when completely filled with water.

5.6.4 Salvage equipment

A variety of smaller complementary equipment is used for personal protection in emergency response operations in offshore rigs. The main items are life jackets, work jackets, buoys, autonomous breathing equipment, fire-resistant outfits, among others. In addition to the international maritime safety requirements and the maritime authority requirements of the country where the rig will operate, designers and risk management experts need to consider the characteristics and the possible accidental scenarios specific to each project. When necessary, personal protective equipment and special equipment should be included in the project and operational routine of offshore rigs, independently of being required by international standards and local maritime authorities. Offshore projects often evolve their operational practices at a rate of speed that revisions to standards are not always able to keep up with. To make up for that the designer and risk management expert involved in offshore activities should have a sense of responsibility with regard to the specification of individual and collective protection resources that are compatible with the risks generated by each associated project and offshore operation.

5.7 Lessons learned

5.7.1 SOS: emergency in FPSO

A state-of-the-art FPSO rig will soon have to deal with an emergency that will mark the professional lives of 110 people who work on it. The modern rig is operating in deep waters, about 240 km from the coast, which makes rescue from on-land teams slow and limited. By sea, hours of navigation would be required, and by air, aircraft would have limited autonomy due to the high fuel consumption for the return trip alone.

Like any offshore rig, the FPSO is divided into a hazardous area and a nonhazardous area. Some of the 110 people are in the process plant at locations with significant hydrocarbon inventories. Another group works in areas considered nonhazardous, such as offices and living quarters, where hydrocarbons are being processed.

Given the recent the project and construction of the facility, the FPSO is equipped with the most modern emergency response resources such as deluge and carbon dioxide fire protection systems, detection and alarm systems, all approved by the platform's classification society. The operating company's escape and abandonment strategy adheres to strict internal operating standards. In the event of abandonment decision, the FPSO may rely on four conventional lifeboats, each of which with 55 seats, equally split between port and starboard lifeboats.

Lifeboats are always available, docked in fixed davits and positioned over the sea, in the direction and location defined for contact with the water. The sea condition in the region is very severe, with frequent storms that produce big waves that make it very difficult for vessels to navigate and approach the rig in case of abandonment.

Among the teams that are on duty, five will have a significant influence on the emergency that is about to happen. The first team is in the main control room where all information about the platform's operation is centralized. It is a team trained for the routine work in the main control room and must know how to act in any possible emergency scenario. Most of the team's activities in the control room are related to the alarms and indicators on the process computer screens that controls the rig. The main control room has a few windows that allow a very limited and partial view of the process plant. The second team is also in the superstructure, but in another part of the same living quarters, where the main control room is. This second team works in a smaller control room, called the remotely operated vehicle (ROV) control room. ROV is sophisticated equipment capable of performing tasks in ocean depths where divers could not survive. ROVs have robotic arms, high definition cameras and a control room from where they can be operated. A third group of operators is where in the FPSOs is called the engine room, located inside the FPSO hull, below the superstructure where the main control room and the ROV local control room are located. It is a confined space, with no windows to the outside environment and therefore unable to follow what is happening in the process plant and living quarters, except by radio, telephone and emergency alarms. The fourth group of operators is in an area considered one with the highest risk of accidents: the process plant. Despite being in open area, operators circulate around pipelines and equipment containing significant quantities of oil and gas. For this reason they must take extra care operating equipment that may generate sparks and work under strict safety procedures and under the approval of a permit to work (PW), in which all risks and safeguards need to be evaluated in advance. Finally, the fifth group is working isolated at the bow, having to cross the entire process plant in case of an emergency, to reach the nonhazardous area and the abandonment resources that are located near the superstructure. As seen in the membrane diagram (Fig. 3.12), all systems that make up a technological enterprise as a FPSO rig have their own safety requirements that function as barriers to prevent accidents. But in addition to the failures that are detected and corrected in each system, there are other types of failures difficult to detect, specifically those associated with human factors (error-inducing environment) and to the safety culture. These are the types of failures, generally related to subjective and behavioral problems, which may cause a potentially catastrophic accident in a rig. And if that happens, there are systems dedicated to emergency response that act as mitigating barriers limiting the possibility of accident cascading effect. Such systems are fire fighting, escape and abandonment systems. In the membrane diagram they appear as barriers that act to mitigate accidents to avoid further degradation of the FPSO safety: the beyond design-basis accident scenario. In this extreme scenario, safety systems are unable to respond. Harm to people, damage to environment and property can lead to a catastrophic accident. The emergency that the FPSO will need to deal with will occur due to failures in preventive barriers, but if operators do not give the emergency the due attention at the right time, mitigating barriers may also fail and a looming catastrophe could change the lives of POB the FPSO.

The operating company and the society are experiencing a time of economic crisis and the organization's objective is to maintain the rig production while ensuring the safety of people, environment, and property. Operators are trained by the company to act in the event of an accident by assessing the situation and coordinating mitigation actions.

According to company's operating culture the rig manager should only initiate the escape and abandonment procedures if the situation becomes irreversible.

It is 3:07 p.m. and a normal working day on the rig, with all the operational parameters in accordance with the set points and within the safety limit ranges. The control room conducts communication tests with the ROV control room, the engine room team on duty, the process plant operators and with the team that performs a task at the bow of the FPSO. During the communication test, each team confirms reception of loud and clear audio signal and takes the opportunity to confirm that all activities underway are normal. All communications between the control room and teams of operators use radio. However, communication in the control room is under the responsibility of a production coordinator, who works as manager assistant and is responsible for exchanging information directly with operators.

Operators in the area may request information to the control room at any time. Prompt response and ability to provide solutions to the problems posed is required by the control room. A phone call from the company's headquarters places the company's general manager in contact with the rig manager. The general manager reiterates the strategic importance of maintaining the rig in operation, given the economic crisis period experienced by the company and society, and the interruption of production under these circumstances could significantly aggravate the company's situation.

At exactly 3:12 p.m. an alarm on the naval console in the main control room indicates that the rig is listed by about 0.5 degree. It is such a small slope that it is impossible to be perceived by operators who are working in the area. The operators activities in the area proceed normally and all safety systems continue to be available. The ROV control room staff requests authorization to the main control room to start scheduled subsea activities and, almost simultaneously, the team working at the bow also contacts the main control room to report the beginning of the routine maintenance activity, authorized by the PW-021.

In the meantime the rig continues to tilt and at 3:17 p.m. the list reaches 1 degree. For the people in the area, though, the slope cannot be noticed, due to the rough sea condition, which causes strong sway motion. The slope does not affect any ongoing activities and operations continue normally, with all safety systems available for use if necessary. In the main control room, operators are trying to figure out what is happening with the FPSO stability while the naval team still has doubts about the list and wonders if it could be a false alarm caused by a faulty panel indicator. Some of the main control room operators continue working normally and notices a banner on the company's intranet announcing a world record for productivity and availability achieved by the rig—a source of pride for the entire company. The operator shares the news with colleagues and soon the entire main control room celebrates the recognition of team's work in times of economic crisis. The work continues smoothly in the area with all systems safety features up and running. During a routine check, ROV control room operators call the main control room to ask if everything is working normally. The control room gives affirmative answer and relays no information about the list alarm, as it is still being analyzed to determine whether or not the rig is actually tilted. The team that is working at the processing plant informs the control room about the completion of the task to be carried out per PW-022, also requesting permission to go to the main control room to close the PW.

It is now 3:22 p.m. and the instruments indicator in the control room remains with the information that the rig the list of 1 degree. Although the indicator remains stable at 1 degree, all operations are proceeding normally and all safety systems continue available, some operators in the control room feel that the rig is actually tilted. The naval team at the control room remains unsure about the reason for the rig list alarm, and whether the rig is actually leaning. But as the situation has been developing for a few minutes already, the naval team suggests that the rig manager make an assessment of the conditions and the need to generate an alarm for abandonment preparation, since, although not confirmed, it is suspected that some problem is affecting the FPSO stability. Contrary to what was experienced in the main control room, operators in the area realize clearly that something is wrong and the rig appears to be leaning. Safety systems continue ready for use if necessary, but the team in the ROV control room realizes through the instruments that the rig is really tilted and gets in contact with the main control room to report work interruption and its reason. In the engine room, the team working in the closed environment also perceives the slope and gets in contact with the main control room to ask for explanation, already concerned about the safety of the rig. The team at the bow carries on the tasks authorized by PW-021.

It is 3:27 p.m. when the telephone in the main control room rings again. This time around, the same general production manager after seeing the banner on the intranet wants to congratulate the entire team on the productivity and availability record. The rig manager and his entire team are proud. Seconds after the call ended, a big jolt suddenly hit rig all at once. Slope rig indicators remain at same 1-degree angle even after the impact caused by the sudden motion of the rig. This leads the naval system operators to conclude that there is a serious problem in the control and instrumentation start making desperate contacts with the room main control panel seeking information and to report problems. The team in the ROV control room warns about smoke in the superstructure, apparently coming from the kitchen. In the superstructure are also located the main control room, meeting points (muster stations), living quarters, access to the engine room besides the ROV control room. A person arriving from the kitchen at the moment warns the ROV control room staff that the jolt caused by the tilt caused heated oil spill on the kitchen floor that started a fire, currently out of control. The ROV control room staff ROV relays the information to the main control room and requests guidance. The atmosphere is tense, the machine room operators also request information about the situation and guidance for how to proceed. Over across the rig, the team working at the bow far away from the superstructure is unaware of the fire in progress. They are isolated, and despite the rig slope try to carry on with the activities but realize that it is impossible to do their work. That is when they discover that it has also become impossible the return to the superstructure due to the rig tilt.

Finally, at 3:32 p.m., the slope of the rig reaches 16.5 degrees, making it impossible to launch lifeboats and the operation of the fire-fighting water system. It is unfeasible to perform the escape and abandonment operation, and should the fire spread in the rig, the only option for the 110 POB will be to jump into sea in their quest for survival. Computer simulations demonstrate that for up to 16.5-degree slope it takes about 20 minutes to gather the 110 people at the meeting points (muster stations) and assess the need for abandonment. However, a prerequisite for this type of operation is that safety systems be operational, especially the lifeboats, which under these conditions has become impossible.

Basically, a malfunction in the control of the ballast system allowed the rig to tilt at 0.5 degree. Upon reaching 1-degree slope, the data acquisition by the control computer "froze," and as a result the ballast system maintained with the incorrect alignment, worsening the stability conditions. Finally, the severity of the accident worsens, causing a jolt, oil spill in the kitchen, fire and loss of safety systems when the slope exceeded 16.5 degrees. Since 3:22 p.m., about 15 minutes after the first alarm, the platform manager already had all the information at hand and the favorable conditions to start the preparation procedures for abandonment, but the productivity culture prevailed, fueled by phone calls and the intranet banners. The bad habit associated with heroism (in this case, the determination to meet the production goal at all costs) disturbed the crisis management activity and caused the postponement of the abandonment decision, to a point when it was too late to safely abandon the FPSO. The platform manager, responsible for making abandonment decisions, influenced by heroism, insisted in keeping the platform production, thus trying to preserve the good reputation of the team, and as a hero, he tried to overcome the crisis without halting the operation. Unfortunately his strategy was not successful and now he has a team of 110 people left with no alternative to abandon the platform. Life rafts can no longer be launched easily either, the only option remaining being to jump into the sea and survive with the life vests amid the risks of hypothermia, malnutrition and shark attacks, until rescue from sea or air is able to overcome the distance and adverse sea conditions.

This scenario is didactic and is not considered in real world studies due to the remote possibility of sudden tilts like the one described. The objective of the text is to emphasize didactically how the lack of a solid safety culture can prevent the right attention at the right time for safety-related issues. In an actual emergency the sequence of occurrences leads those responsible for the abandonment decision-making process to consider, several times during the emergency, whether or not to order abandonment. The assurance of the timely abandonment decision is dependent on issues related to human factors and safety culture that need to be previously worked on starting in the design phase, thus reducing the consequences of possible human errors.

5.8 Exercise

5.8.1 Crisis scenario simulator

Based on the text in this section, it is possible to formulate crisis simulation exercises without the need for a computer or a physically built simulator. For this type of exercise two rooms are required, namely, one for the team that will act as the main control room operators and another for the team that will act as operators in the area. Both teams should not have knowledge of the text in Sections 5.7 and 5.8 before the exercise, or conversely they should be presented a new scenario, similar to this one. Each room should be equipped with a radio and some tasks need to be previously assigned. For ambient sound, each room should play music at medium to high volume. Suggestion: for the group of operators in the main control room, the music and sounds should convey stress, like electronic music alternated with alarm sounds, thus creating a confusing atmosphere for the team that will need to communicate with others, including by radio, with such disturbing background sounds. As for the environment of the operators who will simulate the work in the area, ambient music should be calm and lullaby-like style, to minimize the stress but, at the same time, cause some confusion due to the disconnect between the ambient sound and the information to be announced.

The exercise should be accompanied by monitors (one in each room). They will be responsible for observing the trainees and to distribute the information about the evolution of the scenario in 5-minute intervals. The exercise sets a time limit for the abandonment operation to be initiated by the platform manager.

Independently of the exercise outcome, participants should have some available time at the end to identify the principles of safety culture, human factors and risk management (Sections 2.1, 2.2, and 2.3, respectively) that had greater weight in the final result of the exercise, and then share it with the other team.

Next, we will present the organization and information distribution sequence for the exercise to be executed. Evidently this is a didactic exercise example for training purposes. Changes can be made and other texts can be created and customized according to the purpose of application of the exercise and the need for adaptation of the training to real work condition.

5.8.2 General instructions

- Trainees are divided into two teams, one responsible by the main CONTROL ROOM of an offshore platform and another responsible for the work in OPEN AREA.
- Each team will occupy a different room and the communication between them can only be done by radio. For didactic reasons, the room used by the CONTROL ROOM team will also be considered the place designated as the meeting point (muster station) in case of abandonment operation.
- The CONTROL ROOM team must choose a MANAGER (responsible for analyzing alarm information, opinions from the team and

make the final decisions) and a COORDINATOR (the only team member who can use the radio to send commands and directions to the OPEN AREA team).

- The OPEN AREA team will be subdivided into four smaller groups within from the same room: bow operators, plant operators, engine room operators, and operators in the living quarters. Any member of the OPEN AREA team can call the control room by radio and receive calls.
- As in real life, THE AVAILABLE INFORMATION IS INCOMPLETE.
- Although not essential, a preparatory presentation is recommended with photos showing the different FPSO locations where each team will supposedly be positioned during the exercise.

5.8.3 Instructions about scenario evolution

The instructions should be presented by each monitor every 5 minutes, in a synchronized way between the two rooms. Below are the instructions, considering the scenario described in this section.

Control room team	Open area team
Time line: 0 minute	
All information will be provided using slideshow for both teams. The company and the society are going through a moment of crisis and the objective of the organization is to keep the platform production safe for people, the environment, and property. In the event of an accident/emergency, the control room must assess the situation and coordinate mitigation actions and, if necessary, escape and abandonment. The platform is in normal operation and no alarms. The control room must conduct a communication test with the ROV control room, engine room, plant operators, and bow operators. The communication test consists of calling each team to confirm that they are receiving a loud and clear audio signal.	All information will be provided using slideshow for both teams. The company and the society are going through a moment of crisis and the objective of the organization is to keep the platform production safe for people, the environment, and property. In the event of an accident/emergency, the control room must assess the situation and coordinate mitigation actions and, if necessary, escape and abandonment. The platform is in normal operation and no alarms. Operators in the area must answer calls from the main control room. If there is a need to relay relevant information to the room, radio communication must be used.

(*Continued*)

(Continued)

Control room team	Open area team
Time line: 5 minutes	
Alarm indicates that the platform is listed (0.5-degree slope). It is such a small slope that it is impossible to be perceived by the people in the area.	There are no abnormalities on the platform, all activities proceed without any problems.
There are no consequences for platform operations.	All the safety systems remain operational.
All the safety systems remain operational.	ROV control room staff must request authorization from the main control room to initiate scheduled subsea activities.
A routine phone call from the headquarters reiterates the strategic importance of keeping the platform operation	
The Naval team in the control room does not know the root cause of the 0.5-degree slope.	BOW staff must report the start of routine maintenance authorized by PW-021.
Time line: 10 minutes	
Platform slope increases to 1 degree. It is still too small to be noticed by the people in the area.	There are no abnormalities on the platform. All activities proceed without any problems.
There are no consequences for the platform's operations.	All safety systems remain available.
All safety systems remain available.	ROV control room staff needs to ask if the platform production proceeds normally.
An intranet banner announces world record broken by the platform in productivity and availability in the offshore market and everyone is proud.	The team at the PROCESS PLAN must report the completion (success) of routine work authorized by the PW-022 and inform of their intention to return to the control room to close PW-022.
Naval team in the control room is unable to explain the platform 1-degree slope.	
Time line: 15 minutes	
Control room instrumentation on the platform tilt continues to indicate 1 degree. Such a slope cannot be perceived by the personnel in the area.	Everyone realizes that the platform is listing (tilted).
There are no consequences for the platform's operations.	All safety systems remain available.
All safety systems remain available.	ROV control room staff must report the slope to control room and stop activities.
People in the control room can notice the platform tilt.	The MACHINE ROOM team must report slope and request information about the problem.
Control room Naval team is unable to explain the reason for the platform to remain tilted and suggests assessment of the need to generate the escape and abandonment alarm.	BOW Team should carry on PW-021 activities.

(*Continued*)

(Continued)

Control room team	Open area team
Time line: 20 minutes	
After the banner notification on the intranet, a new headquarters phone call congratulates the platform team for productivity and availability. A big jolt suddenly tilts the platform to a slope currently less than 16 degrees. Fire alarm in the living quarters. Storm of alarms in operating systems. Control room Naval team discovers and reports that the automation systems that control the ballast and stability were "frozen" at 1 degree in the indicator and this defect caused the accident	A big jolt suddenly tilts the platform to a slope currently less than 16 degrees. All safety systems remain available. ROV control room team warns of smoke in the living quarters from a fire in the kitchen. Team also explains that as a result of the sudden inclination, heated oil fell and ignited. MACHINE ROOM team asks control room for explanation one more time. BOW team continues to be held in the same place with difficulty to return to the living quarters due to steep slope—team does not submit any report to the control room (PW-021).
Time line: 25 minutes End of exercise	
Slope exceeds 16.5 degrees making all safety systems unavailable. Escape and abandon operation should have been performed before this slide. The computer simulation indicates an escape time of 19 min 37 s for the didactic scenario used in the exercise. This time does not include the displacement of the lifeboat meeting points. Basically, a malfunction in the control of the ballast system allowed the platform to tilt 0.5 degree. When 1-degree slope was reached, the data acquisition by the control computer "froze" and the ballast system continued with the wrong alignment causing a jolt, oil drop in the kitchen, fire and loss of safety systems when the slope exceed 16.5 degrees. This scenario is didactic and is not considered in real world studies due to the remote possibility of sudden tilts like the one described.	Slope exceeds 16.5 degrees making all safety systems unavailable. Escape and abandon operation should have been performed before this slide. The computer simulation indicates an escape time of 19 min 37 s for the didactic scenario used in the exercise. This time does not include the displacement of the lifeboat meeting points. Basically, a malfunction in the control of the ballast system allowed the platform to tilt 0.5 degree. When 1-degree slope was reached, the data acquisition by the control computer "froze" and the ballast system continued with the wrong alignment causing a jolt, oil drop in the kitchen, fire and loss of safety systems when the slope exceed 16.5 degrees. This scenario is didactic and is not considered in real world studies due to the remote possibility of sudden tilts like the one described.

5.9 Answer

The objective of this exercise is to develop the perception of the importance and the influence of the principles of safety culture, human factors, and risk management defined in Chapter 2, Fundamentals of Risk Management, in emergency and crisis scenarios. The final result of each group is not the most important aspect, but rather the discussion of the Chapter 2, Fundamentals of Risk Management, principles and the exchange of information between groups that needed to deal with the same theoretical scenario in different environments. If by 25 minutes into the exercise the escape and abandonment operation is called, the simple fact of meeting at the muster station (room of the operators from the main control room) does not mean the conclusion of the exercise. The platform manager will still need to make the explicit decision of the abandonment to the sea. It is necessary that everyone understands and accepts such a decision and then effectively execute the abandonment to the sea. Based on the results of the computer simulations performed for the exercise's theoretical scenario, the travel time required to the lifeboats meeting point is about 10 minutes. To this time, another 10 minutes should be added for boarding and launching lifeboats overboard. Therefore, to have enough time to complete the operation before the slope reaches 16.5 degrees and the lifeboats become nonoperational, the decision of the effective abandonment to the sea must be communicated to the POB (gathered at the meeting point) no later than 5 minutes after the emergency occurs. Considering to the didactic nature of this exercise, we can say that the ideal action expected from the teams is that at the time of first information announcement (first 5 minutes) all the POB should be gathered at the meeting point (muster station). Also during the 5 minutes, the platform manager must make the call on the effective abandonment to the sea. After this time, even if the abandonment order is communicated to the POB, while the lifeboats descent process is underway the "jolt" would happen and the lifeboats would be stuck on the side of the platform before the descent into the sea could be completed. However, if the abandonment order were given between 5 and 15 minutes, although it would avoid the lifeboats' unavailability for launching, at least the operators working at the bow of the FPSO would not be blocked and would be able to get to the meeting point, improving their chances of survival. Therefore as part of the evaluation of the exercise result the following points also need to be verified:

- Did the manager order the travel to the meeting point (muster station)?

- Did the manager order the actual abandonment to the sea?
- Did the order for the travel to the meeting point happen within 5 minutes?
- Did the order for the actual abandonment to the sea happen within 5 minutes?
- If the travel to the meeting point and the order of actual abandonment did not happen before 5 minutes, did it at least happen within 15 minutes of the exercise?

If all orders were given within 5 minutes from the beginning of the exercise, it was performed in a timely manner for the safe abandonment to the sea. If the order to travel to the meeting point was given between 5 and 15 minutes, the team working at the bow was spared from the situation of blocked access to a nonhazardous area. If the order for actual abandonment was given between 5 and 25 minutes from the beginning of the exercise, lifeboats would become unavailable, preventing or interrupting its launch into the sea (life rafts and jump to the sea would remain as the only options for abandonment). If no order for travel to the meeting point and no order of actual abandonment to sea were given or understood by the POB within 25 minutes, the exercise gets concluded.

5.10 Review questions

- What is the most important component of the Risk Management Strategic Line to save human lives?
- What is the basis for the choice of the term "hazard escape and abandonment scenario system"?
- What does computer simulations of escape and abandonment allow us to study beyond the conventional theoretical methods and planning?
- Explain the principle of human factors associated with the importance of the escape and abandonment system to save lives.
- What attitude of the survivors of the Piper Alpha accident made the difference between life and death? In what procedure was it planned?
- Why is fighting large-scale fires on an offshore rig limited?
- Give examples of the interaction between people and the platform during the perception of an emergency and during an escape and abandonment operation.

- What is the escape operation?
- What is the abandonment operation?
- What are the differences between primary and secondary escape routes and what are the recommended dimensions for escape and abandonment routes on projects?
- What are the destinations of the secondary escape routes and the primary escape routes?
- Who is responsible for the abandonment order?
- Explain the meaning of the term "time window" in the context of escape and abandonment strategies.
- Can the escape and abandon operation be considered completed after the arrival of the entire POB to the meeting points (muster stations)? Why?
- Give examples where additional means of travel from the physical boundary of installations at shore/in shore are required in escape and abandonment operations.
- Describe the sequence of the escape and abandonment operation with an intermediate task.
- Describe the sequence of the escape and abandonment operation without an intermediate task.
- Describe the sequence of the direct escape and abandon operation.
- Describe the dimensions and basic recommendations for escape routes.
- Describe the meaning of the acronym EERA.
- What are limited access spaces and machine rooms?
- What aspects should be considered in escape and abandonment projects for limited access spaces and machine rooms?
- Under what conditions can plastic materials be used in escape routes?
- Define and describe the characteristics of the meeting points (muster stations).
- What is the main equipment for survival at sea?
- What does the acronym TEMPSC mean?
- What is a davit?
- What are the differences between life raft and lifeboat?
- Explain the concept of launching of conventional rescue boats.
- What is a free-fall lifeboat and how does it work?
- Which concept of launching of rescue boats is most closely aligned with Principle 2 of Efficiency in Risk Management? Why?
- Compare the advantages and disadvantages between the conventional and free-fall concepts for launching lifeboats.

- What is POB and how does it influence the configuration of lifeboats on fixed, SS and FPSO offshore platforms?
- Name three types of davits and their operating characteristics.
- Under what scenarios can a free-fall lifeboat be launched by the davit using cables?
- How often should lifeboats (conventional and free fall) be launched into the sea (for testing)?
- How are life rafts stored and launched?
- How many life rafts are required on fixed platforms? And in other types?
- Explain the characteristics and functionality of the rescue boat in escape and abandonment operations.
- What is lifesaving equipment and how should they be specified?
- What are the limitations for launching lifeboats in the case of listing?
- What is the difference between meeting point and abandonment point?
- What is the procedure should an operator fall overboard while working in an offshore rig?
- What are the options for the means to abandon an offshore rig?
- What are the main risks of a hasty and unnecessary abandonment operation?
- What are the main risks of a delayed abandonment operation?

CHAPTER 6

Emergency control

In the fourth level of priority and importance, component of the risk management strategic line is emergency control. In order to avoid an accident and its consequences, the top requirement is technical and operational knowledge, which means technical expertise of ongoing activities. Second, if a threat is established, generating unacceptable risks to the facility, immediate reduction of hazards becomes the priority, as much as possible. Hazards reduction will contribute to higher likelihood of success in the effectiveness of the third component, which is the removal of agents from the accidental scenario. In this chapter, we will cover the fourth component of the risk management strategic line: emergency control (Fig. 6.1). The systems associated with emergency control aim to provide a response to the accidental scenario, avoiding the escalation of the severity of the event

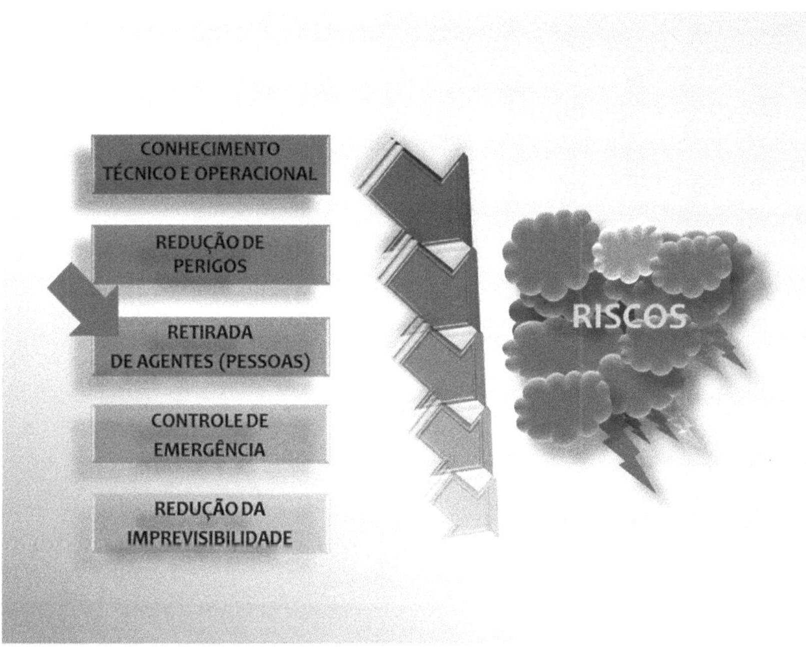

Figure 6.1 Emergency control.

Risk Management in the Oil and Gas Industry.
DOI: https://doi.org/10.1016/B978-0-12-823533-1.00006-X

beyond the control capacity supported by the facility project. When the systems associated with emergency control fail, the facility loses the ability to provide a response to the accident, which establishes the maximum degradation of the accidental scenario: the beyond design-basis accident.

Each system and technical discipline in an oil and gas industry facility contains components and subsystems associated with emergency control. In addition, other systems are designed exclusively to respond to the accidental scenario. In this chapter, we will present the main requirements and technical characteristics of the systems associated with the emergency control component that is part of the risk management strategic line.

6.1 Power generation systems

There are many possible configurations and strategies for the design of power generation and supply systems. As part of an emergency control strategy, the power generation and supply system for oil and gas facility can be subdivided into three circuits: main, auxiliary, and emergency. Each circuit is associated with an independent source of power generation designed to meet specific load demand.

The main circuit supplies power to the process and utilities plant, the auxiliary circuit supplies power to the superstructure (living quarters) and auxiliary equipment. Lastly, the emergency circuit meets the load demands deemed essential and safety related.

For risk management purposes, it is important to characterize emergency power generation systems through the establishment of some fundamental technical requirements:

1. Emergency power generators need to be autonomous and independent. Autonomy is characterized by the generator's ability to provide all the resources required for its full operation (fuel, start-up air, cooling, etc.) separated from the rest of the facility, dedicated exclusively to the emergency power generator. Independence means that postulated accidental events from the project's safety studies that result in the unavailability of the other power generation systems (main and auxiliary) cannot simultaneously cause the unavailability of the emergency power generation.
2. Start-up of emergency power generation systems should be automatic. The automation and control and power protection systems need to

sense the state of emergency that requires the emergency generator to start and send the start-up signal to the emergency power generator without the need for manual actions.

3. There are various types of emergency power generators such as turbo generators and diesel-engine generators. In the case of rotating machines with greater start-up inertia, as for example, with diesel-engine generators, the system should reach the nominal rotation specified in the project within 45 seconds after the loss of main power generation. The detection and recognition of the emergency scenario and all mechanical stages of the start-up operation are also included in this time window.
4. The power generation system needs to be designed to maintain the autonomy and independence necessary for an 18-hour continuous operation in the midst of an accidental scenario without the need for refueling. There are situations where the project requires an extended autonomy greater than 18 hours, and in other projects, on the other hand, analyses and safety studies may justify a reduction of sustained autonomy to values below 18 hours. These options depend on the standards adopted as reference by the project, as well as on safety studies that demonstrate that the autonomy time suggested in the project is technically adequate.
5. With respect to the independence of emergency power generation systems, special attention must be paid to the physical location of the associated equipment in the installation to be protected. Emergency generation equipment (turbo generator, battery pack, and circuits) should be located such that in the event of a fire, flood, or other accident affecting the main power generation it will not cause the unavailability of emergency power generation.
6. Also related to the location, emergency power generation equipment cannot be positioned in areas considered of risk regarding the buildup of an explosive atmosphere (classified areas—Section 6.1.4). Even if the power generation equipment and emergency electrical circuit are located in a room separated by walls, these rooms cannot be adjacent to classified areas. If that is not at all possible, the walls should be built using special high-strength bulkheads in order to preserve the emergency power generation and supply in the accidental scenarios considered in projects. Walls of this type of room should be compatible with the bulkhead classification standards adopted by the project (Section 6.8.2). In this situation, the design of such rooms cannot only consider only forced ventilation as sufficient protection to classified areas.

In addition to the main, auxiliary and emergency power generation sources, the facilities need to be equipped with a battery pack with a DC/AC power inverter, known by the acronym UPS (uninterruptible power supply). The UPS unit needs to be permanently powered (for recharging the batteries) by either the main generation system or the emergency generation system. UPS units should have sufficient power capacity to meet the demand of all energy consumers that need to remain available after an emergency power generator is shut down. These energy consumers are called ESSENTIALS because they ensure the conditions for the execution of escape and abandonment operations.

6.1.1 Essential consumers

Those are the systems directly associated with the protection of people and the equipment considered essential for the operation of escape and abandonment. Essential consumers should continue to be powered by UPS even after the highest level emergency shutdown (ESD), namely, ESD LEVEL 4.

Each project needs to define in advance as a strategy the classification of essential consumers for the related circuit and UPS to be properly specified. Following are examples of consumers classified as essential for different types of offshore rigs:

6.1.1.1 Essential consumers common to fixed and floating platforms

1. Loads defined as essential by the modular production and completion units supplier.
2. Davits for lifeboats and rescue boats.
3. Essential lighting.
4. Helideck lighting.
5. Obstacle signals and markings for aircraft approach.
6. Ventilation and exhaust of rooms containing essential consumers (AC and DC).
7. Flare ignitor (only for units with pilot flame on).
8. Battery chargers.
9. Controllers and auxiliary systems for essential consumers such as well control, fire-fighting water pump (FWP, water, and foam deluge systems), emergency power generators, air compressors, and others, provided that they are defined in the project as such.

10. FWP (water and foam deluge systems)—whenever applicable, for example, in the case of electric pumps. Note: electric FWP must have dual power supply (according to item FWP).
11. Water mist system, when required.
12. Lighting equipment in the lifeboats launch area.
13. Divers support systems.
14. Search and rescue lighting equipment.
15. Battery bank or UPS.
16. Air conditioning system for the main control room, radio room and telecommunications room.
17. Flare purge nitrogen system.
18. Container/turbine room ventilation system.

6.1.1.2 Essential consumers on semisubmersible floating platforms

1. Ballast and sewage pumps (the emergency power generator needs to be specified considering the simultaneous operation of the number of ballast pumps corresponding to 50% of the required ballast system capacity).
2. Hydraulic valve control unit, waterproof doors and hatches, and dampers.
3. Vessel control system (power supply through the emergency circuit and UPS).
4. Flood monitoring system (columns, structural elements, voids, pump rooms, elevator shaft, etc.).
5. Column elevators.
6. Electric inclinometer for the trim and band (power supply via the emergency circuit and UPS).
7. Ballast ring adjustment system.
8. Essential naval systems as defined by the classification society.

6.1.1.3 Essential consumers on FPSO/FSO floating platforms

1. Hydraulic valve control unit.
2. At least one ballast pump, one sewage pump and one for general engine room services or more, according to the design criteria.
3. One or more sealing pump on the deck (for inert gas system), according to the design criteria.
4. Essential naval systems as defined by the classification society.
5. On platforms equipped with instrument air compressors powered exclusively by an electric motor, their power supply source by either emergency generation or auxiliary generation needs to be considered.

Only one of the compressors should be considered for the definition of the power requirement of the emergency power generator.

6.1.2 Safety consumers

Those are the systems directly associated with the people's protection and the rig integrity, which cannot be subjected to power outage during the transfer of the main power generation to the emergency generation. Consumers should remain energized, even in the event of loss of emergency power generation, the power needs to be supplied the UPS unit. Following are some example of consumers that can be classified into this category:

1. With 30-minute autonomy
   a. gas and fire detection system,
   b. fire-fighting system,
   c. closed-circuit TV,
   d. ESD system,
   e. emergency lighting,
   f. telecommunication and intercom,
   g. auxiliary panels and auxiliary power generators,
   h. emergency power generator control panel,
   i. FWP control panel,
   j. flare ignition panel for units with no pilot (closed flare system), and
   k. control and supervision systems by the main control.
2. 12-hour autonomy
   a. Emergency lighting associated with the safety equipment operation, as for example, panels of emergency and auxiliary power generators, meeting points, abandonment points and the main control room.
3. 96-hour autonomy
   a. Equipment associated with navigation required also in offshore rig projects such as navigation lights, fog horns, etc.

6.1.3 Special requirements for cables and lighting

In order to avoid the emission of toxic fumes during fires in the superstructure of offshore rigs or in enclosed spaces with a large concentration of people in the on land facilities, power cables should be halogen free. When the power source is external, to overcome the distance the cable routing should take place preferably through safe areas, where occurrence of accidents and damages is less probable. When it is not possible, for example, for electrical cables that supply essential and emergency services in "Hazardous Areas," the recommended strategy is the redundant routing through two different routes. Moreover, the routing design should consider that a fire risk scenario does not reach both routes simultaneously. As an alternative to the use of different routes, a fire-resistant electric cable can also be used. Cables that supply normal and essential services and that are installed in stabilizing columns in semisubmersible (SS) offshore rigs may pass through the same route, as long as the essential cables receive passive protection, that is, fire-resistant type.

The essential lighting is powered by the emergency generator through the essential panels, and the luminaires must be suitable for operating in classified areas (Section 6.1.4). Even when located outside classified areas, essential lighting equipment needs to meet additional safety requirements (enhanced safety luminaires). Exception is made for luminaires located inside mechanically ventilated rooms or within superstructures, buildings and living quarters. The emergency lighting equipment should be powered by the emergency generator during ESD-4 and, in the event of a power outage from the emergency generator, it should be powered by the UPS unit.

Some electrical equipment needs to be installed in areas exposed to rain, wind and other important influences to safety. Therefore electrical equipment can be classified according to the degree of protection required for each area and the associated risks. Electrical and electronic equipment installed in areas with fire protection provided by water spray systems, or by water mist, as well as adjacent equipment should have an Ingress Protection degree, as required by the manufacturer, according to the standards adopted by the project and, in the case of offshore rigs, also by the standards of the classification society of the installation.

6.1.4 Area classification

Electrical equipment installed in hazardous areas is designed with specific protection characteristics to prevent electrical equipment from becoming an ignition source in an area where a potentially explosive atmosphere can be formed. Such characteristics are compatible with the acceptance of an explosive atmosphere in the area where the electrical equipment will be installed. In order to know the protection characteristics required for each electrical equipment related to its installation location, the areas with the highest expectation of formation of explosive atmospheres are classified through lists, drawings, and design documents. Accurate information on Area Classification is recorded in internationally adopted standards that serve as a reference for the preparation of area classification documents (e.g., API 500/505, IEC 60079, and others cited in Chapter 9: Risk managements systems).

It is very important to understand that the classification of areas is neither considered a risk analysis technique nor a formal study of quantitative or qualitative risk analysis. The area classification drawings show regions demarcated according to the class associated with the explosion risk. It is also important to realize that the classification of areas in fact generates volumes that are representative of the explosive atmosphere buildup region and not strictly areas, as the name of the technique seems to suggest due to its well-established use. These demarcations, though, are only references and are not completely associated with the actual regions of formation of explosive atmosphere, which are dependent on more complex influences such as the direction and intensity of winds at the moment of emission, the degree of confinement, pressure of the sources of emission and leakage, among others. None of these influences are fully considered in the preparation of area classification documents. The objective of area classification is not exactly to accurately reproduce a region of possible cloud formation with an explosive atmosphere. Rather, it is a coarse approach to define areas adjacent to equipment and components of hydrocarbon installations that deserve special attention regarding the type of protection to be required for the electrical equipment installed in them, based on the experience accumulated by experts and operators.

This experience is recorded in internationally adopted standards that serve as a reference for the preparation of area classification documents. The use of such standards makes it possible to specify the energized equipment in the processing area, even at an intermediate stage of the project, without the need for more sophisticated studies, which can only be carried out with technical quality later at more advanced stages, as the project gets wrapped up. Through the classification of areas, even at the early stage of the project, it is possible to define in a conservative way (prioritizing safety) the areas of theoretical risk of formation of explosive atmosphere. Each classification level is associated with its respective safety requirements for the electrical equipment to be installed in a given region. In summary, the classification of areas is not a safety study but rather a helper tool for designers to use as part of the work related to the specification of materials for electrical design of facilities with hydrocarbons. The demarcations in the area classification drawings are not equivalent to the documents of the analyses and studies of gas dispersion, fire propagation, and explosion (Chapter 7, Reducing unpredictability). These types of studies are much more accurate and closer to the real scenarios, and they can be used in risk analyses, unlike the classification documents of areas that are not applicable for this purpose.

The classification of areas topic is important for all experts in risk management, but for those who need to work directly in the specification of electrical equipment and in the preparation of technical documents and drawings for area classification, it is necessary a deeper level of understanding, through a specific course, which is outside the scope of this book. The main parameters that influence the classification areas are: rate of flammable material release, the lower limit of flammability, ventilation, and relative density of gas or vapor. Next we will present some basic concepts and general information about classification of areas and related documents, with a focus on risk management in facilities of the oil and gas industry. As a note, Section 3.3.2 also covers some of these concepts with complementary information.

6.1.4.1 Concepts, physical, and chemical phenomena

1. *Ignition temperature:* lowest temperature at which an explosive atmosphere can ignite. The ignition temperature of a substance, in a solid, liquid, or gaseous state, is the minimum temperature required to start or cause combustion, regardless of an igniting element. The ignition temperature is also called the autoignition temperature or apparent ignition temperature. The determination of the ignition temperature value varies according to the dependence between several factors that directly interfere in the final result. Some of these parameters are: composition of the gas and air mixture, shape and size of the space where the ignition occurs, heating duration, ignition source and temperature, and catalytic effect of the materials involved.
2. *Explosive atmosphere:* stoichiometric mixture of air and flammable vapors in the ideal proportions for an explosion to occur in the presence of an igniting source.
3. *Combustion:* chemical reaction in which a substance is combined with an oxidizer resulting in the release of energy in the form of heat and/or light (flame).

4. *Combustion speed:* based on the type of flammable vapor, source of ignition and the ratio between the amount of vapors and the amount of oxygen, the combustion reaction can occur at different rates of speed.
5. *Deflagration:* combustion reaction that occurs at a speed below the speed of sound (subsonic) and may generate pressure waves that move at subsonic speeds.
6. *Detonation:* combustion reaction that occurs at speeds above the speed of sound (supersonic) generating shock waves that move at supersonic speeds.
7. *Explosion:* sudden expansion of gases, generating a shock or pressure wave. It can be confined (in an enclosed space) or not confined (in a sparsely occupied open space).
8. *Dust explosion:* combustion chain reaction of tiny solid particles such as dust and fibers.
9. *Boiling liquid expanding vapor explosion (BLEVE):* explosion resulting from the external heating of a vessel or tank containing liquid hydrocarbons in which the external heating causes the internal pressure of the vessel or tank to increase past its mechanical strength causing its rupture releasing vapors and liquids to a lower-pressure external environment. The thermodynamic transient produces a vapor cloud that generally gets ignited in contact with the same external heat source that caused the event or it can be ignited by any other external ignition source that reaches the vapor cloud.
10. *Flash point:* lowest temperature at which a liquid releases enough vapors to form a flammable mixture.
11. *Vapor pressure:* it is a measure of the tendency of a liquid to evaporate. The higher the vapor pressure the more volatile the liquid is and the lower is its boiling temperature relative to other liquids with lower vapor pressure at the same reference temperature.
12. *Flammable liquid:* liquid having a flash point below 37.8°C (100°F). Flammable liquids are classified into:

    Class I: with a flash point below 37.8°C (100°F) and vapor pressure that does not exceed 2068.6 mmHg (40 psi) at 37.8°C (100°F). Flammable liquids are further subdivided into:

    Class IA: with a flash point below 22.8°C (73°F) and a boiling point below 37.8°C (100°F).

    Class IB: liquids with a flash point below 22.8°C (73°F) and a boiling point $\leq$ 37.8°C (100°F).

Class IC: liquids with a flash point equal between 22.8°C (73°F) and 37.8°C (100°F).

Note: classification according to ASTM D 323-Standard Method of Test for Vapor Pressure of Petroleum Products (Reid Method).

13. *Combustible liquid:* liquid that has a flash point ≤ 37.8°C (100°F) classified into:

Class II: with a flash point between 37.8°C (100°F) and 60°C (140°F).

Class IIIA: with a flash point between 60°C (140°F) and 93°C (200°F).

Class IIIB: with a flash point above 93°C (200°F).

Note: classification according to ASTM D56-Standard Method of Test for Flash Point by the Tag Closed Tester.

14. *Flammability limits:* the flammability range defines the lower flammability limit (LFL) and upper flammability limit (UFL) within which the concentrations of the substance establish an explosive atmosphere. Concentrations of the substance below the LFL are considered to be a nonflammable "lean mixture." Concentrations of the substance above the upper limit of flammability are considered as "rich mixture," in which the amount of oxygen is insufficient for the substance to burn.

6.1.4.2 American and international standards

There are two lines of approach for preparing area classification documents and for the specification of the respective electrical equipment. One line of approach is based on the National Electrical Code (NEC), an American national standard, and the publications of the American Petroleum Institute (API). The second line is based on the international standards of the International Electrotechnical Commission (IEC). There are no conceptual incompatibilities between the standards, but the international trend is to adopt the international (IEC) standards. European countries and Mercosur countries (Brazil included) adopt IEC standards. American (NEC) standards recommend in their texts that the approach to the theme in new projects be based on international (IEC) standards as well.

6.1.4.3 Electric and nonelectric ignition sources

The explosive atmosphere can be ignited by energized equipment due to sparks, arcing, corona discharge, and also if the surfaces of electrical equipment are overheated and get in contact with the explosive mixture. Deficiencies of electrical insulators can generate sparks and corona discharge. A spark is a sudden discharge of electrons in a high electrical field, while the arc is a sustained flow of electrons. The Corona discharge occurs due to the rupture of the dielectric strength in the region affected by a very high voltage. However, this voltage is not sufficient to cause arc or sparks, but enough to ionize the air in the surrounding area. The closing and opening of electrical contacts cause frequent sparks and arcing, depending on factors such as the material of the contacts and electrodes, voltage and intensity of current, AC voltage frequency, and the speed of the open–close contact reversal cycles. Not all sparks have sufficient high energy level to cause ignition and for an explosive atmosphere reaction. The minimum ignition energy is only reached at higher voltages (above 127 V AC). Control and communication equipment in general operate on low-power electrical signals: typical 24 Vcc supply voltage and standard transmission current between 4 and 20 mAcc. The use of microprocessor chips requires low voltages around 5 Vcc; arcing occurs at voltages above 300 Vcc.

High-temperature operations such as welding and cutting of metal sheets, or other forms of metals, performed improperly, are frequent sources of explosions. Open flame or exposed welding arc is the source of ignition. The high temperature of plates during these processes can also cause the ignition of an explosive atmosphere. A typical accident is one caused by welding or cutting an "empty" tank (tank used to store liquid combustible). An inefficient or lacking purge can generate an explosive atmosphere and create the latent risk of explosion. Only services control through a strict permit to work system can be effective to prevent accidents of this nature.

Another source of ignition is electrostatic discharge, characterized by the release of accumulated electrical charges from one charged material to another that has lower resistance to the ground potential. Even without being powered by a source, electrical sparks can be produced by static electricity. Therefore this type of source needs to be considered including in nonelectrical equipment. Another spark source is parasitic current that occurs due to the potential difference between two points (voltage point and ground) and the intrinsic electrical characteristics of the materials, mainly the electrical resistance. It is necessary to have a highly conductive connection for all conductive parts of the equipment, so that the electric potential difference (voltage) is reduced to a safe level. Radiation energy should also be considered among the possible sources of ignition. The main sources of radiation that can cause ignition of an explosive atmosphere are ultrasonic—density and level meters; electromagnetic—radio waves; electromagnetic—infrared (IR), ultraviolet (UV), and visible light; and ionizing radiation—radioactive sources. Some equipment produces radiation and the control the emission parameters needs to be included as part of their design. Thus this type of equipment can be safely used, since the radiation produced is limited on a permanent basis. Moreover, they need to be tested and certified before they can be operated in potentially explosive areas.

There are other nonelectric sources of ignition. Furnaces, ovens and boilers are sources of ignition because they operate at high temperatures. Furnaces or ovens can burn fuel oil, thus always generating risks of formation of explosive atmospheres in its surroundings. During the start up or shutdown of the furnace, the risk may be the highest when there is an interruption of in the air supply, which enables the formation of mixtures of atmospheric air and flammable vapors within the flammable range. Well-designed and installed furnaces and ovens are not an uncontrolled source of ignition under normal operation because the safety of this equipment is ensured by the correct control of the proportions of the mixture through the adequate air flow in relation to flammable vapors, which maintains the atmosphere below the lower flammable limit.

The large volume of air that passes through the gas turbines and the combustion chamber allows in some cases the turbines not to be considered sources of ignition. The turbines installation locations are normally not classified as hazardous, based solely on the equipment. But despite that, external electrical components should have some special level of protection. It is not unusual for the turbine, combustion chamber, and boiler to be located in the same space or to lack adequate separation. In such cases, careful analysis is required, considering the possibility of an explosive atmosphere, that is: it is necessary to classify the area and use equipment and devices suitable for this type of location. The boiler control is often guided by dependent limits, in which air and fuel flow rates are controlled in a fixed ratio. When the boiler load increases, the air flow is increased first, and when the boiler load is reduced, the fuel flow is decreased first. Thus the formation of excess fuel is avoided.

Flames, gases and hot particles may be found within the combustion machines, either during normal operation or during equipment failure. Safety requirements need to be met to prevent hot flames and gases from being released from the enclosures. Another source of ignition is hot surface. If the temperature on the surface exceeds the ignition temperature of the local area atmosphere, it can be considered as an ignition source. Other considerations need to be made for special situations, like for example, in the case of a flammable mixture, as turbulent flow or in the form of a jet, traveling toward a heated surface. Even at high surface temperatures, due to the travel speed of the mixture, it will not be sufficiently heated to reach the hot surface temperature and, therefore it will not be ignited. If the hot surface area is small in comparison to the volume of the moving mixture, the hot surface will not be able to heat up the mixture to the temperature required for ignition to be started. This is the most common situation for luminaires and motors installed in hazardous, open, and well-ventilated areas. The hot surfaces are the result of energy losses of systems, equipment or components during normal operation. It is not recommended the use of equipment whose operating temperature is higher than the ignition temperature of the explosive atmosphere that can be formed in the environment. The operating temperature should not exceed 70% or 80% of the ignition temperature.

A source of nonelectrical ignition is the mechanical spark that may occur during the collision of two surfaces or when they maintain constant contact, as for example, in grinding activities, in which a beam of sparks is formed, or in abnormal conditions, when two parts accidentally come into contact with one another.

Mechanical sparks can be generated during normal cutting and finishing equipment operation, and therefore they must be excluded from hazardous areas. Structural failure of rotating parts, insufficient lubrication in sliding parts, and similar situations can also generate such instant sparks when the equipment is defective or it fails. The use of nonsparking metal, such as bronze and some aluminum alloys, and nonmetallic materials, can reduce the likelihood of formation of a mechanical spark.

Adiabatic compression and shock waves within pressurized tubular structures can also become a source of ignition. An example is the fracture of a tubular fluorescent lamp, which is filled with an atmosphere of hydrogen and air (in addition to some other metals in much smaller quantities).

6.1.4.4 Degree of risk source

A source of risk is an equipment or a process plant site where a substance can be released to form an flammable/explosive atmosphere. The sources of risk can be classified according to the frequency and duration of the substance release. For example, storage tanks containing liquid hydrocarbon have permanent volumes of vapors above the liquid surface. Or, during operation when the internal pressure in a vessel exceeds some normal limit value, the relief valves can release vapors to the environment. Thus the sources of risk can be classified as:

1. *Source of risk of continuous degree:* when the release of a substance occurs continuously for long periods of time or frequently during short periods. An example is the surface of a flammable liquid located in a fixed roof storage tank without inertization. It can also be mentioned the surface of flammable liquid that is open to the atmosphere, either continuously or for long periods, as is the case with water and oil separators.
2. *Primary risk source:* when the substance is released periodically or occasionally, under normal operating conditions. It may be caused by repair operations, frequent maintenance, and predictable conditions. Some examples: pump seal, compressor seal or valve seal, as long as the release of flammable product is expected under normal operating conditions; water drains in vessels that contain flammable liquids and that may release flammable product to the external environment during water drainage under normal operating conditions; sample collection locations where flammable product is released under normal operating conditions.
3. *Secondary grade risk source:* release that is unexpected during normal operation that occurs infrequently and for short periods. The release of the substance occurs under abnormal operating conditions or due to a rupture/failure in a process equipment. Albeit being abnormal they are postulated as accidental scenarios. Some examples: pump seal, compressor seal or valve seal, flanges, connections, and pipeline accessories, where the release of flammable material to the external environment is not expected to occur under normal operating conditions; sample collection points, relief valves, vents, and other openings where the release of flammable material is not expected under normal operating conditions.

6.1.4.5 Ventilation types

Ventilation is an essential parameter in the classification of areas and may justify the declassification of a supposedly hazardous area. Environments can be classified with respect to ventilation as follows:

1. *Adequately ventilated environment:* naturally or artificially ventilated rooms, buildings or equipment enclosures.

2. *Environment with natural ventilation:* where the air motion and fresh-air renewal occur due to wind effect and/or a temperature gradient. The following are considered naturally ventilated: environment open to the exterior environment in all directions; environment protected by a roof, wall, or screen with free areas (open on side or upper walls) $\leq 60\%$ of the area calculated as the environment perimeter times 2.5. For gases or vapors heavier than air, the free areas should consider the lower regions, and for gases lighter than air, the upper regions.
3. *Limited ventilation environment:* it has obstacles that make it difficult, but do not block the natural air circulation.
4. *Environment with blocked ventilation:* No air movement and probable accumulation of flammable gases or vapors forming an explosive atmosphere.
5. *Environment with artificial ventilation:* an artificial system of fans or exhaust fans is used to prevent the formation of an explosive atmosphere. The artificial system should either be able to perform at least 12 air changes per hour or to provide an air flow of 0.46 m^3/minutes/m^2 of the area of the environment in consideration, whichever is greater, under the conditions of atmospheric pressure and temperature between $-10°C$ and $40°C$.

For any type of ventilation, another very important factor should be considered: the degree of ventilation. This qualitative parameter is representative of the ventilation efficiency at the site, which makes it possible the verification of the ventilation adequacy for reducing the degree of risk of an specific area. This parameter is related to the wind speed and the number of air changes per unit of time.

6.1.4.6 Group and zone classification

Area classification is a project requirement, used in the construction and processes operation in industries that handle with flammable substances. The main criteria for area classification of industrial plants or locations with the possibility of formation of an explosive atmosphere are related to the probable characteristics of the explosive atmosphere and the likelihood of the actual formation of such atmosphere.

The classification of an area is associated with its mapping into a zone.

From the area classification of an industrial plant, it is possible to specify the equipment with compatible electrical classification, adapting the equipment specification to each location. The criteria for equipment classification are, for the most part, the maximum spark energy that can be produced and the highest surface temperature that can be reached. International standards classify equipment electrical in:

Group I—manufactured for operation in underground mining and group II—manufactured for operation in other industries. Group II is subdivided further into IIA, IIB, and IIC, according to the substances present in the explosive atmosphere that may be formed. With regards specifically to the area to be classified, international standards classify them into zone 0, zone 1, and zone 2.

The classification by zones is mostly influenced by the degree of risk source and ventilation parameters. The zones are classified as zone 0, zone 1, and zone 2.

Zone 0—for an area where an explosive atmosphere consisting of a flammable mixture of gas, steam, or mist and air is continuously present, for long periods or high frequencies. Zone 1—for an area where the occurrence of an explosive atmosphere composed of a flammable mixture of gas, vapors, or mist under normal operation is occasionally probable. Zone 2—for an area where an explosive atmosphere consisting of a flammable mixture of gas, vapor or mist is probable during short periods of time under abnormal situations. Fig. 6.2 shows an example of a drawing with area classification by zones 0, 1, and 2 in an offshore project.

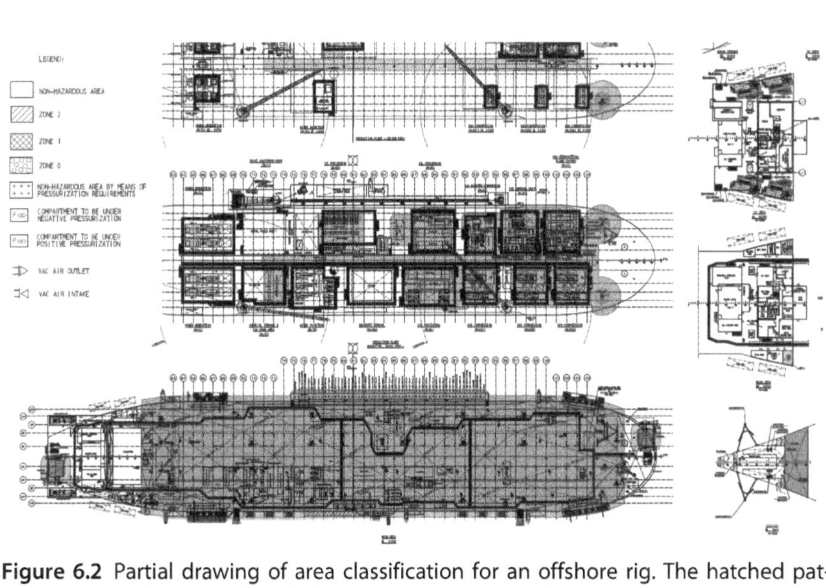

Figure 6.2 Partial drawing of area classification for an offshore rig. The hatched pattern indicates the areas classified as zone 0, zone 1, and zone 2. The electrical equipment installed in the classified areas should meet the specific safety requirements of each type of zone.

6.2 Heating, ventilation, and air conditioning systems

The design of these systems include the choice of air intake and exhaust locations, the associated air currents and airflow, calculation of losses, duct design, thermal insulation, cleaning and treatment equipment, air replacement and air recirculation subsystems, suspension particles control, fans, filters, humidity controllers, measuring instruments, and cooling machines of various types. Forced air circulation also helps the maintenance of the adequate temperature by reducing thermal stress for people and equipment. Adequate ventilation systems also make it possible to keep the concentrations of toxic contaminants at acceptable levels, especially in confined spaces. Even in the office and living quarters areas, ventilation is important for the control of odors and fumes harmful to health and for the functioning of equipment. Through an efficient ventilation system it is possible to control the presence of microorganisms, dust and other particles. Another important contribution of ventilation systems to safety is to limit the concentration of carbon dioxide, maintaining adequate air quality for people. But this type of application becomes even more critical when there are carbon monoxide emissions as it is the case with machines such as diesel-powered pumps and generators. Only the handling of exhaust and the release of the combustion gases in the proper place can ensure an efficient protection for certain environments subject to contamination by carbon monoxide. Below, we will present some important safety recommendations related to heating, ventilation, and air conditioning systems in facilities in the oil and gas industry:

1. For closed or partially open environments, air conditioning and ventilation systems should have minimum capacity of 12 changes per hour of the volume of the space maintained by it or sufficient air flow for sustained concentration of gases and vapors below 20% of the LFL, when the space to be protected is subject to flammable gases or vapors.
2. Battery rooms in particular require special care due to hydrogen vapors buildup caused by batteries recharge. As a result rooms reserved for this equipment need to comply the requirements of international standards and other standards applicable to the related type of facility. One of the possible means to ensure hydrogen dilution is to keep its concentration below 4% of the LFL and have at

least 100% exhaust fan system redundancy, that is, at least two systems, each of which with the ability to meet 100% of the demand. Battery rooms with forced-air ventilation systems should be designed with capacity of at least 30 air changes per hour. The location of the air intake and exhaust points needs to be carefully defined. The air intake of the battery room should be located near the floor level, and the exhaust vent location should be located at the highest level near the ceiling, in order to avoid the formation of hydrogen gas pockets in the highest room areas.

3. If a classified area loses its main ventilation source, the system should be designed for the automatic startup of the backup ventilation system, without relying on any manual intervention. Rooms housing essential equipment (Section 6.1.1) should be designed to ensure continuous ventilation operation even in the event of failure of the main ventilation system.
4. Ventilation systems might transfer air conditions from one location to another. For example, in the event of a smoke-generating emergency, gases can be enter an area through the ventilation system and unnecessarily affect the space with smoke, heat, etc. One way to reduce this risk is to avoid some interconnections of the air conditioning ducts system, for example: the exhaust fan system of battery rooms and laboratories with risk of contamination must not have interconnections with other compartments. Also areas with different classification levels with the respect of electrical equipment installation should not be connected by the ventilation system.
5. Some rooms may have windows or another type of opening to the external environment. In such cases, if there is a classified area located less than 3 m from the window, door or any other type of opening, it is necessary to maintain a positive pressure in the room that directs the air flow from the room to the outside and never the reverse. Besides, the pressurization needs to be monitored to prevent the system failure without the knowledge of the operators in the main control room. Conversely, closed rooms with sources of emissions of gases and vapors that can potentially form an explosive or flammable atmosphere will need to maintained at a pressure level lower than the pressure at any of the surrounding environments. Thus if there is a leak of vapors or gases in any these rooms, the air flow will be from the outside toward the room, instead of the other way around.

6. The operating parameters of mechanical heating and air conditioning ventilation systems need to be continuously monitored. Fluctuations and most importantly failures need to generate signals and alarms in the main control room so that operators can take actions to respond accordingly.
7. Rooms, offices, workshops, accommodations and any other closed environment considered within the safe area should have all outside air intakes equipped with devices that prevent gas flow to protected areas. These devices may be filter mist eliminators with filter and/or water-tight or fire dampers. In the event of a confirmed gas leak, the fire dampers needs to be closed automatically to protect the enclosed areas that are part of the safe areas.
8. The ventilation, heating and air conditioning systems have an exterior air intake vent that performs exchanges with the purpose of maintaining the air quality circulating in the system. The location of this outside air intake point is very important for the system safety and efficiency. The air intake needs to be done from a location in the safe area, at least 3 m away from any classified area, and 4.5 m from the exhaust vent of the own ventilation system (to avoid undesirable air recirculation) and also 4.5 m from the combustion gas discharge point and vents.
9. Some special rooms can be protected by CO_2 fire-fighting systems (Section 6.6.5). For the efficient operation of these systems during a fire, the volume of CO_2 should be contained within the area to be protected, without leakage through the ventilation system. For this reason, areas protected by CO_2 need to have dampers to ensure air tightness in the ducts and/or ventilation slots in the rooms, in order to prevent CO_2 leakage.
10. Oil and gas industry facilities often contain bulkheads, partitions and fire walls. But for the uninterrupted routing of the ventilation system ducts they need to pass through these special bulkheads without reducing the level of protection at what is characterized as a vulnerable point. To solve this problem when the ducts pass through a bulkhead, the duct should have a fire damper installed exactly at the point of intersection between the duct and the bulkhead. This damper, when closed, needs to ensure the same protection category as the rest of the bulkhead. The fire damper panels should be installed so that the damper elements are located on the side of the bulkhead in the region of lower risk.

6.3 Flushing, purging, and inerting systems

Vessels, tanks, pipeline segments, and other spaces containing hydrocarbons can be emptied to meet operational needs. Residual volumes of hydrocarbons in these regions can generate an explosive atmosphere. It is essential for the protection of such spaces from fire and explosion that flushing, cleaning and inerting operations be performed. This is even more so during a process plant startup for the first time or after a period of maintenance and repairs, to carry out an efficient flushing operation. Through this operation, undesirable particles and substances can be removed, preventing damage and contamination from occurring during normal operation, besides accidents. When it is necessary replacing the fluid in circulation in a section of a process plant or to interrupt the use of equipment and pipelines, even if just for a short period of time, there can be no residual quantities of hydrocarbons, otherwise an explosive or flammable atmosphere can be formed from these residues inside the equipment, vessels, tanks, and pipes. Purge is the complete removal of the fluid present in the system to maintain the safety conditions of the process plant.

Another related operation is inerting that is applied to ensure inert atmosphere inside equipment, vessels, tanks, and pipelines to prevent fires and explosions. Inerting is widely used in the oil and gas industry, an important example being the storage tanks for FPSO offshore platforms, shuttle tankers, etc. These tanks are not always in operation for transfer of the oil produced, but they need to be standby and ready. However, due to their large volumes, such tanks need to be subjected to an efficient inerting operation that can ensure protection against fires and explosions resulting from explosive/flammable atmospheres formed by residual volumes of hydrocarbons. The purpose of inerting is to create either poor or rich atmosphere, outside the range of flammability limits. With this purpose, some operators elect to use the natural gas widely available in this type of facility to inert these tanks, by creating a rich gas mixture inside the tanks. However, it is not recommended. A failure to monitor these tanks or the failure of their structural containment can allow air into the equipment and eventually form an explosive/flammable atmosphere. That is why it is not recommended that the inerting of cargo tanks be performed using fuel gas. In fact, whenever possible, inerting should be achieved by a poor gas mixture (concentration below the LFL), incapable in terms of hydrocarbon content to cause an explosion or fire, rather than by a rich mixture that can be altered due to a containment or monitoring failure.

6.4 Gas detection system

The presence of gaseous hydrocarbons in an area with potential sources of ignition characterizes an accidental basic scenario for which oil and gas industry facilities need to be protected against. Control actions such as operation halt, shutdown of power supply, diversion of volumes, and depressurization must be initiated as soon there is a risk of gaseous hydrocarbon emissions forming flammable atmospheres. For efficient protection, it is essential to have a safe and timely detection of gaseous hydrocarbons so that automatic signals can initiate control and mitigation actions. The presence of toxic gases also needs to be monitored as these gases can put in harm's way the life and health of people present in the oil and gas processing facilities, depending on the type of product and plant. Oil and gas facilities with large hydrocarbon inventories need to be equipped with an accurate and efficient gas detection system.

There are various types of gas detectors and analyzers, based on different technologies. These instruments are major cost components and should be located at strategic points in the facility, aiming to optimize costs without compromising detection efficiency. A wide array of manufacturers and suppliers develop detectors and complete detection systems alike. The diverse technologies require specialized knowledge in instrumentation, control, and automation. Adding to that, products get continuously redesigned. New detectors based on new principles are released frequently. Regardless the particularity of each specific equipment made by a given manufacturer, some specific precautions must be observed by the risk management experts who supervise the work of designers and technicians of instrumentation and automation of gas detection systems. Below, we will present some recommendations for gas detection systems of relevance to the quality of risk management in facilities in the oil and gas industry.

6.4.1 Flammable gas detection

The number and location of detectors in a system should be determined based on previous projects experience and from the experience of operators with technical knowledge of the installations processes and routines. In addition, gas dispersion studies (Section 7.2.2) need to be conducted for new situations, tailored specifically for each project. These studies consider the influence of prevailing wind currents, include some of the effects of bulkheads and obstacles, postulate the probable volumes of emissions caused by leaks that may happen at the plant, and in addition consider some specific emergency situations. The results of these studies contribute to the refinement of the document for the location of detectors, correcting possible unprotected areas or removing detectors when it can be justified. Gas dispersion studies in general are conducted using computational fluid dynamics (CFD) tools and rely on a variety of data and assumptions, which make them very vulnerable to distortions if they are not performed under the judgment of experts with solid operational experience in oil and gas processing plants. In the lack of the operational and practical vision, gas dispersion studies will be limited to academic scope, that is: they can yield results using an appropriate scientific method, but these results can be excessively theoretical and, therefore represent an inaccurate approximation of the much more complex reality of operational situations.

As part of a broader strategy, risk management experts need to check the monitoring of all air intakes in the heating, ventilation and air conditioning systems. The ingestion of flammable gases through the air intakes must be prevented.

Methane gas detectors should be installed at each air intake with this purpose, to detect gas intake hazards and immediately generate the signals for mitigation actions. A confirmed gas signal at the air intakes should initiate actions such as: generating an alarm in the main control room, performing the closing of fire dampers, shutdown the ventilation system in the affected area, and also activating the level 3 emergency shutdown system ESD-3 (Section 4.3).

The most suitable types of detector for monitoring flammable methane gas (CH_4) leaks are the IR sensors, which can be point sensor (Fig. 6.3) or open path sensor (Fig. 6.4). The IR sensor is based on the principle that most heteroatomic molecules (formed by two or more different atoms) have the property of absorbing a significant part of the IR radiation that passes through them at certain wavelengths or frequencies. IR detectors employ an IR radiation frequency which is absorbed by virtually all hydrocarbons. This radiation is not absorbed by moisture or other air components, which ensures the detection of hazardous products that may normally leak in oil and gas processing plants. The operation of such detectors becomes more stable with the use of two different wavelengths, namely, one for absorption and the another for reference. In point sensors, flammable gases and vapors must penetrate an internal chamber containing IR radiation. On the other hand, open-path sensors generate an

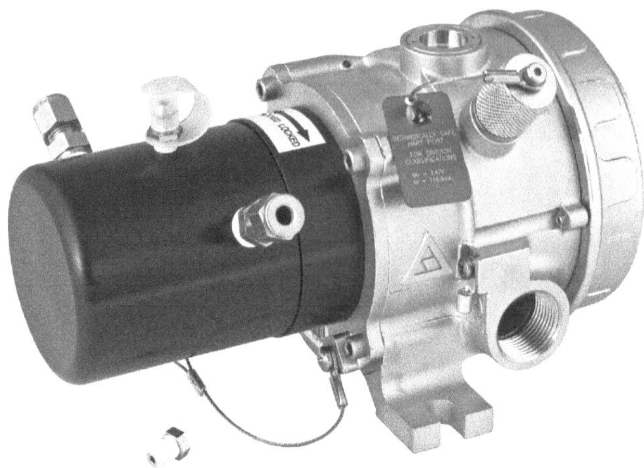

Figure 6.3 Infrared point gas detector. *Courtesy Det-tronics.*

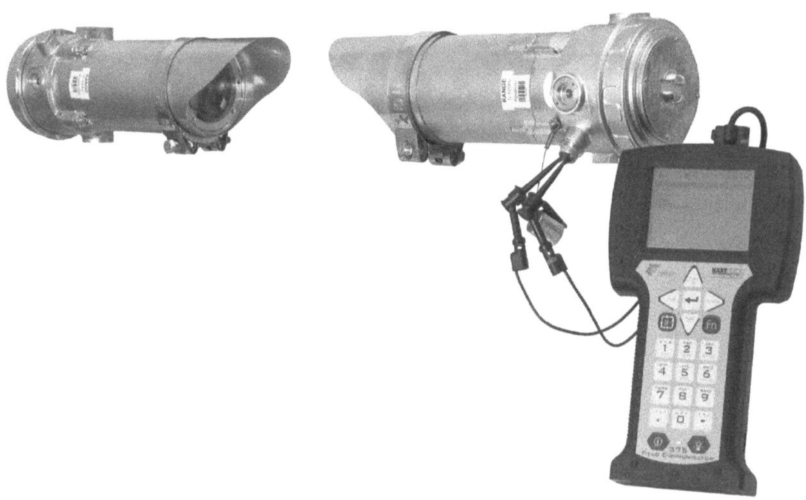

Figure 6.4 Open path infrared gas detector. *Courtesy Det-tronics.*

IR radiation beam that travels outdoor distances between tens and hundreds of meters.

Open-path detectors are not a replacement for point detectors, which are capable of accurately monitoring a specific point. They are complementary; open-path devices are useful, for example, to monitor perimeters of storage tanks areas, perimeters of processing plants, pump seals, or geometrically aligned flanges.

They also serve to detect clouds from a leak that disperses and gets diluted quickly. In this type of leak, the cloud is unable to stabilize in sufficient concentration to be detected by the point detector, but the open-path sensors can detect the leak due to the integration effect along the path of the IR radiation beam. The integration effect concept considers the lower explosivity per meter limit values traveled by the radiation beam. The detection open-path sensor, therefore, depends on the radiation absorption along the entire beam path, considering the integrating effect. It is worth noting that open-path detectors do not work properly in congested areas or areas subjected to vibration and, obviously, gas detectors cannot become sources of ignition themselves, as they need to be installed in locations with higher likelihood of occurrence of an explosive atmosphere. Therefore IR detectors should be specified with adequate degree of protection for operation in hazardous areas.

The monitoring of hydrogen gas (H_2) can be performed by catalytic detectors (Fig. 6.5), with an suitable level of protection to operate in classified areas. Catalytic detectors explore the effect of a catalytic material that is sensitive to any flammable gas or vapor coming in contact with the sensor. The working principle is based on a sensor (also called pellistor) composed of a metallic wire resistor (or a thermistor), which is coated with a catalyst that promotes the oxidation of the combustible gas or vapor by the oxygen in the air, generating a localized combustion. Combustion raises the temperature, which in turn changes the electrical resistance of the resistor or thermistor. The pellistor is connected to a bridge circuit along with the other similar element, but the latter element is coated with an inert, noncatalyst material, which compensates for variations in the ambient temperature. The pellistoris heat controlled so that it does not become a source of ignition causing a fire or explosion. The pellistor is mounted in an explosion-proof enclosure, and the product to be detected penetrates the sensor through a flame arrester device (sintered metal or ceramic insert that work as heat exchanger for the gases generated in an eventual internal combustion to cool down before escaping to the external environment, preventing the spread of combustion). Catalytic detectors are suitable for combustible and flammable gases, such as hydrogen.

The flammable gas detectors (CH_4 and H_2) should be designed for safe failure, that is, in the event of a failure or operation halt, the detectors can be considered activated, as if they had detected flammable gas in the environment. For greater reliability, the detection systems use systems

Figure 6.5 Catalytic gas detector display calibrated for methane gas detection. *Courtesy Det-tronics.*

with voting logic 2 out of 3, where for every 3 detectors that monitor a certain risk area, 2 detectors must be activated (detecting flammable gas) to characterize the conditions to trigger the gas leak confirmation signal.

6.4.1.1 Post-CH_4 methane gas confirmation actions

Safety actions must start after the confirmation signal of combustible gas methane CH_4 detection by two (point and/or open path) sensors, within the same monitored area or in air intake areas. The lowest concentration to trigger the detectors should be 20% of the LFL, considered "low concentration," and 60% of the LFL, considered "high concentration," both for point detectors. For open path detectors, the activation concentration should be 1 LFL linearmeters, considered "low concentration," and 2 LFL linear meters, considered "high concentration."

The control actions to be generated in case of confirmed gas in monitored areas are: alarm in the control room; power shutdown of electrical equipment installed in open areas, which are not suitable for operation where gas has been detected; activation of the ESD-3. The alarm in the control room should go on when the gas is detected by the first detector (a single activated detector indicating 20% LFL concentration will suffice). In the case of gas confirmed in the air intake points, the following safety actions should be generated: alarm ringing in the control room; closure of dampers and shutdown of the ventilation system of the affected area; operation of the ESD-3.

Instrument air compressors, when installed in an open or partially open area, need to be monitored by methane gas detectors. The gas detection by of any sensor should generate an alarm in the control room. The voting logic for the confirmed gas signal in the area of instrument air compressors should be 2 of n (where $n \geq 3$). The confirmation by two detectors should generate the machine shutdown signal and/or stall the instrument air system start up.

6.4.1.2 Posthydrogen gas confirmation (H_2) actions

An alarm should ring in the control room when a gas detector (a detector at a concentration of 10% LFL) is activated. The confirmed gas signal (detected by two detectors with 10% of the LFL) should initiate further control actions, such as backup exhaust system startup. In addition, the gas signal confirmed for 15% gas concentration (detection by two detectors of 15% of the LFL in the same zone) must generate a signal to halt the full batteries recharge operation. Hydrogen gas detectors should be installed near the exhaust system intake points in the battery rooms. The exhaust fan system should be installed at the highest location in the battery room. The battery room design should prevent gas build-up (pockets).

6.4.2 Toxic gases detection (H_2S)

In most cases, toxic gases leaks such as H_2S (hydrogen sulfide or sulfidic acid) are monitored by a type of detector with an electrochemical functioning principle. The electrochemical cell is like an incomplete galvanic cell, in which one of the reagents required to complete the chemical reaction is the gas to be detected. Presence of gas causes the galvanic battery to be complete, which results in the potential difference between the electrodes combined with an electric current as the sensor's detection signal. Gas dispersion studies should be considered aiming to correct the detector's location design in order to optimize costs associated with the quantity of equipment. If studies confirm the need for certain regions to have additional detectors, it should be considered in the project. H_2S detectors (Fig. 6.6) need to be suitable for operation in explosive atmospheres, that is, they cannot become sources of ignition in the event of a gas leak in the area where they are located. Safe failure is intrinsic to the technical characteristics of toxic gas detectors. In the event of operation halt or failure, the detection system needs to interpret the event as presence of toxic gas that was responsible for their activation. If additional detectors are active in the same area, it means that a confirmation signal based on the design voting will generate all the emergency actions considered in the project.

Figure 6.6 Electrochemical gas detector display calibrated for H_2S toxic gas detection.

6.4.2.1 Post-H_2S gas confirmation actions

The confirmation signal of large toxic gas leaks is composed by voting logic 2 of n, where $n \geq 3$ detectors. This means that the confirmation signal is obtained based on at least two detectors activated at a concentration of 20 ppm in an open area, where dilution conditions are more favorable. The main response actions following the confirmation of toxic gas are: alarm ringing in the control room; ESD-3.

Lower concentrations can be used as a reference, especially for air intake monitoring. In this case, the alarm should be triggered in the main control room at a concentration of 8 ppm, even if only one detector gets activated. If two detectors in the same area are activated at 8 ppm concentration, it should also generate the confirmed toxic gas signal and additional actions be initiated such as closure of the dampers associated with the activated detectors. H_2S concentrations of up to 10 ppm are below the level that causes eye irritation, and so the reference value of 8 ppm for the activation setpoint is very conservative in terms of people's protection. It is worth noting that a leak can generate a low-concentration cloud initially and suddenly rise as the accident evolves, which justifies the conservative limits adopted for the concentration setpoints for activation of the H2S detectors.

6.4.3 Monitoring gas contamination (H_2S/CH_4)

Some equipment inventories such as expansion vessels for cooling water systems may be contaminated by the presence of gases and therefore need to be monitored by methane (CH_4) gas detectors capable of identifying excessive concentrations, for example, in the vents area.

As part of this strategy, the activation of any detector with the related function should generate an alarm signal in the control room, indicating contamination in the water circuit. A gas (CH4) monitoring system should be provided for the water replacement circuit of the heating water system, subject to contamination. Areas containing equipment or pipelines where stale water (water as a byproduct the produced oil) may occur should be monitored by H_2S detectors.

6.4.3.1 Monitoring asphyxiating gas (CO_2)

Some oil and gas facilities may have to cope with CO_2 inventories. In these types of installations, the need to install CO_2 detectors must be confirmed by a gas dispersion study for the assessment of CO_2 releases at the level of risk of the associated accidents. Air intakes and turbines exhaust hood of the air generation system do not necessarily need CO_2 monitoring, except in cases where specific analyzes confirm for the need for such monitoring.

The monitoring of asphyxiating gas (CO_2) leaks can be performed by point type IR detectors. These detectors need to be specified with the level of protection compatible with the plant area classifications. In the event of an operation halt or failure, the detection system must interpret the CO_2 detectors as if they were activated by the presence of asphyxiating gas. If additionally other detectors are activated in the same area, it means that a confirmation signal based on the voting design will generate all the emergency actions considered in the project.

6.4.3.1.1 Post-CO_2 gas confirmation actions

Safety actions should be initiated following the confirmed CO_2 gas signal. This signal is formed by the activation of two detectors in the same area (voting logic 2 of n, $n \geq 3$), but the detection by only one detector should generate the alarm notification in the main control room of the facility.

CO_2 concentration setpoints should be used as a reference to initiate specific response actions such as:

1. Concentration of 3900 ppm should produce an audible alarm in the main control room.
2. Concentration of 30,000 ppm should initiate control actions such as alarm in the main control room and in the affected area, besides the ESD-3.

Some oil and gas facilities can process mixture streams with variable concentrations of CO_2 and CH_4. The design of the gas detection system in these installations needs to consider the possibility of leak detection of any of the gases. Therefore in some cases detection may be more effective using only CO_2 detectors and CH_4 in others. The need for detection based on both gases is only required in specific cases identified by safety studies.

6.4.4 Specification and location of gas detectors

Table 6.1 shows a summary to aid in the specification of gas detectors. However, it is worth mentioning that the use and location of gas detectors should be based mainly on the operational experience, the experience with previous projects, the standards applicable to each project and the thermodynamic characteristics of the processes of each installation. In addition, a supplementary verification can be performed based on gas dispersion studies, fire propagation, and explosion applying CFD analyses to previously selected scenarios.

Table 6.1 Specification of gas detectors.

Gas and detector type	Monitored area	ESD voting logic	Alarm voting logic
Natural gas, CH_4, open path (IR)	Risers area	2 of n ($n \geq 3$)	1 of n
Natural gas, CH_4, point, or open path (IR)	Process plant and wellhead Turret Air intakes outside living quarters	2 of n ($n \geq 3$)	1 of n
	Superstructure air intake (living quarters)	2 of n ($n \geq 2$)	1 of n
Natural gas, CH_4, open path (IR)	Expansion vessels of cooling/heating water system	ESD not applicable	1 of n ($n \geq 1$)
Toxic gas, sulfidic acid, H_2S (electrochemical)	Equipment and pipelines with stale water	ESD not applicable	1of n ($n \geq 1$)
	Process areas and wellheads Risers area Turret Air intake	2 of n ($n \geq 3$)	1of n
Hydrogen H_2 (catalytic)	Battery room	ESD not applicable	1 of n ($n \geq 2$)
Asphyxiating gas, carbon dioxide, CO_2 (IR)	Process areas containing equipment or pipelines with potentially significant CO_2 inventory	2 of n ($n \geq 3$)	1 of n ($n \geq 1$)

Notes: Air intakes located adjacent to one another can be commonly monitored by open path detectors. In such cases, at least three detectors should be installed in the joint monitoring area ($n \geq 3$). ESD, emergency shutdown; IR, infrared.

6.5 Fire detection systems

Flame, heat, smoke, thermal, and thermal-velocity detectors provide the ability for diagnosis, confirmation, and initiation of countermeasures for fire scenarios.

Detection and response time has a major influence on the consequences of early stages of a fire. A prompt detection and confirmation of the fire improves the chances of the event being controlled in the "incipient stage fire." Especially when there are large hydrocarbon inventories, as is the case in the oil and gas industry facilities, safety systems are designed based on the strategy of fighting the incipient stage fire. The objective is to avoid the growth of the incipient fire to a fully developed fire scenario fueled by large hydrocarbon inventories. The amount of energy stored in hydrocarbon inventories justifies accurate and highly reliable detection systems. Next we will introduce the main types of detectors and their characteristics for use in oil and gas installations.

6.5.1 Flame detection

The main technologies employed in flame detection use UV detection, IR detection and the combination of both radiations (UV/IR). The burning of gases during combustion generates a flame that can be visible or invisible to human eyes. The function of flame detectors is to generate a signal from the exposure of their sensors to the radiant energy of a flame. The detection system should be designed to generate signals for the alarm notification in the control room and confirmed fire signals, adhering to a voting logic. Thus spurious alarms are reduced as much as possible.

The radiation emitted by a flame travels at the speed of light and the flame detectors are fast acting devices (reaction time in the order of milliseconds). Flame detectors can be of the open path type (they monitor a region and not just a point), but for this application its efficiency needs to be evaluated. It is necessary to consider environmental conditions such as the influence of obstructions by structural elements and equipment, or the possibility of presence of dense smoke or gases. Due to their fast action, flame detectors are generally the choice for locations with significant risk, such as fuel storage and transfer areas, industrial processing areas, and in situations where explosions and fires with fast and catastrophic escalation can occur.

6.5.1.1 Ultraviolet detector

The UV sensor responds to invisible radiant energy on the UV scale (below 4000 Å wavelength). It is a fast acting (UV) detector. The presence of solar radiation, electrical discharges (lightning bolts) can mask the detection and, therefore, the UV flame detectors (Fig. 6.7) are designed with sensitivity only in the range of 1800–2500 Å, to reduce false alarms. The intrinsic limitations of UV detectors can generate spurious alarms when there is interference caused by radiation sources such as X-ray machines, arc welding and natural electrical discharges (lightning). Another source of interference is the presence of smoke capable of filtering UV radiation. The efficiency of the detector is limited in areas where smoke may be present before the flame forms. The use of UV detectors relies on previous technical studies on the adequacy of the environment and associated accidental

Figure 6.7 Ultraviolet flame detector. *Courtesy Det-tronics.*

scenarios, considering that the described limitations can make unfeasible the use of such equipment.

6.5.1.2 Infrared detector

The IR sensor operating principle is based on a photovoltaic or photo resistive cell, with a filter and lens system. The sensor responds to invisible radiant energy above 7700 Å. Like the UV detector, the IR detector (Fig. 6.8) is also fast-acting and responds to many heat sources, being susceptible to false alarms, even when equipped with discrimination devices protection. IR detectors can also generate spurious alarms under high humidity levels. One means of reducing the generation of spurious alarms is the use of detectors of triple IR sensor (IR3) type. IR3 detectors have a high level of rejection of false alarms and some are equipped with features capable of detecting sensor obstruction by dirt, snow, or other elements that may make affect the sensor sensitivity. The use of UV detectors relies on previous technical studies on the adequacy of the environment and associated accidental scenarios, considering that the described limitations can make unfeasible the use of such equipment.

Multispectral IR detectors can also be used to detect invisible flames, formed during the combustion of hydrogen gas.

6.5.1.3 Infrared/ultraviolet detector

These detectors (Fig. 6.9) are designed where UV or IR detectors used individually can generate excessive spurious signals. UV detectors can respond to sources other than flames, such as lightning, X-rays, or

Figure 6.8 Infrared flame detector. *Courtesy Det-tronics.*

welding operations. Similarly, IR detectors can generate spurious signals to various hot objects that produce intermittent radiation such as electric heaters or ventilation and heating system pipelines. The combination of a UV sensor and a single frequency IR sensor in a single detector is a strategic solution where the simultaneous response of the two sensors is the only way to achieve the confirmation and triggering of signals and alarms. These two (UV/IR) detection elements monitor different parts of the radiation spectrum, and there is virtually no source of false alarm that is common to both sensors. This allows the detector ignore radiation sources that affect only the UV sensor or the IR sensor, without limiting the ability to respond to a fire even in the presence of potential sources of false alarms, such as welding, X-rays, or very hot objects.

6.5.2 Heat detection (fusible plug)

The term heat detector can be used for a specific type of detection system whose sensor element is a metal that normally melts at 70°C/158°F. This

Figure 6.9 Ultraviolet and infrared flame detector. *Courtesy Det-tronics.*

system is known as fusible plug system, based on its operating principle that includes a compressed air line in which the fuse plugs are installed to monitor a specific area. If sufficient heat is generated to melt the fuse plug metal, the compressed air line undergoes sudden pressure drop as the air gets released by the plug itself after fusing. This pressure reduction is detected by the sensor and a signal is generated for alarms and postfire confirmation actions. However, the use of this type of system has been progressively decreasing due to the many advantages and superior reliability offered by other types of electronic detectors. Operators often complain about operational problems and false alarms generated by fusible plug systems. The compressed air lines are built with small diameter pipes (tubes) and in most case need to reach long lengths, causing the compressed air to travel significant distances. They can be exposed to shocks resulting in leaks that generate spurious alarms. There is also a risk that the fusible plugs will receive some improper masking and as a result will not melt at the expected temperature. There is evidence of interference caused by dirt or paint residues that can inappropriately mask the entire detection system. It is the opinion of many operators that the system shows some fragility due to these vulnerabilities and they are often removed from operation justified by possibility of undesirable interruptions in the operation caused by spurious alarms. Although removing detection systems without additional safeguards is a serious failure, the designer needs to consider that the use of fusible plug systems can increase the chances of errors in risk management, which is in fact a justification for the adoption of a more reliable heat/temperature detection system.

6.5.3 Smoke detection

Smoke can be defined as all the visible and invisible particles resulting from the combustion process that are transported through the air. The visible combustion particles form the thicker, heavier smoke of a blazing fire and correspond to a small density of large-size particles size in the volume monitored. The invisible combustion particles (e.g., transparent smoke from a fast-burning fire), on the contrary, are associated with high density of small particles. When a smoke detector detects the presence of these visible or invisible particles it will generate an alarm signal. The main types of smoke detection technologies are photoelectric/optical and ionization smoke detection.

Photoelectric (or optical) smoke detectors have a sensor that consists of a photocell and a light source. They are usually specified for areas where fire is expected to produce large quantities of visible combustion particles. The operating principle is that the smoke that gets in the light path will cause an obstruction and prevent the light from reaching the photocell, thus generating an alarm signal. Another principle used is based on the interference of the smoke with the light beam that when the light gets reflected in a photocell it will capture the reflection and as a result also generate an alarm signal.

The photoelectric detectors can be of spot or open path types. The spot smoke detector includes all elements in a small unit, while the open path type consists of a light source and a photosensitive receiving device. The sensors are located at opposite ends of the area to be protected, and usually installed near the roof.

Ionization detectors are able to detect the presence of visible and invisible combustion particles, but are more sensitive to invisible particles. The rationale is that its efficiency relies on the smoke containing a minimum density of particles, thus allowing the activation of the ionization sensor. Invisible combustion fumes contain more particles than visible combustion smoke. The sensor consists of a power source, detection circuits and an ionization chamber with a small alpha radioactive source. The DC source and a battery or a rectifier maintains an electrical potential between two plates acting as electrodes. The air molecules inside the chamber are split into ions and electrons under the influence of the alpha radioactive source (ionization). Given that equal charges repel one another whereas opposite charges attract each other, the charged particles between the plates will flow in the direction from the negative electrode to the positive electrode. The smoke particles in the ionization chamber will bond with the charged particles and as a result reduce the flow. The detector's circuits are sensitive to the flow reduction, thus generating an alarm signal.

The installation of smoke detection systems needs to follow some recommendations regarding the location of the detectors, for example:

1. Areas subjected to high levels of combustion particles such as garages, boiler rooms or designated smoking areas should not be monitored by smoke detectors to avoid spurious alarms.
2. The installation of smoke detectors in the adjacent to air supply diffusers should be avoided because of the possibility of reduced sensitivity due to the air flow and its interference with the smoke.
3. The use of dual-chamber ionization detectors helps to avoid problems related to factors such as: humidity, barometric pressure and temperature, which cause false alarms or smoke detectors sensitivity loss. The two ionization chambers in these special detectors have the following roles: one performs the detection function while the other acts as the reference chamber to compensate for variations in humidity, barometric pressure and temperature.
4. For smoke detectors installed in open areas, protection against the elements, such as rain, snow, air currents, as well as dust, and other contaminants need to be considered.

The combination of different sensor technologies increases the detection capacity and improves its efficiency. Smoke detectors using combined ionization detection sensors and photoelectric sensors are sensitive to both visible and invisible combustion particles. These detectors utilize the advantages of both types of sensors to provide comprehensive coverage.

6.5.4 Thermovelocimetric detection

Two types of thermal detectors are available, namely, fixed temperature, or thermovelocimetric. There are also other types of detectors that are variations of these main types, such as the thermovelocimetric detector with compensation and the combined detector. Thermovelocimetric detectors generate an alarm signal when the air temperature rises above a predefined rate (e.g., 15°F/minute). These detectors are faster to respond than fixed temperature detectors, because thermal delay is not a relevant factor in the operation of thermovelocimetric detectors. But thermovelocimetric detectors are not immune to false alarms, and it can happen when they are exposed to rapidly rising temperatures, other than from a fire. Thermovelocimetric detectors should not be installed near ovens, devices with open flame, freezers, frequently open doors to other high-temperature environments, or under conditions where sudden changes ambient temperature variations may occur. Thermocouple detectors may fail when exposed to a gradual slow-growing fire. This vulnerability can be overcome through the use of a combined system of thermocouple detectors and fixed temperature detectors simultaneously.

6.5.5 Fixed temperature heat detection

The working principle of sensors of fixed temperature detectors is based on the temperature of the internal detection element, which must reach the nominal activation setpoint. One type of sensor element uses thermocouple, which consists of two metallic elements with different coefficients of thermal expansion. When heated, the elements expand at different rates, creating deformation of the bimetallic element, which causes the mechanical closure of the electrical contacts. Another type of sensor uses fuse elements that fuse after a certain temperature, causing the mechanical closure of electrical contacts. The bimetallic elements return to their original position when the normal temperature is restored and, therefore, they are reusable, but the fuses are nonreusable and need to be replaced each time they get activated.

Temperature detectors, like other types of detectors, can be spot type or open path type.

6.5.6 Specification and positioning of fire detectors

Table 6.2 contains a summary to aid in the specification of fire detectors. However, it is worth mentioning that the use and location of fire

Table 6.2 Fire detectors specification.

Detector configuration type	Monitored areas	ESD voting logic	Alarm voting logic
Flame (IR3—triple infrared)	Offloading equipment, riser connections, riser input areas, turret internal spaces, storage of flammable and/or combustible products, wellhead, process plant, "coamings area" on the main deck of offshore platforms, cargo pump rooms, shuttle tankers	2 of n ($n \geq 3$)	1 of n ($n \geq 3$)
Smoke (optical)	Essential electrical panel rooms, main electrical panel rooms, transformers	2 of n ($n \geq 3$)	1 of n ($n \geq 3$)
	Control rooms and electrical equipment (batteries, battery chargers, electrical panels), radio and telecommunication rooms, spaces under floor and above ceiling in control rooms	ESD not applicable	2 of n ($n \geq 2$)
	Lodges, cabins, staircases, wardrobes, cafeterias, gym and recreation rooms, access to machine compartments, offshore platforms pontoons, offices, living and entertainment rooms, auditoriums, crane booths, videoconferencing rooms	ESD not applicable	1 of n ($n \geq 1$)
	Air intakes for ventilation of control rooms, muster stations, and meeting points to prepare for abandonment	ESD not applicable	1 of n ($n \geq 1$)
Temperature (thermovelocimetric)	Deposits and warehouses, pantries, workshops, laundry room	ESD not applicable	1 of n ($n \geq 1$)
	Infirmary, laboratories, paint deposits, kitchen, crane equipment rooms	ESD not applicable	2 of n ($n \geq 2$)

(Continued)

Table 6.2 (Continued)

Detector configuration type	Monitored areas	ESD voting logic	Alarm voting logic
Flame and fixed temperature (electric)	Turbines hood (room or compartment) for power generation or for gas compressors, closed compartments containing internal combustion engines or diesel tanks	2 of *n* ($n \geq 2$)	1 of *n* ($n \geq 2$)
Fixed temperature (electric)	Saunas and similar venues	ESD not applicable	1 of *n* ($n \geq 1$)

ESD, emergency shutdown.

detectors should be mainly based on operational experience, experience with previous projects, standards applicable to each project and the thermodynamic characteristics of the processes of each facility. Furthermore, complementary verification and corrections can be performed based on studies of gas dispersion, fire propagation, and explosion using CFD simulations applied to previously selected scenarios.

6.6 Automatic fire-fighting systems

When it comes to safety in the oil and gas industry, the first system that is usually come to one's mind is the fire-fighting water system. But, considering the large inventories of hydrocarbons that are part of the production chain of the industry, much more than powerful pumps and sophisticated automation systems is required to give the right attention at the right time to matters related to safety. Equipment and automation systems are not sufficient to provide an adequate response to prevent or mitigate accidents with fire and explosion scenarios. It is necessary to establish a response strategy associated with equipment and automation systems.

We can make a didactic comparison (Fig. 6.10) between the equipment that is usually specified in projects and the severity of the postulated accidental scenarios for oil and gas facilities. The example presented below helps to understand that a good risk management strategy is much more important than the excessive elevation of "brute force," through the

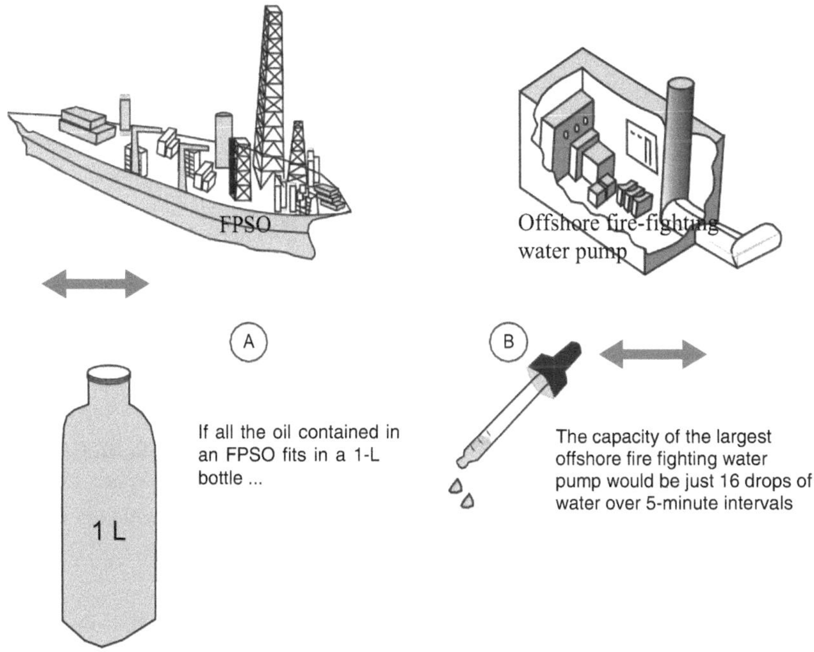

Figure 6.10 Didactic comparison that illustrates the technical limitations of equipment to fight large-scale fires in installations with large hydrocarbon inventories. Higher in importance than the capacity of the equipment and the level of automation is the definition of an efficient strategy to prevent and respond to accidents. (A) If all the oil contained in an FPSO fits in a 1 L bottle ... (B) The capacity of the largest offshore fire-fighting water pump would be just 16 drops of water over 5-minute intervals.

overspecification of equipment, and automation abuse. Taking a FPSO offshore rig as a reference, the largest fire-fighting water pumps (FWP) are supplied with a flow rate of 2500 m^3/hour. This means that during a 5-minute operation, these large pumps can supply around 208 thousand liters of water. A modern FPSO has a storage capacity of 1.6 MMbbl of oil, which is equivalent to 254 million liters of oil. It would take more than 100 hours for the largest offshore pumps to be able to deliver the same volume in fire-fighting water. Comparatively, this would be the same as trying to extinguish a fire in 1 L of oil with just 16 drops of water, in 5-minute intervals. This task although is not impossible, it requires a very well-designed strategy! Statistics provided by the Brazilian Navy show that performing fire-fighting procedures in less than 1 minute after the fire starts has a 90% chance of being extinguished in the first 5 minutes. Conversely, a 5-minute delay can result in 90% chance that it will be extended to more than 2 hours until the fire is completely extinguished.

6.6.1 Water spray fixed systems (deluge)

The water spray system is better known as the "deluge" system. The function of this system is not exactly to fight the flames directly, but rather to provide sprinkled water (as a deluge) to be applied in bulkheads, pipelines, vessels, etc., with the objective of cooling the equipment surfaces and, thus delay the escalation of the severity of the event. It is mainly applicable to equipment that contains inventories of liquid and gaseous hydrocarbons, since these inventories can be affected by the heat released by a fire in its vicinity. This could provoke the increase in pressure and temperature, with releases of hydrocarbons and possible equipment rupture with loss of containment and aggravation of the fire scenario.

Deluge systems are designed to maintain a certain flow rate compatible with each protected equipment. For example, the well area should receive cooling water at the flow rate of 400 L/minute per well. The risers entry area is protected, considering the flow rate of 10 L/minute m^2 on the connection flanges, valves, and other connections.

These data are specified by the standards applicable to each project (Section 9.11). The project also needs to consider that the system is capable of accessing the nozzle located in the areas of most limited access, to ensure their proper operation in all accidental scenarios considered in the project. These data are also specified by the standards applicable to each project (Section 9.11). The deluge system basically includes pumps, valves, distribution lines, and nozzles. The lines need to be designed as short as possible to reduce the risk of damage and costs. In addition, a hydraulic balancing of the main water distribution system is carried out. This hydraulic balancing allows to adjust the flow rates for all the consumers of a facility, within the permitted tolerances, ensuring the water application rates established in the specifications of a project.

There is a special type of valve used in deluge systems called "automatic deluge valves (ADV)." The function of these valves is to open the main lines immediately in response to a triggering event, causing the start-up of fire pumps. In the event of a confirmed fire in a given area, the detection system will send a signal to open the deluge valve associated with the affected area and, additionally, trigger the FWP start up. The automatic or manual (local) opening operation of the deluge valves triggers the start-up sequence of the FWP. Moreover, the system is normally maintained pressurized, and when water is demanded by a consumer, for example, due to the opening of a hydrant, the resulting pressure drop is sufficient to cause the deluge valves to be actuated by the fluid it carries, in this case water. The ADV deluge valve set is supplied in skids or cabinets, including auxiliary equipment and bypass valves required for emergency maneuvers. Each set needs to be properly identified and associated with the area to which it protects, but it cannot be installed within the area which it is designated to protect, so that the fire that it is intended to protect against does not prevent its safe operation. The assemblies are installed in safe and easily accessible areas for manual operation, which may become necessary in the case of another failure, such as automation failure (Fig. 6.11).

All equipment that contains inventories of flammable and/or combustible liquids needs to be protected by the water spray deluge system (Fig. 6.12). However, exceptions are equipment installed in areas protected

Figure 6.11 Automatic deluge valve assembly installed in a cabinet with identification and operating instructions.

Figure 6.12 Construction and assembly phase of the water spray deluge system. The equipment receives water spray on its external surface as protection against heat from a fire in the adjacent areas.

by other fixed systems or installed in areas where a possible flood may aggravate the accident.

Additional exceptions, to be justified on a case-by-case basis, are the specific situations in which studies and analyzes of fire propagation (Section 7.2.1) can justify the noninstallation of a deluge system.

Some equipment can pose certain difficulty in interpretation for less experienced designers because, although they contain hydrocarbons, the quantities in which they are found are either minimal or are associated, for example, with the presence of oil or gas in water. Gas flotation units, hydrocyclones, deaerators, and hot water expansion vessels are not required to be protected by sprays. Spray protection through a deluge system in pipelines containing flammable and/or combustible fluids should be assessed considering fire propagation analyzes, as well as the rules and standards applicable to each project. In the case of offshore rigs, protection by the deluge water system needs to include equipment that contains inflatable and/or combustible liquids located inside the turret, in the case of installations containing this equipment. In these cases, it may be necessary to install a swivel specifically for the deluge water system, which allows the passage of the main pipelines to supply the system inside the turret. An alternative is the installation of an external water monitor cannon (Section 6.7.3) aimed at the turret to provide protection, but a specific assessment of the efficiency of this alternative solution is necessary.

Pressure vessels and pipeline segments containing predominantly gas and whose liquid inventory is not significant after automatic depressurization (ESD) do not need to be protected by water sprays systems (deluge systems). However, in these cases, the automated depressurization should be directly associated with the fire detection system of the affected area. Areas considered less hazardous do not require a deluge system, such as living quarters, cafeterias, and workshops. These locations should be protected by manual fire-fighting resources, or other applicable systems.

The projects may differ as to the criteria for opening the deluge valve, that is, the water deluge system start up. Some commonly adopted criteria are:

1. *Pneumatic actuation:* automatic opening occurs by depressurization of the fusible plug detector's pipes, when fusible plug systems are included in the project. The manual opening can be performed by opening the manual valve for instrument air supply in projects where the deluge valve assembly is equipped with this type of device.
2. *Mechanical manual actuation:* the opening of the ADV assembly can be performed directly through its handwheel or by the bypass valves.
3. *Electric actuation (manual remote):* the opening of the ADV assembly can be performed remotely, for example, from the main control room. The opening command should be of the "energize to open" type, so that in the event of loss of the power supply it does not cause the improper opening of many deluge valves at the same time. In these cases, the generation of the ADV opening signal must be preceded by a warning about the possibility of the command causing an output demand above the design value for FWP. During a fire, only the ADV valves associated with the affected area should be opened. If other ADV valves are opened improperly, FWP may not be able to meet the high flow demand. Therefore ADV valves should be very well identified with respect to the fire zones covered by them.

   Before the operator decides on remote manual opening, he should ensure that no ADV valve is opened unnecessarily.
4. *Electric actuation (automatic):* the activation of two flame (IR3) detectors in the area protected by the deluge valve assembly should generate the automatic opening signal of the associated ADV valves.

6.6.2 Foam-water spray systems

One of the main fire-fighting systems for large liquid hydrocarbon inventories is a foam generation system (Fig. 6.13). There are many types of foam generation equipment and several types of foam. The selection of the system and the adequate foam should be based on the particular type of hydrocarbon. The most commonly used systems are: foam monitor cannons, foam deluge, and portable foam generation systems. The basic principle is to suppress combustion through a foam layer that aims to reduce the oxygen supply from the air that feeds the fire. Depending on the type of hydrocarbon, an inadequate foam can be diluted by the hydrocarbon itself during the application, which makes it ineffective for fire fighting. Therefore it is important to check the foam-hydrocarbon compatibility. Factors such as wind interference and dust can also affect the fire-fighting efficiency.

When the foam is applied by foam cannons, the coverage area of each cannon should be analyzed so that possible shadow areas can be identified and protected by an intelligent arrangement of cannons. They should be positioned to avoid interference with equipment and structural elements. Interference can obstruct the foam jet and reduce the operation efficiency. Another factor that can compromise foam fire-fighting operation is the relative displacement between the foam and the burning hydrocarbon. If, for example, the foam is launched over a liquefied natural gas (LNG) pool

Figure 6.13 Foam fire-fighting fixed system in test operation, installed to operate in the containment basin built to protect spills from pressure vessels. *Courtesy J.C. Melchior's personal archives.*

fire on the main deck of an offshore rig, the wind and the installation motion caused by the sea waves, the foam can move relative to the burning hydrocarbon pool, and if there is also a slope, the foam can even drip away from the installation and be lost.

In addition, depending on the hydrocarbon characteristics and viscosity it can also move, further complicating the fire-fighting operation. Some resources like containment basins and other coaming devices can be installed to contain liquid spills and retain foam in the proper location for the operation. The basins are designed to contain leaks and LNG pools formed as the result of loss of containment in tanks, vessels, and equipment. But the basins also contribute to the efficiency of the fire-fighting foam, since they work also as a container to the foam itself. The use of coaming is recommended for large areas subject to leaks resulting in LNG pools. In this type of solution, some dividers can be added to the floor or surface to limit the displacement of oil and foam pools. These fixed dividers can be created using building materials such as metal flat bars to subdivide the floor or surface subject to leakage into smaller areas. Thus LNG pools of hydrocarbon and the foam can be contained within the polygonal boundaries formed by the bars, preventing fluids from flowing to other areas and from expanding the risk areas. The drawback is that the components that form the coamings hinder movement on the floor, something that needs to be considered by the designers.

Mixed types of foam fire-fighting systems can be used simultaneously, for example, combining monitor cannons with a fixed foam spray system (foam deluge). This makes it possible a more efficient operation in case of fires in large storage equipment, LNG pools either static or in motion. Fixed systems can be designed to be triggered automatically, remotely from the main control room, or by local manual action.

6.6.3 Fire-fighting water pumps

The fire-fighting water systems for oil and gas facilities should be supplied with a configuration of autonomous FWP and designed to meet 100% of the maximum demand calculated in the project, considering all safety requirements established by international standards and other standards adopted in the project.

The autonomy of FWP may vary starting from the minimum values required by international standards, according to the characteristics of each project. A conservative recommendation for autonomy rating for FWP in offshore rigs is 18 hours.

FWPs should be designed to allow operational testing without leaving the facility unprotected during testing. FWP assemblies can assume various configurations regarding the number of pumps, but in all projects at least one backup FWP is required. FWPs need to be installed in strategic locations based on the possible accidents and need to be protected from the weather. Therefore FWPs should be installed outside of hazardous areas, such as process plant area, areas with tanks, pressure vessels, etc.

One of the important aspects to safeguard the autonomy of the FWP is the minimum distance to be kept between the pumps that make up the assembly configuration. They should be located as far apart as possible. Preferably, they should be in symmetrical and opposite locations. For example, in the case of offshore rigs, on opposite edges of the deck or on different decks. In case of limitations imposed by layout and physical space FWPs need to be installed in locations adjacent to risk areas, FWP rooms or enclosures should be isolated by firewalls. The choice of the FWP installation location should be defined so that, in any accident scenario postulated by the risk analyses (Chapter 7, Reducing unpredictability), the FWP configuration is able to maintain the availability of 100% of the maximum design flow. This means that if one of the FWPs is affected by one of the postulated design accidents, becoming nonoperational as a result, the physical arrangement, as well as the independence and operational autonomy of the other unaffected FWPs should ensure that the system continues to deliver 100% of the maximum design flow, through the other unaffected FWPs.

FWP need to be designed for robustness and reliability. They are part of the last protection systems for oil and gas facilities and should operate even under severe conditions, including those unfavorable for the integrity of their components. The interruption of a fire-fighting operation is one of the last options in crisis management, and should only occur in rare situations such as fires where water can aggravate a scenario, for example, causing floods in floating installation compartments, damage to electrical equipment that is essential during the emergency or, when, in the case of very particular fires in which the presence of water may increase the intensity of the flames (fire in cryogenic hydrocarbons—Section 6.10), or even special materials fires in in which the water steam generated by the operation can extend and spread the contamination by toxic substances (e.g., radioactive materials). Therefore FWP shutdown during the operation should be avoided as much as possible and preferably performed by manual action. Automatic shutdown should be provided only for overspeed (risking destruction of the pump) or short circuit.

Indicators of the operational status of the FWP in the main control room should include the "FWP operating/stop" indicators, "FWP selected for local/remote control" indicators, "shutdown due to equipment failure" indicator. FWPs need to be designed to ensure sufficient discharge pressure to meet the needs of the consumer located in the most unfavorable location of the facility. Another important aspect is the use of safety and emergency dedicated equipment. FWPs should have specific and exclusive use for fire fighting. At most, some auxiliary systems such as smaller pumps (jockey pumps) for pressurizing fire ring water lines (Section 6.6.4) can be replaced by other shared resources. However, this is not applicable to all types of facilities, such as floating offshore rigs. In this case, the pressurization of the fire ring can only be done through a branch shared with the conventional seawater intake system, provided that some conditions are met:

1. The interconnection between the systems should be done in the stretches outside confined environments.
2. A check valve and a shut-off valve with automatic shutdown and simultaneously with the FWP start-up signal should be installed in the interconnection branch to prevent partial FWP flow loss through this interconnection during an emergency.

6.6.3.1 Types of fire-fighting water pumps

Designers should consider the international technical standards and other standards adopted by the project to specify the best FWP type for each situation. The analysis for choosing the FWP needs to take into account the characteristics of the designed facility, the availability of power sources, the availability of water intake points and sufficient volumes, the facility physical isolation that may hinder external help, the availability of physical space for assembling the pump system, and among other factors.

Basically, centrifugal pumps are specified in the design of most fire-fighting water systems. Depending on the size of the facility, a simple assembly of centrifugal pumps, that includes at least one backup unit, may be sufficient to meet the project's safety requirements. But some systems may require much more sophisticated and complex FWP. Booster pumps may be required to increase the flow rate and also diesel power generation units to cover vulnerabilities regarding the availability of external electricity sources. The choice of the pump itself is a matter of hydraulic facility design, in which the pump system flow rate must meet the requirements of the fire-fighting water system. In this case, the flow is determined by calculating the fire-fighting water demand (we will deal with this aspect later in this section).

The FWP use electric motors, diesel motors, steam turbines, or the combination of all of these sources. Diesel or electric engines are most commonly used in projects. In situations where the water supply source is located below the center line level of the discharge flange and the pressure is insufficient to supply the pump's water inlet, a vertical shaft turbo pump may be specified. In some projects, it may be appropriate the use of submerged pumps to enable water intake in wells (land), or at sea in offshore rigs, for example. It is possible to design a configuration that combines a submerged pump with a booster pump.

One of the most important FWP types is the offshore hydraulic diesel pump (Fig. 6.14). Its name is due to the dual-pump assembly, namely, a booster pump powered by a diesel engine, and a lift pump powered by a hydraulic unit which is also powered by the diesel engine. The booster pump and all auxiliary equipment are supplied in an assembly in which most of the equipment are mounted inside a container (enclosure) composed of fire-resistant bulkheads. The main equipment of the assembly are a fuel tank with enough inventory to complete an operation, control panel, air cooling unit, hydraulic unit for activation of lift pump (hydraulic power pack), diesel engine, a booster pump, operational transient

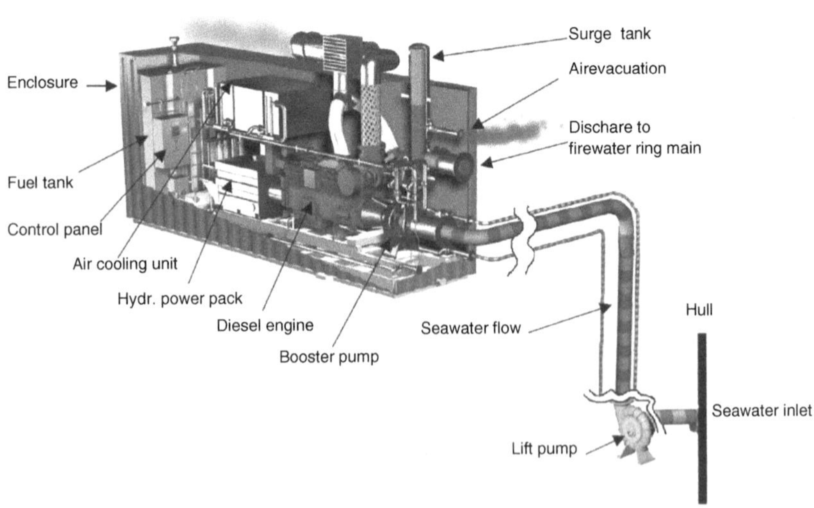

Figure 6.14 Diesel hydraulic fire-fighting water pump for applications in offshore rigs with a dry lift pump, positioned in the lower areas inside the hull. *Courtesy Framo–Frank Mohn do Brasil ltda (www.framo.no).*

accommodation tank (surge tank), motor gas exhaust vent (air evacuation), flange for connecting the discharge pump with the fire firewater system (discharge to firewater ring main), water seawater intake pipelines (seawater flow) and a lift pump.

Fig. 6.14 shows a configuration considering that the project supports the installation of the lift pump inside the offshore rig, in the lower areas of the interior of the hull. This is only applicable when the specific conditions of the project allow the elevation pump to be located close to the hull, with access to a seawater inlet.

There are projects, though, for which this strategy is unfeasible, being then necessary to adopt another configuration in which the lift pump needs to be equipped so that it can submerged and locally on the exterior of the rig hull.

Such configuration is shown in Fig. 6.15, where the main difference is a protection caisson to house the submerged equipment, which includes an internal protection pipe (pipe stack with hydraulic pipes) for the hydraulic circuits of the water fire-fighting lift pump. The submerged lift pump configuration is the most widely used in recent projects. Care must be taken with respect to the depth of the submersion of the lift pump, as if this depth is insufficient, there may be operational failures and deficiency in the fire-fighting water supply. The calculation of the appropriate

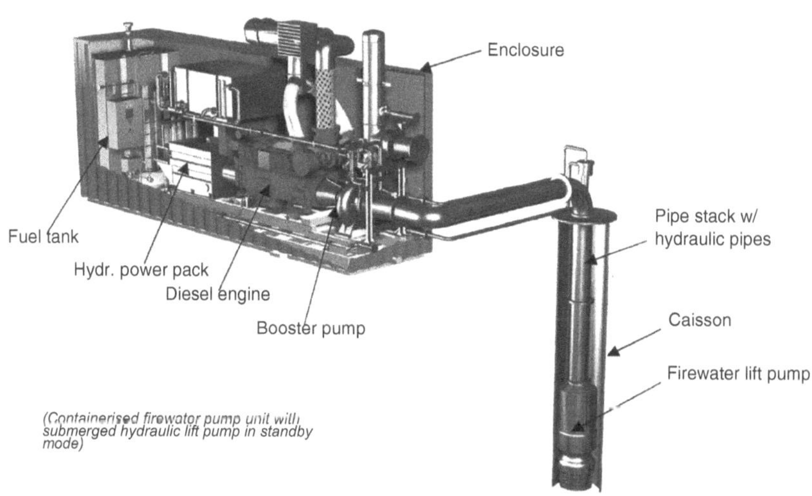

Figure 6.15 Diesel hydraulic fire-fighting water pump for applications in offshore installations with submersible lift pump, positioned in a Caisson (protection enclosure outside the hull). *Courtesy Framo—Frank Mohn do Brasil ltda (www.framo.no).*

submersion depth needs to take in consideration the 100-year design wave (the largest wave that can occur at the rig location, determined as part of the project's safety studies). It is also necessary to consider the maximum angles of listing of the installation, in the case of naval damage postulated by the safety studies of the project. In all postulated accidental scenarios, sufficient number of FWP needs to be fully operational to meet 100% of the fire-fighting water demand. During the offshore rigs, transport operations between the coast and the offshore location, the fire-fighting water system also need to be available, however, considering that the hydrocarbon inventory is minimal during transport, the estimate of the demand for fire-fighting water to be considered in these specific circumstances can be reduced.

With respect to projects in which 100% of the fire-fighting water demand depends on pumps powered by electric motors, these should have dual power source, namely, the main power generation and the emergency power generation sources. However, this is not required when it is powered by a diesel-electric system, in which a dedicated diesel-generator supplies the electric power to the FWP totally independent from the other installation power supplies.

FWPs need to reach their nominal performance level within 45 seconds, including those that depend on diesel engines start up to become operational. This is an important requirement, considering that diesel engines have significant inertia that needs to be compensated for so that FWP can be fully operational within this time frame. FWPs need to start up automatically following the confirmed fire signal (automatic), or by local manual actuation (on site), or by remote manual actuation (control room). All available FWP must start at the same time on demand (including available backup pumps), to avoid that precious time be lost due to detection/diagnosis should one of the pumps fail during start up, and due to the narrow 45 seconds time window in the case of diesel-powered pumps. After the start of the operation and the fire-fighting water supply stabilized, the crisis manager may at his discretion turn off the backup pumps. Whenever possible the a FWP should be turned off by local manual actuation, which implies care with the design of the FWP shutdown interlocks, so that the FWP do not get turned while in full operation due to reasons of secondary importance, contributing to the aggravation of the accidental scenario. The exhaust gases from the FWP diesel engines should be discarded through an independent exhaust duct, attached directly to the diesel engine, and with an outlet vent at a previously defined location in the outdoor area.

Project limitations, especially those related to the power supply and costs, may require designers to specify pumps powered by different energy sources for the same set of FWP. When necessary, some safety requirements should be observed. Table 6.3 presents such requirements.

6.6.3.2 Important notes about Table 6.3

FWP electric pumps can be powered by the main power source. But in this case, the physical location of the main generation equipment needs to be in the area of least risk, that is, in a place where a fire in the process area does not cause the main generation to be unavailable. It is considered as an independent source of electrical power for FWP the whole set formed by actuator, generator, distribution panel, control systems, and transmission cables through the essential loads panel. The electrical power supplies for FWP are considered independent when an accidental event that affects the availability of one of these power supplies does not affect the availability of the others.

6.6.3.3 Fire-fighting water demand

The design, redundancy configurations, and technical specification of FWP are dependent on the calculation of the fire-fighting water demand. This calculation is similar to the design of a hydraulic system, in which the flows of all consumers need to be added, considering the pressures required for the most distant and high consumers during operation. As with any hydraulic project, pressure losses need also to be considered, as well as the types of materials used in the pipelines. In addition, the overall fire-fighting design strategy of the facility needs to be considered in the calculation.

Regarding the strategy, the facility is divided into zones subject to deluge, where the water supply for all consumer equipment in the same area is controlled by a single ADV. If any equipment in a given zone requires water supply from the deluge system due to an emergency, all other equipment in the same zone will also be served by the system, even if not all equipment is affected by the accident.

This is the side effect of the use of a single ADV valve for all consumers in the same area. Therefore designers need to calculate the total demand for each zone of the facility and then identify the zone with the highest demand for fire-fighting water, to be used as the initial reference for FWP design and specification. But it is still necessary to consider that when a fire occurs in a facility with significant hydrocarbon inventories,

Table 6.3 Safety requirements for fire-fighting water pumps.

Configuration of FWP redundancies	Power source	Electric power requirements
2 FWP × 100% each (only 1 FWP is sufficient to meet the total demand requirement)	2 electric	Both electric FWPs should be connected to the emergency circuit, however, the emergency power generator should be designed to power only one FWP. Therefore both FWP should also be powered by the main circuit, as an alternative source.
	1 diesel-pow-ered1 electric	The electric FWP should be connected to the emergency circuit and should also be powered by the main circuit, as an alternative source.
3 FWP × 50% each (2 FWP are sufficient to meet the demand requirement)	3 electric	All electric FWPs should be connected to the emergency circuit, however, the emergency power generator should be designed to power only two FWPs. All FWPs should also be powered by the main circuit, as an alternative source.
	2 diesel1 electric	The electric FWP should be connected to the emergency circuit and should also be powered by the main circuit, as an alternative source.
	1 diesel2 electric	All electric FWPs should be connected to the emergency circuit, and the emergency generator should be designed to power two FWPs. All FWPs should also be powered by the main circuit, as an alternative source.

(Continued)

Table 6.3 (Continued)

Configuration of FWP redundancies	Power source	Electric power requirements
4 FWP × 33% (3 FWP are sufficient to meet the demand requirement)	4 electric	All electric FWPs should be connected to the emergency circuit, but the emergency generator should be designed to power only three FWPs. All FWPs should also be powered by the main circuit, as an alternative source.
	2 diesel2 electric	All electric FWPs should be connected to the emergency circuit, and the emergency generator should be designed to power only two FWP's. All electric FWP's should also be powered by the main circuit, as an alternative source.
	3 diesel1 electric	The electric FWP should be connected to the emergency circuit, and the emergency generator should be designed to power an FWP. The electric FWP should also be powered by the main circuit, as an alternative source.
	1 diesel3 electric	All electric FWPs should be connected to the emergency circuit, and the emergency generator should be designed to power the three electrical FWPs. All electric FWPs should also be powered by the main circuit, as an alternative source.

FWP, fire-fighting water pumps.

the flames generated in the accident area can produce temperature elevations sufficiently high to activate the detection system of the adjacent areas and generate the opening signal of other ADV valves, which considerably increases the flow rate required during the emergency. Therefore it is not sufficient to obtain the flow rate values of each zone individually and compare it with each other to identify the highest value to be used in the design. It is necessary, however, to add the water flows of the adjacent zones that may require water during the emergency to each zone's flow rate. Based on studies of fire propagation (Section 7.2.1), it is possible to assess which zones are affected by each accidental fire event postulated in the project. Thus for each system, water flows from all zones with the highest likelihood of being affected by an actual fire need to be considered.

The maximum design demand is the one required by the largest consumer system (sum of the water flows required by the affected areas in a given fire scenario). The fire scenarios and the zones affected in each system are determined considering the information obtained by the fire propagation studies. "Adding to that the water demand by two hose lines of 1½ inch," that is, the water flow required by the fire brigade to have the ability to use a fire hydrant during the emergency, in addition to the deluge system.

Table 6.4 shows a summary of the calculation (basic design phase) of fire-fighting water demand for a SS offshore rig (didactic example). The systems are arranged in columns (A–K) while the zones (1–11) are arranged in rows forming a spreadsheet in which the underlined numbers are the absolute water flows of each zone and the other numbers in each column represent water flows from other affected areas, which therefore also need to be taken into consideration during calculation of the demand for each system. The total consumption line is the direct result obtained in the calculation of each hydraulic system (considering pressure losses, flow rates, application rates, and required pressures). But it is necessary to add to the table lines other complementary flow values such as the margin (30%) for hydraulic balancing. This is necessary to account for adjustments to the water flow values may occur throughout the subsequent phases of project, construction and assembly. It is also necessary to include the aforementioned addition of a fire hydrant for use during the emergency. The grand total obtained up to this point is in line with the values of the demand required by the system. But in order to determine the value of the water demand for the design of the FWP, a margin of safety still needs to be added. The value of this margin can vary for each project (in the example, it was considered 8%). Finally, using the table in the example, we can identify the highest demand among all systems in the design.

Table 6.4 Summary of calculation of water demand for fire fighting of offshore rig type SS (m^3/hour flow rate values).

	Consumer systems										
	A	B	C	D	E	F	G	H	I	J	K
Zones	1	2	3	4	5	6	7	8	9	10	11
1	51.51										
2		516.72									
3			554.52								
4				133.80							
5					633.23	633.23	633.23	633.23	633.23	633.23	
6					709.57	709.57			709.57		709.57
7					499.25		499.25	499.25		499.25	
8					493.50	493.50	493.50	493.50	493.50	493.50	
9					366.56	366.56		366.56	366.56		366.56
10					280.80	280.80	280.80	280.80	280.80	280.80	280.80
11					280.80	280.80			280.80	280.80	280.80
Total consumption	51.51	516.72	554.52	133.80	3263.72	2764.47	1906.79	2273.35	2764.47	2187.59	1637.73
Hydraulic balancing margin (30%)	15.45	155.02	166.36	40.14	979.12	829.34	572.04	682.00	829.34	656.28	491.32
Fire hydrant	40	40	40	40	40	40	40	40	40	40	40
Demand required by the system	106.96	711.74	760.87	213.94	4282.83	3633.81	2518.82	2995.35	3633.81	2883.86	2169.05
Margin of safety (8%)	8.56	56.94	60.87	17.12	342.63	290.70	201.51	239.63	290.70	230.71	173.52
Design flow rate	115.52	768.68	821.74	231.06	4625.46	3924.51	2720.33	3234.98	3924.51	3114.57	2342.58
Configuration of 3 FWP × 50 % each				Flow	2312.7 m^3/h		3 × 50%				
Pump commercially available				Flow	2350 m^3/h		3 × 50%				

FWP, fire-fighting water pump; SS, semisubmersible.

This demand value is important for pump specification purposes. The system identified (E) as setting the design value is highlighted, with flow rate value of 4625.46 m^3/hour. Taking into consideration that the configuration adopted by the project is 3 FWP × 50%, it means that we would need three pumps with a flow rate of 2312.7 m^3/hour. But on the market, suppliers only offer pumps with a nominal flow rate of 2350 m^3/hour, and for this reason the project needs to specify a conservative flow rate of 2350 m^3/hour for each of the three FWP.

Some techniques can be adopted as a strategy to reduce the demand for fire-fighting water. Experience in projects has shown that distances of 15 m or more between equipment and pipelines containing hydrocarbon inventories can desensitize some areas that would be affected by heat if such distances were shorter. The use of firewalls and the depressurization of equipment normally included in ESD strategies can also be used as a demand optimization resource.

Some areas deserve special attention by designers. Helideck, if included in the installation can be located at high elevations and need to be supplied with a significant water flow, regardless of any difficulty. Another important consideration is excess pressure in the system. During an operation, fire-fighting water systems have their pressure regulated by a control valve that diverts part of the flow to another location. In the case of offshore rigs, the water surplus can be flowed back to the sea. Excess pressure, in addition to compromising the sprays efficiency and the integrity of the installation, can also create difficulties for fire fighters, since lines with excess pressure can get out of the control of the fire crew due to the accumulated kinetic energy, causing accidents where the nozzles themselves can hit people and equipment. The operating pressure of the hydrant nozzles, in the proximity of and at the same level as the area being designed, should be at least 490 kPa (5.0 kgf/cm^2), taking into account the losses beyond the hydrant valve with two 15 m × 38 mm hoses (1½ inch) in diameter and an estimated flow rate of 20 m^3/hour per nozzle. However, the design should also consider that for the safety of the system and the fire crew, water pressure should be limited to 7.0 kgf/cm^2 at these locations.

6.6.4 Fire-fighting water distribution system

The means of fire-fighting water distribution is an important part of the design and risk management strategy in oil and gas facilities. The definition of the water distribution system includes the evaluation of the pipeline routes that ensure the most protected and least irregular paths and with fewer direction changes to reduce the risk of damage during the emergency. Fire scenarios can evolve into explosion scenarios and the intensity of the heat from the flames can contribute to reduction of the mechanical strength of support, which in turn can cause damage to the fire-fighting water lines.

A widely used strategy is the distribution by fire-fighting water ring. This type of distribution is commonly referred to by the simplified term "fire ring" (Fig. 6.16). The fire ring is composed of (as its name suggests) a closed loop distribution circuit, and as result it allows the isolation valves included in the circuit to be shut off in the event of damage or rupture of a section to block the loss of water and pressure at the affected location. Thus even if a rupture or large leak occurs, the affected section can be isolated and the rest of the ring remains available for water distribution to consumers. Obviously this requires an accurate study of the best location

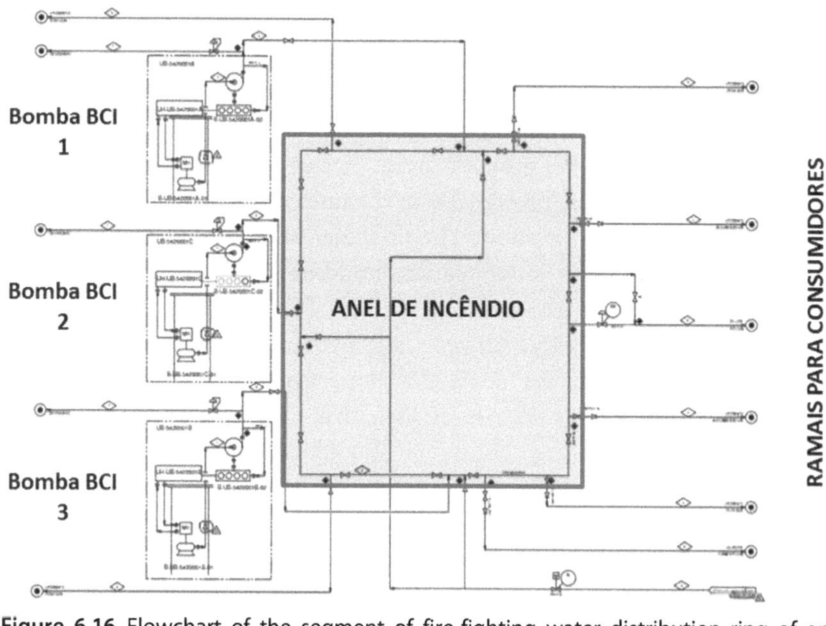

Figure 6.16 Flowchart of the segment of fire-fighting water distribution ring of an offshore rig. The "fire ring" itself is located in the highlighted area (continuous pipeline represented by a rectangle from which the branches and connections to the pumps originate). Consumers, shut off valves, and connection points to the main supply lines from fire-fighting water pumps are also included in strategic locations on the "fire ring."

of the shut off valves and the connection points of the ring with the main supply lines originating from the FWP. The number of valves in a fire-fighting water line should be kept to a minimum to avoid improper shut off and system failures during the fire-fighting operation. Therefore a thorough analysis of the fire ring is so important regarding the physical location of the valves, consumers, and points of interconnection with the main supply water system from the FWP.

The fire-fighting water distribution system reaches nearly all locations of the facility, mainly in the processing areas where the main consumers are concentrated. But is through the fire ring as a segment of the fire-fighting water distribution system that the final branches for consumers depart and where the main valves of the system are installed. For this reason, the fire ring routing should be protected, located outside areas with clear risks of mechanical damage, and should be arranged so that the physical arrangement itself along with the structural elements of the installation can protect it. Dedicated equipment directly associated with safety systems operation should be of exclusive use by the ring and therefore it should not supply other consumers that are not for fire-fighting purposes. It is from the fire ring that the branch lines are originated to supply the various systems considered in the demand calculation (see summary table in Section 6.6.3) for fire-fight water (deluge). And the water for each of these systems should be supplied by an independent branch. The ring should be kept full and pressurized to a minimum pressure of 1 kgf/cm^2 at its most unfavorable location. The flow rate of the system used for pressurizing the fire-fighting water system should be at least 20 times the consumption flow rate of the different points of the system, which can be estimated at around 0.15 L/minute per hydrant, however, it does not need to be greater than the demand for one hydrant. The pressure in the fire ring should be kept at a level such that the pressure drop resulting from the opening of any consumer should automatically start up the FWP.

As mentioned previously, the fire ring, its branches and secondary branches should be isolated by shut-off valves that may become necessary to avoid damaged pipes during a fire-fighting operation (the main objective of these shut-off valves is not only to "facilitate maintenance"). The design strategy needs to provide the assurance of water supply availability to areas of the facility in the event of maintenance, leakage, or rupture of a segment of the ring and/or its branches. In such cases, it needs to be ensured that on each floor/deck of the facility the water supply for at least

50% of the hydrants remains available, even after damage to the integrity of the fire ring. Regarding maintenance, the minimum number of shut off valves that needs to be installed is determined so as to allow operational flexibility in performing maintenance activities without causing the system to become unavailable. However, as mentioned earlier, the greater the number of valves the greater the chances that some of them will be in an inadequate position (closed) due to an operational failure during the fire-fighting operation. This means that the designer should not consider the convenience of continuous operation during maintenance tasks, but rather should optimize the number of shut off valves in the fire ring to the minimum number of valves required. The supply branches destined to consumers located at high elevations should be protected by design to avoid hydrodynamic transients and disturbances such as water hammer, especially during start-up of FWP. Care needs to be taken to prevent valves installed on the fire ring from being left in the wrong positions, including the identification of all valves, especially those responsible for blocking the ring of fire so that they can be easily visible considering adverse conditions such as the presence of smoke, water mist and steam formed during the fire-fighting operation. The valves should be easily accessible and preferably be operated from the facility floor/deck. The manual operation of these valves and access to them should be designed taking into consideration ergonomics and human factors aspects. Colors, signs, mechanical locks, access stairs, mezzanines, physical effort for access and operation, temperature, and visibility conditions, besides all human factors should be considered in the design of the valve maintenance stations for each valve or set of valves that are part of the fire ring.

Different materials can be used for the fire ring pipelines. Facilities subjected to severe corrosion conditions can be built using cupronickel pipes to replace carbon or stainless steel pipes. The fire ring can alternatively be built using pipes made of fiberglass-reinforced plastic materials, as long as it is kept permanently filled with water and the conditions established in international standards are met (Section 9.11).

6.6.5 Carbon dioxide fire extinguishing system

Fire-fighting water cannot be used in some environments, either to protect electrical and electronic equipment from damage, or due to the nature of the burning material. One of the most widely used options is the deluge using carbon dioxide (CO_2), which costs slightly less than other gases but has the disadvantage of the risks of asphyxiation to people in the accident area and also during the CO_2 units maintenance when installed indoors. CO_2 deluge systems are designed according to internationally adopted safety standards (Section 9.11). The system concentrates all the CO_2 required for fire fighting in a central high-pressure CO_2 cylinder set ready for use. Some projects may define a backup set so that, in case the system is triggered and the operation is successful, the facility can restore its normal operating condition more quickly, using the backup set. However, this needs to be carefully assessed, because all CO_2 cylinders need to be subjected to periodic tests and refills, which demands significant continuous logistical effort, whereas the replacement of postoperation depleted cylinders is a single operation and with no product shortage in the market. However, in some specific installations, in remote locations or offshore, a CO_2 cylinder backup set may be justified.

The design of cylinder sets does not consider the CO_2 inventory required to supply all consumers simultaneously, but instead to fulfill the requirements of the largest environment since, from risk management standpoint, independent accidental events are not expected to occur simultaneously in all consumer environments of the system. A distribution system equipped with directional valves should allow the gas to be discharged only in the environment that is part of the accidental event. Rooms housing electrical equipment with installed capacity ≤ 1000 kVA, and rooms that house internal combustion machines, with installed power greater than 375 kW should be protected by a CO_2 deluge system.

Control rooms floors and/or false ceilings that include high voltage electrical cables should be protected by a local, standalone dedicated CO_2 cylinder set. In this case, a specific directional valve should also be available for each compartment served by the system (floor and ceiling), and the stored gas inventory should meet the demand of the largest compartments.

Another local set of CO_2 cylinders can be specified to protect the atmospheric vents, vent post, filters, and kitchen range hoods. When these local sets are specified, a backup CO_2 cylinder set should also be considered. Firing up the set of vents and vent post should be done automatically, given the low risks of suffocation.

Rooms that house CO_2 sets and dry transformers do not require CO_2 deluge protection. CO_2 cylinders concentrate energy through the gas internal pressure. If the cylinders get heated during a fire, overpressure can make them an additional hazard. Therefore the installation of CO_2 should be avoided in environments located in columns and pontoons of SS offshore platforms and in areas of difficult access in other types of offshore platforms. It is also not recommended to protect confined environments with CO_2 deluge in remote locations of the installation such as rooms in the lower hull of offshore platforms.

Safety requirements for populated areas are established to address the danger of CO_2 asphyxiation, especially with respect to the instant when the system is activated. If the system is activated improperly, people may be caught off guard by the sudden oxygen reduction in the air and not have enough time to travel the complete environment escape route. Therefore the activation of the CO_2 deluge system should comply with some specific safety requirements:

6.6.5.1 For local manual activation (buttons)

The opening of the directional valve of the room to be protected (valve that aligns the environment to be supplied with CO_2) and the pilot cylinder valve (which prepares the system for operation) should be done through manual electric actuators installed externally to all the accesses to the room to be protected and to the CO_2 cylinder set room.

The manual electric actuators need to be of the type "lift the lid and press the button" to reduce the error-inducing environment and identified by markings and safety signs. The contacts should be of the normally open type, that is, the trigger should be energized to activate the deluge in the environment to be protected.

6.6.5.2 For mechanical manual activation (valves)

The opening of the directional valve and the valve(s) of the pilot cylinder(s) should be by manual activation directly on the CO_2 cylinder set associated with the room to be protected. The reset (repositioning to the ready position) of the pilot valve and the directional valves of the CO_2 cylinder set should always be done manually.

6.6.5.3 CO_2 deluge alarm

The discharge of CO_2 into the environment should be preceded by an audible and visual alarm signal in the room. In the adjacent external areas, visual alarm signal should be installed next to the access of each area, using red intermittent light indicators. The actuator should produce a 30-second delay in the effective deluge of the environment to be protected, before the CO_2 gas is released. To prevent the delay from being bypassed, the actuator system should be designed for the delay to be generated through a pneumatic device, preferably operated by the CO_2 gas itself. An alarm announcement should also be issued in the control room, confirming the gas discharge completion after the delay. The CO_2 deluge confirmation signal should be generated for the control room from the pressure switch located downstream of the directional valve.

For ambient noise level above 90 dBA, visual alarms should be included in the room subject to flooding as warning of the upcoming CO_2 gas discharge, for example, revolving red warning lights or strobe lights. The location of visual alarms should be such that it can be seen from anywhere in the environment. The audible alarms for CO_2 flooding should be pneumatically actuated by the CO_2 gas itself, in compliance with the applicable international standards (Section 9.11). The audible alarms should be designed based on ergonomics and the maximum noise level limit of 90 dB (A). The frequency ranges of the environment background noise should also be assessed, so that the alarm sound frequency is perfectly audible under the ambient conditions.

Access doors to rooms protected by CO_2 should have a warning sign in the exterior. An example of a warning signs should read: "Do not enter while the light is flashing—Area Flooded with CO_2." During the discharge, a red light signal located above the warning sign should indicate that the system has been activated. In addition, autonomous breathing equipment, properly conditioned, should be installed close to all access entries to CO_2-protected environments.

Carbon dioxide fire suppression system CO_2 as fire suppression gas is still widely used but is becoming an out of date technology. Several specialized companies have developed other gases and suppressor products, with trademark, being able to obtain detailed technical information about "cleans agent fire suppression systems" with each supplier. However, carbon dioxide fire suppression system CO_2 is still the most widely employed in much of the offshore industry (Fig. 6.17).

6.6.6 Water mist fire suppression system

Environments protected by a CO_2 flooding system may be protected, alternatively, by a water mist system. CO_2 can cause asphyxia when people remain inside the protected environment. This does not happen in the water mist system because the fog produced is not asphyxiating, and so it is possible for people to remain in the environment to be protected. The water mist system generates a cloud of atomized water from high pressure

Figure 6.17 CO_2 room with cylinder set of a carbon dioxide fire extinguishing system.

source and special nozzles. The water droplets are sufficient to stop the chain reaction that takes place in the flames, and this is the main role of the water mist systems. This effect characterizes the water mist as a fire suppression system, which uses pressurized water distributed through a pipeline to be sprayed in the protected environment. Such protective effect is achieved by the design of the special nozzles so that the water turns into a mist. The mist transforms to vapor in the proximity of the fire, extinguishing it very quickly via an accelerated cooling process. This system has the ability to lower the temperature from 716°C to 136°C in less than 10 seconds.

Each liter of water nanodroplets in contact with heat expands into 1700 L of water vapor. This vapor causes a significant increase in the environment humidity, preventing the fire from creating an air current with a fresh oxygen supply to feed the flames. Through the system, each water drop is "broken down" into 8000 nanodroplets, which means that a water mist system uses 100 times less water than a traditional spray or deluge system to extinguish a fire. Good results are achieved by nanodroplets in class A and B fires, and equipment is not affected, unlike large volumes of water that can cause damage to equipment. It is possible to protect electrical equipment with water mist systems, as long as such equipment manufacturing characteristics are compatible with the system.

The environments for which the water mist protection system can be recommended as an alternative solution are: turbo machine rooms, FWP rooms, auxiliary power generator room (without electrical panels in the area of influence), emergency power generator (ditto), as long as the water mist system does not depend on electrical power, and limit-access environments.

The type of flooding of the environment (localized in some areas of the environment vs entire environment) should be conditioned to the level of protection required for the electrical equipment installed in the room. Some electrical equipment has a level of protection that allows exposure to rain and moisture. The water mist system needs to follow the requirements of international standards (Section 9.11) and should act automatically triggered by the confirmed fire signal generated by the detection system, besides the remote manual activation option (from the control room) or locally activated. Protected environments cannot be exposed to wind or air currents because they reduce the system efficiency. Access to protected environments should be marked by an external warning sign that reads: "area flooded with water mist." When the discharge is completed, a red light indicator above the warning sign should indicate that the system has been activated. Audible and visual alarms should also announce the flood to people within the environment. In external accesses, the announcement should be issued 15 seconds before the water mist is released, through an intermittent red light signal. The 15-second delay between the announcement and the flood should be achieved by interlocking with the confirmation signal from the fire detection system. That is, between the detection/alarm and the actual flooding of the environment there should be a 15-second delay.

6.7 Additional fire protection systems

In addition to large fixed systems, other smaller fixed equipment and portable systems are used to complement the operation of the main fire protection systems. Some of this equipment can be transported and positioned to protect a specific activity that is being carried out at a location that needs complementary safety features. Other smaller equipment helps clear the way for escape routes and to rescue victims. There is also equipment that is located at strategic points for response to specific scenarios considered in the safety studies of the projects.

6.7.1 Fire hydrants

The hydrants should have two 2½ inch diameter outlets, along the periphery of the complete facility. The main strategy to define the location of the hydrants is to prevent that a fire in the area protected by it hinder its operation. To ensure the control conditions by the fire brigade members, the maximum hydrant operating pressure should be 1373 kPa (14 kgf/cm^2). All hydrants should be equipped with quick-coupling connections. In the ideal situation any point of the facility can be protected by at least two water jets from different hydrants. One of these jets should come out from a 15-m hose and the other from a two 15-m segment hose.

Fire-fighting equipment should be enclosed in cabinets installed adjacent to each hydrant in open spaces. The internal areas should be protected by hydrants supplied with a nozzle outlet of 1½ inch in diameter and 15-m long hoses, in strategic locations such as along corridors. In the proximity of external areas, fire hydrants with two 1½ inch outlets should be installed. The same type of configuration with two 1½ inch outlets should be adopted for rooms within the hull of offshore rigs, such as engine rooms and pontoons.

6.7.2 Mobile foam generating equipment

In process plants and in areas where equipment operates with flammable and/or combustible liquids, portable fire-fighting foam systems should be installed as specified in the applicable standards (Section 9.11), with minimum autonomy of 30 minutes. Foam protection system should be available to protect alcohol tanks. Foam-generating liquid (FGL) drums should also be available, as well as all equipment for any specific application and operation. Storage areas for combustible products should have a containment dike. Thus in the case of leaks or foam release, the dike will retain both the combustible product and the foam itself, preventing the fire from spreading.

In the case of facilities with helidecks, the refueling station should be protected by a foam fire-fighting system. The minimum operating pressure upstream of the helideck foam generation system should be 687 kPa (7 kgf/cm^2), based on the simultaneous use of two launcher guns. Means should be provided so that the foam system can be activated directly from any of the three fire-fighting positions of the helideck, where the foam cannons are located. The design of foam generating electric pumps should consider two independent power sources isolated from each other.

6.7.3 Fire-fighting monitor cannons

Fire-fighting water and foam monitoring cannons should be provided as part of the protection strategy and for compliance with regulatory requirements (Section 9.11). For example, in production probes, fixed monitor cannons should be installed with an estimated flow rate of 2000 L/minute per cannon. The positioning of the cannon should ensure that any point can be reached simultaneously by water jets from two different guns.

6.7.4 Fire extinguisher

Fire extinguishers are equipment used in all areas of industrial facilities for manual fire fighting. The fire extinguisher type needs to be specified according to the needs of each area to be protected, meeting the requirements of the standards applicable to each project (Section 9.11). The distance from any point in the facility to the access of a fire extinguisher should not be greater than 10 m, except for indoor areas where the distance can be 15 m. Fire extinguishers located in open areas should be equipped with weather protectors.

6.7.5 Auxiliary equipment

Whenever required, the facilities should be provided with strategically positioned cabinets containing fire-fighting support equipment, such as protective clothing, special shoes, personal protective equipment, and tools, always in compliance with applicable standards.

6.8 Passive fire protection

The term passive protection is used with reference to means of delaying the spread of fire through specific material coatings and fire resistant bulkheads. Passive protection is employed to protect equipment, pipelines, and structural components from exposure to fire. The application of such means are generally defined based on safety studies and analyses (Chapter 7, Reducing unpredictability) and they contribute, for example, in the prevention of the failure of steel structural components due to the loss of their mechanical strength. In addition to fire protection coatings that delay the effects of fire, passive protection may consist of bulkheads classified into different levels of protection, according to their fire resistance ratings.

Bulkheads are used as firewall partitions and should be installed in high-risk areas to isolate them from areas of lower risk. Thus areas that need to be protected include frequently populated areas, areas that house important safety-related equipment in compliance with the requirements of the applicable standards and the recommendations of the fire and explosion propagation studies.

The strategy for the adoption of passive protection consists of grouping the different spaces of the facility based on their similarities, generating a classification regarding the risk requirements. As an example, a space classification strategy adopted for offshore rigs is as follows:

1. Control stations
   - **a.** Radio and communication room;
   - **b.** process utility systems control room;
   - **c.** room housing critical equipment such as programmable logic controllers, fire and gas detection panels, among others;
   - **d.** battery room of the emergency power supply system;
   - **e.** panel battery charger room and transformer room for the emergency power supply system;
   - **f.** ballast and cargo system control room;
   - **g.** equipment room for navigation and dynamic positioning systems; and
   - **h.** telecommunication equipment room.
2. Vertical accesses

   Interior stairs, except those located in spaces where machines and/or equipment are installed. A stair that interconnects two closed spaces and that only has a fire door on one of its ends should not be considered an independent space, but rather part of the space it is connected directly to without a fire door.
3. Corridors

   All internal spaces for circulation in offices and accommodations.
4. Accommodations
   - **a.** cabins,
   - **b.** offices,
   - **c.** infirmary,
   - **d.** bathrooms

   - **e.** lavatories,
   - **f.** recreation rooms
   - **g.** auditoriums,
   - **h.** meeting rooms,
   - **i.** movie theaters, and
   - **j.** libraries.

5. Service areas (high risk)
   - **a.** Kitchens,
   - **b.** cafeterias containing warming equipment,
   - **c.** warehouses containing flammable material,
   - **d.** warehouses with areas $\leq 4\ m^2$, and
   - **e.** laboratories.

6. Service areas (low risk)
   - **a.** Warehouses with areas less than 4 m^2,
   - **b.** laundries and drying rooms, and
   - **c.** refrigerators.

7. Open spaces

Spaces in open areas, excluding hazardous areas with process activities.

8. Well areas

Spaces in open areas where wellheads and Christmas trees are located.

9. Process area, hazardous areas, and cargo pump room
   - **a.** Process area: deck areas where process equipment such as those listed below is located:
     - **i.** manifolds and production lines,
     - **ii.** production separators,
     - **iii.** test separators,
     - **iv.** launchers and receivers,
     - **v.** oil and gas measuring station,
     - **vi.** heat exchangers and vessels with indirect heating,
     - **vii.** electrostatic treaters,
     - **viii.** oil and gas treatment equipment,
     - **ix.** gas treaters,
     - **x.** oil transfer pumps, and
     - **xi.** gas compressors.
   - **b.** Hazardous areas: those where the presence of flammable atmosphere is likely such as:
     - **i.** FPSO and FSO decks located on the cargo tank and

   ii. storage area for aviation kerosene and other liquid fuels with ignition points below 60°C.

   Cargo pumps room: spaces where cargo pumps are installed (only applicable to FPSO, FSO, and similar installations).

10. Machine spaces category "A"

   All spaces containing oil or gas boilers or internal combustion engines used for:

   a. main engines driver and

   b. other purposes in addition to driving motors, if their combined power is greater than 375 kW.

11. Other machine spaces

All spaces containing internal combustion engines, actuators, boilers, ventilation or air conditioning equipment, power generators, main electrical panels, and similar spaces.

6.8.1 Determination of the type of partitions

The criteria for selecting classified partitions or bulkheads should be based on the standards applicable to each type of project (Section 9.11). The International Code for the Application of Fire Test Procedures establishes the reference criteria that are adopted internationally. Basically, the bulkheads are classified based on tests defined by heat flux and test duration time.

Class "A" bulkheads are made of steel or material with equivalent strength, having a stiffened structure, being efficient in preventing the passage of smoke and flames when subjected to an hour long fire test. They should have insulation with approved noncombustible materials, so that the average temperature increase of the unexposed face do not rise more than 139°C (250°F) above the initial temperature, and the temperature at any point, including any joint that may exist, does not increase more than 180°C (325°F) above the initial temperature in the following time intervals:

1. class A—60–60 minutes,
2. class A—30–30 minutes,
3. class A—15–15 minutes, and
4. class A—0–0 minutes.

Class "B" partitions are made up of bulkheads, floors, ceilings or ceilings built to prevent the passage of flames, at least during the first half hour of a fire test. They need to provide insulation such that the average temperature of the unexposed face does not rise more than 139°C (250°F) above the initial temperature and that the temperature at any point does increase by more than 225°C (405°F) above the initial temperature during the following test times:

1. class B—15–15 minutes and
2. class B—0–0 minutes.

Class "B" partitions should be built using approved noncombustible materials and all materials used in their construction and assembly, except in cases that are made explicit in international standards.

Class "H" partitions should meet structural stability and integrity requirements after 2 hours of exposure to hydrocarbons and resistant to greater heat flux and higher temperatures than class "A" partitions.

Aiming to simplify the specification of partitions, designers can use tables based on criteria established by international standards. Table 6.5 allows the classification of partitions that work as a floor/ceiling. First, the spaces to be separated by the partitions are classified based on the 11 definitions of spaces described previously. Then using the table, the intersection between the column that defines the type of space above the partition and the row that defines the type of space underneath is the type of partition to be used.

The same method allows the identification of vertical bulkheads that separate adjacent spaces. Table 6.6 allows the identification of the specification for vertical bulkheads.

6.8.2 Observations cited in the tables

1. The bulkhead should be made of steel and, if there are openings they should be sealed, being unnecessary for it to be classified as category "A." If covers are used to close the openings, the project should then take in consideration aspects related to area classification, explosion, and fire. When separating two process areas, all implications due to communication between the separate areas should be evaluated, such as area classification, calculation of water demand for fire, etc.
2. When an area contains an emergency power source or a component of the emergency power system and the adjacent area contains the main power generator or a component of the main power system, the bulkhead that separates these areas should be class A60.
3. The bulkheads of the accommodation corridors, including class A/B doors should be extended from floor to ceiling. In places where the

Table 6.5 Fire resistance of decks (floor or ceiling) that separate adjacent areas.

	Top area / Lower area	1	2	3	4	5	6	7	8	9	10	11
1	Control stations	A0	A0	A0	A0	A0	A0	(1)	H60	H60	A60	A0
2	Vertical accesses	A0	(1)	A0	A0	A0	A0	(1)	H60	H60	A60	A0
3	Corridors	A0	A0	(1)	(1)	A0	(1)	(1)	H60	H60	A60	A0
4	Accommodations	A60	A0	A0	(1)	A0	(1)	(1)	X	X	A60	A0
5	Service areas (high risk)	A60	A0	A0	A0	A0 (5)	A0	(1)	H60	H60	A0	A0
6	Service areas (low risk)	A15	A0	A0	A0	A0	(1)	(1)	H60	H60	A60	A0
7	Open deck areas	(1)	(1)	(1)	(1)	(1)	(1)	(1)	(1)	(1)	(1)	(1)
8	Well areas	H60	H60	H60	X	H60	H60	(1)	[a]	H60	H60	H60
9	Process areas, hazardous areas, cargo pump rooms	H60	H60	H60	X	H60	H60	(1)	[a]	(1)	H60	H60
10	Category "A" machine space	A60	A60	A60	A60	A60	A60	(1)	H60	H60	(1/2)	A60
11	Other machine spaces	A15	A0	A0	A0	A0	A0	(1)	H0	H0	A0	(1/2/5)

Notes: X = configuration not allowed.
[a]To be defined by the project.

Table 6.6 Fire resistance of bulkheads that separate adjacent areas.

	Areas	1	2	3	4	5	6	7	8	9	10	11
1	Control stations	A0 (2)	A0	A0	A60	A60	A15	(1)	H60	H60	A60	A15
2	Vertical accesses		A0 (3)	A0 (3)	A0 (3)	A0	A0 (3)	(1)	H60	H60	A60	A0
3	Corridors			C	B15	A0	B15	(1)	H60	H60	A60	A0
4	Accommodations				C	A0	B15	(1)	H60	H60	A60	A0
5	Service areas (high risk)					A0 (5)	A0	(1)	H60	H60	A0	A0
6	Service areas (low risk)						C	(1)	H60	H60	A60	A0
7	Open deck areas							(1)	(1)	(1)	(1)	(1)
8	Well areas								—	H60	H60	H60
9	Process areas, hazardous areas, cargo pump rooms									[a]	H60	H0
10	Category "A" machine space										(1/2)	A0 (2)
11	Other machine spaces											A0 (6)

[a]To be defined by the project.

class B ceiling is continuous on both sides of the bulkhead, the space vertical bulkhead can be extended just up to the ceiling. The access doors from corridor to dormitories or public areas can have ventilation openings in the lower half area of the door. Vertical accesses such as staircases that connect only two floors should be protected, by the installation of at least one class A door with automatic closing, in order to avoid the rapid spread of fire from one floor to the other. When connecting more than two floors, vertical accesses should be enclosed by class A walls and protected by class A doors with automatic closing on all floors. These doors cannot have devices that allow them to be kept open.

4. The classification of the partitions shown in the tables is only necessary for adjacent areas with mixed purposes. For example, a kitchen adjacent to a paint store requires the bulkhead to be A0, but it is not required in the case of adjacent paint stores.
5. Bulkheads between cargo pump rooms and category A machine compartments can be penetrated by the cargo pump shaft. The penetration should have an airtight seal to keep the gaseous atmosphere contained in the pump room.

6.8.3 Interference of classified bulkhead and penetrations

Oil and gas facilities have a large number of pipelines and cabling that in many cases can interfere with classified bulkheads. Whenever it is necessary to penetrate a bulkhead or partition, whether pipelines, ducts, trays, or cables, the integrity of the bulkhead with respect to its classification must be ensured, including at the penetration point. Approved fireproof sealing materials should be used in the penetrations.

6.8.4 Interference between classified bulkheads and doors and windows

Bulkheads and partitions can also interfere with doors and windows. Doors and windows should follow the classification of the bulkheads where they are located. Fire doors should be equipped with an automatic closing device.

6.8.5 Structural protection

Materials under the effect of high temperatures due to fires may suffer a degradation of their mechanical strengths. To avoid the loss of structural integrity, all structural components (including the turret structure of offshore rigs) that may be exposed to the action of fire and whose failure may compromise the installation's structural integrity to some extent, should receive passive fire protection (PFP) to withstand fire conditions and comply with applicable test standards (Section 9.11). The application of PFP to structural components should also take in consideration safety studies such as those of fire and explosion propagation (Chapter 7, Reducing unpredictability).

6.8.6 Materials for passive protection

The material to be used as a coating for PFP should comply with international standards adopted by the project (Section 9.11). The application of epoxy-based materials in PFP is only allowed in open, normally uninhabited areas, and is subjected to prior technical assessment. The application of PFP in pipe fittings should be easily removable to allow inspection and maintenance. Material used as PFP should be compatible with the characteristics of the environment where it is applied (e.g., do not install a blanket in external areas).

6.9 Protection systems for confined equipment

Confined equipment such as some types of turbines for power generation systems needs to be protected by systems compatible with each type of equipment. The supplier and the designer of the original equipment should be consulted to include all the specific protection systems that are necessary.

The most widely used systems for fire-fighting purposes in these types of space are water mist and CO_2 flooding systems.

6.10 Accidents with cryogenic products (LNG)

A trend in the oil and gas industry is the increase in facilities for processing, transferring and transporting LNG. Besides being a hydrocarbon, LNG is also a cryogenic liquid. Cryogenic substances are in a gaseous state under normal conditions of temperature and pressure, and to be liquefied, they must be subjected to temperatures below $-150°C$. Due to this characteristic, LNG aggregates all hazards related to hydrocarbons and low temperature, both in terms of harm to people and damage to installations. Cryogenics is a very specialized area and some phenomena that occur at cryogenic temperatures may astonish even experts with experience in risk management involving hydrocarbons. The materials used and by extension the related equipment need to be compatible with the phenomena associated with cryogenics. Some materials do not support extreme temperature variations. Material shrinkage and expansions are well-known thermal phenomena, but in cryogenics the variations are extreme and occur at much higher rates of speed, which makes the application of conventional materials unfeasible. These differences affect even the most well-established liquid hydrocarbon fire-fighting concepts. Cooling or fire-fighting hydrocarbon using water is a well-established technique. However, due to the cryogenic temperatures of LNG, water at room temperature becomes a heat source that increases the evaporation rate of the cryogenic liquid hydrocarbon, forming a large gas cloud, and the subsequent increase in the flames intensity. So, risk management activities in facilities with inventories LNG requires basic knowledge of the physical phenomena related to the processing, transfer, and transportation of cryogenic products in combination with traditionally adopted and consolidated practices for the protection of people and facilities with conventional hydrocarbon inventories.

6.10.1 Knowing the cryogenic characteristics of liquefied natural gas

LNG is a colorless, odorless liquid that resembles boiling water in environments close to normal temperature and pressure conditions. LNG is normally stored in temperatures around $-162°C/-260°F$. It is considered cryogenic because its boiling temperature is below $-150°C$, as shown in Table 6.7, in which cryogenic products are highlighted.

The (net) LNG density is 420–480 kg/m^3 (relative density of 0.45). The typical composition is listed in Table 6.8.

The LNG composition changes as the storage time increases. Its components progressively evaporate starting with those with the lowest boiling point (methane). There is an increase in concentration of the highest boiling point/heaviest hydrocarbons. The density and energy of LNG also increases.

Table 6.7 Comparison of boiling points between liquefied natural gas and other materials.

Temperature (°F)	Temperature (°C)	Reference phenomena
212	100	Water boiling point
70	21	Ambient temperature
32	0	Water freezing point
31	−0.5	Butane boiling point
−27	−33	Ammonia boiling point
−44	−42	Propane boiling point
−238	−150	Cryogenic temperature
−258.5	*−161.4*	*LNG boiling point*
−298	−183	Oxygen boiling point
−319	−195	Nitrogen boiling point
−422	−252	Hydrogen boiling point
−454	−270	Helium boiling point

LNG, liquefied natural gas.

Table 6.8 Relative characteristics of liquefied natural gas.

Component	Percentage	Boiling point	Relative density	Molecular weight
Methane	83.0−99.0	−161.0°C/ − 258.7°F	0.422	16.04
Ethane	1.0−13.0	−88.6°C/ − 127.5°F	0.546	30.07
Propane	0.1−3.0	−42.1°C/ − 43.8°F	0.590	44.09
Butane	0.2−1.0	−0.5°C/31.1°F	0.582	58.12

Notes: LNG, Liquefied natural gas.

The expansion rate (from liquid to vapor) is relatively high, on the order of 1/620 (Table 6.9).

The hazards of LNG are associated with its cryogenic liquid characteristic combined with the risks related to hydrocarbons. The greatest risk of LNG−air mixture is flammability. Unconfined vapor cloud explosions (VCEs) are unlikely due to the slow flame spread. Clouds in partially confined areas (congested or partially closed plants) or in totally confined areas may enable explosions with overpressure waves formation. People can suffer injuries due to contact with the cryogenic liquid, cold equipment surfaces, and with cryogenic liquid pipelines, as well as due to contact with cold vapors. LNG vapor is not toxic, but it is a simple asphyxiating gas characterized by the absence of oxygen. It does not cause aquatic toxicity nor for waterbird wildlife. The LNG autoignition temperature is 537°C/999°F, higher than gasoline: 456°C/853°F. The flammability limits in air are 4.7% (LFL)/14.9% (UFL) and are above the

Table 6.9 Comparison of expansion rates.

Products	Expansion rate
Butane	222
Propane	290
LNG	580–620
Nitrogen	642
Hydrogen	788

LNG, liquefied natural gas.

gasoline limits: 1.4% (LFL)/7.4% (UFL). The flames produced by LNG reach 1330°C 2426°F temperature, while gasoline flames reach 1027°C/1880°F. The LNG combustion heat is 50.24–54.43 MJ/Kg (21,600–23,400 BTU/lb), higher than for gasoline: 43.54 MJ/kg (18,720 BTU/lb). The LNG burning rate is 12.5 mm/minute, while is 4.0 mm/minute for gasoline.

The LNG vapor cloud can be become visible by water condensation in the environment. The density of the vapor cloud is directly influenced by its temperature, which can vary between −107°C and −160°C. This means that the vapor cloud can be lighter, have the same weight, or be heavier than air. The initial vaporization rate in the soil (10:1) is 10 m^3 /minute of steam/m^2. A 10 × 10 m LNG pool (100 m^2) in the initial vaporization will produce 1000 m^3 of steam per minute. As a reference, water has an initial vaporization rate in the soil (50:1) 50 m^3/minute of steam/m^2.

6.10.2 Basic accidental scenarios and liquefied natural gas cryogenics

1. Open area spill (without confinement)

   LNG spill in an open, unconfined area with good ventilation forms a vapor cloud caused by the heat transfer from the floor surface or soil to the LNG pool from the spill. The rate of speed of vapor formation due to evaporation is inversely related to the thickness at LNG pool regions, that is, the shallower the region the faster the evaporation. In general, for a surface without depressions, recesses and cavities, the more peripheral LNG pool regions tend to generate steam more intensely because of higher heat transfer rate from adjacent regions.

2. Open area spill (with confinement in soil/floor cavity)

   LNG spill can form a LNG pool in an open area where the LNG accumulates in some kind of depression, recess, or cavity in the ground or floor. In this case, the depression region produces a confinement effect that changes the behavior of the LNG spilled at that location. The behavior of the LNG pool formed by the spill will be greatly influenced by the capacity of the terrain and the floor to absorb LNG, especially in the region where the LNG is retained. If the terrain allows LNG to penetrate its superficial layer, promoting absorption, then evaporation and vapor cloud formation will be much slower. Conversely, if the cavity where the LNG is confined has nonporous surfaces, without capacity to absorb the cryogenic liquid, evaporation becomes more intense, thus facilitating the formation of the vapor cloud.

3. Spill in water (without confinement)

   LNG spill can occur over sea water, a river, lake, or another large body of water. The evaporation rate of a LNG spill in unconfined water (e.g., water not in a retention basin or tank) is much higher than in the case of spills on the floor or soil, because heat transfer from water to LNG is usually more intense. When water is not contained in a limited space such as a retention basin, the water circulation under the spill is more efficient, further intensifying evaporation.

4. Spill in water (with confinement)

   When the volume of water in which the spill occurs is confined to a tank, retention basin, or pool, the initial evaporation rate is as high as in spills in unconfined water. But as the event evolves, the heat transfer rate from water to LNG is reduced. It is caused by water reaching its freezing point, which creates an ice sheet that limits heat transfer and evaporation.

5. LNG pool fire

   LNG spills can get ignited and establish a LNG pool fire scenario similar to those that occur with other liquid hydrocarbons. The characteristics of the flames vary according to the size and shape of the LNG pool, the wind speed, and the composition of the spilled LNG. The horizontal extent of the flames is mainly influenced by the dimensions of the LNG pool and the wind speed. The flame height is about 2–3 times the approximate diameter of the LNG pool and is also affected by wind speed.

   Comparatively, under the same ventilation conditions, the LNG burning rate is 12.5 mm/minute, and the gasoline burning rate is 4 mm/minute. Therefore a LNG pool with the same volume of gasoline produces less instantaneous thermal radiation. In the case of gasoline spills, the total heat produced by the complete LNG pool burning is higher than for LNG, but the heat is generated over a longer period of time because gasoline burns more slowly. Therefore for two LNG pools with the same volumes and ventilation conditions, the gasoline pool generates less intense thermal radiation and flames of smaller heights than a LNG pool because the total radiation produced by the gasoline pool is distributed over a longer period of time.

6. Flash fire (cloud ignition), explosion, and fireball

The flash fire or VCE (Section 3.3.2) is the sudden, intense cloud combustion. Flash fire is not an explosion, because the burning reaction speed is not sufficient to produce a shock wave. It should also not be confused with jet fire, which is a phenomenon that occurs when a pressurized pipeline suffers a loss of containment (leakage) that allows the formation of a pressurized gas jet. Some LNG cloud formations may lead to a flash fire. For ignition to occur, the LNG concentration must be within the flammability limit range.

The likelihood of a flash fire is increased when a LNG cloud becomes elongated and narrow, like smoke from a cigarette, because the LNG concentration varies significantly along this type cloud shape. A flash fire will happen when a cloud region within the flammability limit range reaches an ignition source. Factors such as the speed and variation in direction of the wind have an important influence on the shape of the LNG cloud, as the one described. The higher the wind speed, the less likely it is that flash fire will occur, as more intense winds will disperse the cloud. The characteristics of the spill and the LNG pool also influence the risk of flash fire. If LNG is spilled into a retention basin with raised walls, these walls will act as obstacles, creating a barrier to be overcome by the vapors emanating from the spill. It acts to slow down the cloud formation as well as also its extension and elongation.

A more critical situation can occur if the cloud becomes so long that it reaches confined spaces such as buildings, warehouses and housing. If the cloud is able to penetrate these confined spaces and if the concentration is within the ignition range, then an explosion can occur if the explosive atmosphere in the presence of an ignition source. The explosion generates a shock wave capable of causing severe injury to people and damage to the installation. Another critical situation can occur if downward-moving air currents create a concentrated plume near the ground level, keeping the vapors lighter than air for longer. In this specific condition, the vapor cloud may not be dispersed as fast and this situation can extend the length of time during which the cloud is within the flammability limit range, and as result increasing the risk of ignition. If a plume with such characteristics gets in contact with an ignition source, a large fireball-type combustion will occur, generating a very intense ball-shaped burning with significant heat release.

7. BLEVE

   BLEVE (Sections 3.3.2, 6.1.4) is a well-known phenomenon with high risks for hydrocarbon facilities. But in theory, in the context of processes involving LNG, BLEVE risk is not major. It is unlikely to occur in LNG containers due to design factors such as construction of dual-containment structure tanks (for insulation material containment or vacuum). This construction characteristic makes the outer containment structure a barrier to the flame that prevents it from ever reaching the internal structural component. This mitigates the heating effect and the sudden evolution of the accident. Depending on the material used for thermal insulation between the containment surfaces, it can also contribute for the reduction of heat transfer speed to LNG.

6.10.3 Emergency control and liquefied natural gas cryogenics

LNG emergencies cannot be mitigated by professionals who lack specific fire-fighting training in cryogenic products, that is imperative for the understanding of cryogenics-related phenomena that occur during a LNG fire. As we have mentioned earlier, in the case of a LNG pool fire or in a container with LNG, water fire fighting does not yield the same result obtained in a fire involving conventional hydrocarbons. The LNG cryogenic temperature below −150°C causes the water used for fire fighting to become a heat source, accelerating the evaporation of hydrocarbon, expanding the cloud, fueling the flames and increasing the intensity of the radiated heat. The water can be used in the vicinity of the LNG pool and in unignited cloud to accelerate the evaporation process, as long as there are no risks of ignition in the area of the accidental event and the fire brigade crew is experienced in adopting the technique. Otherwise, the best control action should ignition occur would be to allow the product to be completely consumed by the burning and water cooling should be done only to protect equipment that is not directly exposed to fire, but under the effects of heat radiation from the fire in its vicinity. Water fire fighting can never be done directly over a LNG fire.

In the event of a LNG leak, it should be stored in retaining basins designed in places with high risk of loss of containment, such as underneath tanks and areas with multiple connections, among others. Such retaining basins should be built using materials capable of withstanding the thermal stresses associated with the cryogenic product. Reinforced concrete is not such a material because of its inability to deal with extreme temperature variations. When a facility is equipped with multiple LNG spill retaining basins, the basins overflow can be collected by an impoundment basin (Fig. 6.18). Thus as a result LNG leakage gets stored outside the process area. Along with better ventilation LNG is able to evaporate with less risk of ignition and injury to people and damage to the facility. Retaining basins need to be designed with systems for the purpose of monitoring and removing accumulated water. They should be equipped with a foam flooding system to form an insulating layer between the LNG spill and vapors in the atmosphere. This layer serves to limit the formation of the vapor cloud and to retard the evaporation in a controlled and safer environment. The foam also helps minimize the risk of LNG coming into contact with ignition sources and control the dispersion of vapors. The foam heats up the vapors causing them to rise to the atmosphere more quickly to get dispersed, preventing the formation of a hazardous vapor cloud of at low elevations. Foam also plays an important role in controlling possible fires in the spill area, in the presence of an

Figure 6.18 Foam flood system test in which the liquefied natural gas spill retaining basin is flooded with foam spilling the excessive volume into the channel connected to the impoundment basin. *Courtesy J.C. Melchior's personal file.* LNG, *Liquefied natural gas.*

ignition source inside the basin or in the LNG pool, controlling the burning, and avoiding catastrophic escalation with great heat radiation.

As mentioned earlier, fighting LNG fires requires knowledge of specific phenomena related to cryogenics. In particular, the fire-fighting mechanism in cryogenic fires using high-expansion foam is worth special mention. The foam is formed by the proper mix of water and the FGL. Although direct water fire fighting with LNG is not recommended, water is also present in the foam, and as already mentioned, it acts as a heat source capable of aggravating the fire scenario. The fire-fighting mechanism with foam generation in cryogenic fires (Fig. 6.19) allows vaporization control, heating up the vapors so that they rise to the atmosphere more quickly, and lastly to minimize the risk of flash fire due to vapor cloud formation near the ground. The foam layer also allows LNG to be insulated from the air, reducing the chances of ignition. Nonetheless, constant monitoring of the area under foam combat is necessary, because the most important mechanism of this combat occurs under the foam.

When the foam is applied over LNG, the water in the foam freezes and forms a layer of ice in the bottom area of the retaining basin. This process creates something similar to ice stalagmites, which are like cones

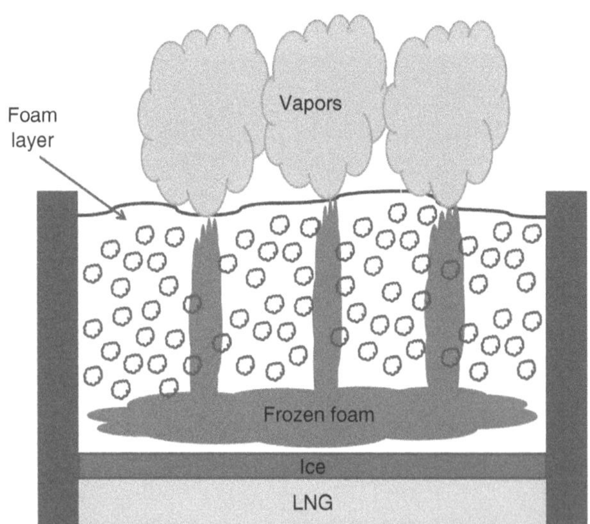

Figure 6.19 Evaporation control mechanism for liquefied natural gas leak during combat using high expansion foam. As the foam water freezes, it forms stalagmites that redirect the vapor to the region above the foam layer.

with the base at the bottom of the basin and a height that extends to the upper layer of foam. These ice formations are useful for channeling LNG vapors, concentrating the gases above their tips. They act as pipes that pass through the foam layer, leaving it intact. This mechanism organizes the formation of vapors and the release into the environment. As long as this mechanism is in effect, the situation remains in a more easily controllable condition. But ice stalagmites may not last throughout the operation, and as a result the stability of the vapor release process may be affected. It is essential to monitor the foam fire-fighting operation so that complementary foam volumes can be applied if necessary, or the control of possible ignition sources as an additional actions.

In the event of a momentary ignition, it can be controlled by chemical powder extinguishers or alternatively keep the foam layer under monitoring, depending on the severity escalation of the event and the associated risks. Burning under the action of high-expansion fire-fighting foam can be an emergency control solution that accelerates the process of safe depletion of the LNG inventory (Fig. 6.20).

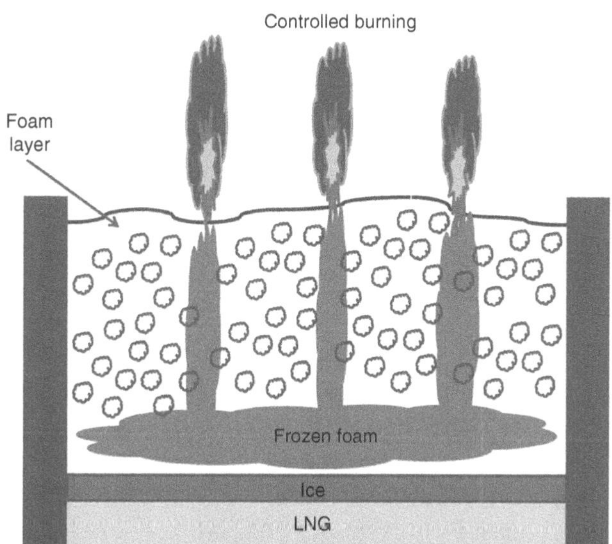

Figure 6.20 Liquefied natural gas spill burn control mechanism during fire fighting using high expansion foam. As the water in the foam freezes, it forms stalagmites that redirect the vapors to the region above the foam layer. These vapors can ignite and the combustion can be kept under control through this mechanism.

Fire fighting can also be performed using portable chemical powder fire extinguishers, in case of flames in small LNG pools, or to control minor ignitions during fire fighting using high-expansion foam. Chemical powder fire extinguishers can also be used to clear the way for escape routes affected by LNG spills. The types of chemical powder fire extinguishers that can be used for LNG fire fighting are sodium bicarbonate fire extinguishers (compatible with class B and C fires) and ammonia monophosphate fire extinguishers (compatible with class A, B, and C fires). Potassium bicarbonate is also compatible for LNG fire fighting, being quite effective, however, an approved product is not yet available in all countries. Considering the effectiveness and market availability, it is recommended to use ammonia monophosphate fire extinguishers in facilities with LNG inventories. But we reiterate the importance of specific training for fire-fighting cryogenic products like LNG. Training on conventional fires, even including hydrocarbon facilities, is not sufficient to train fire brigade personnel and technicians for fire-fighting cryogens such as LNG.

6.10.4 Rapid phase transition and liquefied natural gas cryogenics

One of the most important phenomena related to the safety of cryogenic products (such as LNG) is the rapid phase transition. A liquid can have the beginning of its boiling retarded due to the lack of an energy concentration point. Compared to the boiling water phenomenon in a simple pan, the first bubbles form at the bottom of the pan where there is higher concentration of energy, such as at points of irregularities in the bottom metal surface.

A drop of liquid in contact with a hot surface forms a vapor film between the liquid and the surface. This distance created by the film is sustained by the evaporation rate, and this process causes the drop to be supported by the steam "film." The drop does not actually touch the hot surface. When a drop falls on a very hot surface, the effect of the steam film will make the drop "float" and the drop will remain stable in this situation for as long as the region of the film contains no point of significant energy concentration. In the case of LNG spill in water, due to the temperature difference associated with the cryogenic characteristics of LNG, the water will act as if it were the hot surface in the previous example. So, large drops of LNG will be produced which will be supported above water surface by the LNG steam film. If the LNG composition is predominantly pure methane, no violent evaporation shall occur. However, if the LNG composition contains a large amount of ethane, propane, and/or butane, the methane will evaporate first, increasing the concentration of heavy hydrocarbons. This will cause the LNG boiling point to increase up to a threshold above which the heat transfer is no longer sufficient for sustained steam production to form the supporting vapor "film." At such time instant, the superheated LNG comes into contact with the water, from points of energy concentration and a very fast heat transfer starts. This phenomenon results in a sudden and violent vaporization—called "explosion without flame" or "fast phase transition."

The rapid phase transition can create a localized shock wave, with harm to people and damage to structures within the event's radius of action. The composition of the LNG determines the likelihood of occurrence of the phenomenon. The higher the methane concentration in LNG, the less likely it will occur. Another influencing factor is the water agitation state. The more agitated the water, the more likely the fast phase transition phenomenon occurrence.

6.11 Subsea safety equipment

In offshore rigs, there are emergency situations that require the immediate shutdown of subsea wells. This type of closure impacts the entire offshore production chain and requires the best of designers' skills to prevent unnecessary shutdowns. At the same time, in the event of an emergency, if performed for safety reasons the operation cannot fail. To ensure efficiency in closing offshore wells, the subsea arrangements are equipped with automatic underwater shut off valves whose actuation needs to be assessed based on the consequences analysis studies (fire propagation and explosion). As an initial approach to the problem, the installation of subsea automatic shut off valves in the output and input pipelines and in production gas pipelines of satellite wells should be considered. Similarly, automatic shut off subsea valves should be planned for the gas pipelines for the interconnection of offshore rig with subsea manifolds.

6.12 Fire brigade and rescue crew performance

It is not rare for accidents to be reported related to emergencies with increased number of victims due to harm to fire brigade members and rescuers. There are cases where they are the only victims. It is important that the decision to send the fire brigade and rescuers be made based on a diagnosis of the developing accidental scenario followed by an immediate assessment of the response resources capacity with respect to the scenario in question. When facing limited response means in relation to the emergency severity, the decision should be not to expose fire brigade and rescuers to risks that might increase the number of victims as a result of their actions. They can only be assigned to fire-fighting tasks if the available means of response offer clear chances of success. For this reason, any emergency in high-risk facilities such as in the oil and gas industry should first be recognized and declared. The emergency recognition is obtained through the diagnosis of the facility's highest authority, who needs to be technically qualified for this specialized work. The declaration of the emergency involves gathering all the people in the facility at previously established meeting points and the immediate communication of the emergency situation to all those with responsibility role defined in the emergency communication roster. Once people are assembled and the emergency is declared, the authority must conduct the scenario diagnosis, evaluate the response conditions and call the fire brigade and rescue workers if deemed feasible in terms of minimizing losses. Otherwise, if the resources available are below the minimum required to cope with the emergency severity, the authority must begin the escape and abandonment operation as established in the original facility project.

6.13 Crisis management and decision making

Decision making in an emergency is characterized by time pressure, which unlike other managerial situations requires immediate response. For this reason, the technical and behavioral profile of those responsible for leading the response actions and main decisions during a crisis is distinct from other activities. First, practical operational experience is required and not just theoretical knowledge on the facility operations. One must know the pertinent physical and chemical phenomena and have a solid technical foundation regarding engineering processes related to the operation and the project of the facility. But this is not sufficient. In addition to all the technical knowledge required, the best possible decision during the crisis will be made if the facility's highest authority (responsible for crisis management) has an adequate behavioral profile for this role. Therefore it is essential that technical skills are complemented by behavioral skills so that the best decisions can be made given the time constraints that crisis scenarios impose.

Regular training (at least once a year) in crisis management is essential for professionals directly involved in the decision-making process during emergencies. Complex and mission critical facilities should provide training in simulators for operators involved in crisis management. There should also be psychological monitoring of these professionals so that weaknesses can be identified. Professionals in these conditions should be at least temporarily removed from their activities as prevention.

That happens when psychologists and behavior analysts identify a personal precarious psychological situation that may impair the professional's performance in the midst of a crisis management operation.

The characteristics of each facility need to be considered in crisis management training. Generic trainings are not recommended. The success of actions to be taken during a crisis depends on the combination of technical and behavioral aspects, so that the specific problems related to the technology adopted in the facility need to be taken into consideration. Lack of technical knowledge related to the facility can lead to wrong decisions, slow decision-making process, fear, and blackout. Conversely, the technical expertise of the processes, a good fit of the personality profile and the behavioral training are essential elements in the preparation of competent professionals for decision making and crisis management roles.

A situation that has been identified as a frequent cause of the aggravation of accidental scenarios during crisis management is the high degree of automation of facilities and reduced intervention by operators in the day-to-day operations. We herein coin the term "manophobia" as operators' fears of manually operating systems that are normally operated automatically. This term can also be applied to everyday situations, for example, a driver of an automatic transmission car who has resistance (fear) of driving a car with manual transmission. Many accidents in different areas associated with technology such as aerospace, navigation, operation of process plants, etc., have behavioral factors related to manophobia as their root causes. The general automation trend, from household appliances to large process plants, has developed into a generalized manophobic trend. As systems are most often automatically operated, when automation fails requiring manual intervention, operators can be reluctant to take over the operation influenced mainly by the departure from the their routine activities, which for the most part consists of automation monitoring. Manophobic behavior is a direct threat to the safety of facilities and equipment, which has led to the reflection on the need for reduction of the level of automation under certain situations, even though there is an available technology.

As we have said, manophobia is some type of fear. In terms of risk management, we can define fear as the set of human reactions caused by the perception of the disconnect between a hazard and the technical knowledge required for its diagnosis and to manage its risks. There are always unknown factors about any hazard.

No matter how much science and technology develop, there will never be an absolute assurance that a hazard does not materialize in losses. There is no total guarantee that accidents can be avoided. What risk management engineering can contribute to this situation is with the ability to provide safeguards that keep the risk of the hazard causing losses down to a level that it can be accepted. Fear while facing hazard is established when there is a perception of the incompatibility of the knowledge available to manage its risks. During the handling of a crisis, there is always going to be some knowledge gap related to the scenario, despite the quality of the professionals, equipment and all the associated technology. So, there will always be some component of fear as part of the natural human reaction. The difference between natural fear and the one that leads to a catastrophe is the level of knowledge and solid experience on the operation of the facility or equipment.

6.14 Selecting and identifying accidental scenarios

It is a requirement to have prior knowledge about the main accidental scenarios that threaten the installation to work in emergency control. Every conceptual definition of a scenario has limitations. It is impossible to anticipate all the possibilities of accidents, natural influences, human behavior, and the complexity of their interactions. When we conceptualize a scenario for studies or simulation, in reality we are establishing the limits of our study and simulation. In a real-world situation, these limits do not exist and any unforeseen scenario can happen. This justifies the use of the term risk management, which denotes the hazard management and not its complete elimination followed by total safety assurance. Total safety does not exist in objective terms. Safety is related to the degree of risk acceptance by each person or professional. Risk acceptance level is an individual characteristic. Under the same quantified risk, some people may feel safe while others may not. On the same flight, with the same crew, on the same plane, not all passengers may be feeling safe. But the risk is the same for everyone. Risks can be assessed, calculated and, in some cases, quantified considering the safeguards defined in a project. This activity is more in line with the term to manage risks than with the term to provide safety. Once the risks are managed, the quantified values can be more or less accepted, depending on the subjective influences by whoever evaluates them. When the results obtained from risk management lead to the (subjective) perception of safety by decision makers, usually it means that the risks have become acceptable. On the other hand, when they lead to the perception of lack of safety, decision makers usually lean toward seeking justifications for rejecting the risks, even if they can be corroborated by theoretical quantitative analyses.

The greater the number of scenarios evaluated and the variety of influences considered in the definition of accident scenarios, the greater their contribution in managing the associated risks. Evidently, the increase in parameters considered in the studies, as well as variables and scenarios make the work involved more complex.

Exactly for this reason, beyond a certain degree of complexity, computational tools are the only options for making risk analysis and management viable. Only they make it possible the processing of a large amount of information, logical sequences and interlocks.

Even with the use of computational tools, however, it is essential to define scenarios mainly based on operational experience and historical accident data. Priority should be given to the inclusion of the most general and probable cases in the list of scenarios. The frequency of occurrence of a scenario can never be calculated with absolute accuracy. Despite the great effort for improvement, it is still much more reliable and realistic to choose the scenarios to be studied, based on history and operational experience. Values calculated from databases and statistical analyses can be supported mathematically, but it does not provide any guarantee that these numbers will be in agreement with the operational reality. There is no replacement for operational experience and technical operational knowledge as part of the definition of the best scenarios for risk and safety management. Theoretical calculations should be utilized to identify opportunities for corrections of the perception based on operational experience, when applicable.

The project of oil and gas facilities need to identify the most important and the most likely accidental scenarios, considering not only those with great destruction, but also, for example, scenarios such as the leak of gas without ignition of the released volumes. These scenarios should also consider that the physical damage to the facilities in these circumstances is not significant to affect the effectiveness of the abandonment means planned in the project. Such scenarios may also assume that there are no concentrations of gases capable of intoxicating people.

Scenarios of this kind assume a gas leak due to damage or to human error, which justifies the escape order and, later if necessary, the abandonment. Therefore in this type of scenario, it is considered that all escape and abandonment routes are available and no gas ignition during the accident or intoxication of people is postulated. For this reason, this type of scenario also relates to the order of escape and abandonment during training (standard scenario of escape and abandonment) or for any other reason that is not caused by significant damage to the facility. As a variation in this type of scenario, the toxic gas leak (such as H_2S) can be considered, and its impact or of any other gas on people should be assessed, even without any damage to the facility integrity. Typical scenarios considered in the analysis of gas leaks without a fire are the rupture of pipes with a large gas inventory, loss of integrity of pressure vessels, rupture of risers in the lift gas system, flare failures in which the burning can be interrupted, hard-to-detect small leaks subjected to generating gas confinement, among others.

Fire scenarios should be used to investigate the effect of the temperature generated by the flames on the facility and on people. They also serve to investigate the effect of toxicity and limited visibility due to smoke on people's activities. The number of possible fire scenarios in oil and gas facilities is very large. It serves as a justification for the need to prioritize the scenarios with the highest chance of occurrence and with the most severe consequences for the facility and for people. Pool fire scenarios associated with equipment where significant hydrocarbon inventories are stored are usually considered for analysis purposes. Likewise, jet fires resulting from the loss of the integrity of pipelines and pressurized equipment.

Scenarios involving the occurrence of an explosion are studied separately. In this case, equipment and pipelines most likely to generate explosions are the ones chosen, and the consequences of shock waves that can be generated are also investigated.

Other specific accidental scenarios, such as naval damage in offshore rigs, also need to be studied according to the characteristics of each installation. There are no limits to accidental scenarios that can be postulated, but there are associated technical and cost limits. It is necessary to identify the most important and representative scenarios to be studied in detail. The more significant in terms of the operational reality the scenarios chosen are for the studies, the greater the level of protection to be obtained from the results of the analyses. Most frequent scenarios despite their minor consequences may deserve more attention than extremely catastrophic scenarios with very little likelihood of occurrence. The evaluation of these aspects is essential for a sound definition of accidental scenarios for safety studies.

6.14.1 Design basis accident

This is a concept adopted in high technological level risk management projects. In addition to the scenarios postulated and studied throughout the facility project, a specific accidental scenario is defined as a "design basis accident." It aims to identify the worst and most severe accidental scenario for which the facility has safeguards, that is, for which the safety systems designed for the installation is capable of responding to. The design basis accident definition is valuable for guiding the complete operation, both from the standpoint of understanding the safety systems limitations as well as for checking the consistency of the operational procedures. It becomes a definitive reference for the whole facility life cycle, from conceptual project to its decommissioning. For example, the project of an offshore rig may define as a design basis accident scenario a "naval malfunction with tilting the rig to the maximum angle at which it is possible to launch a lifeboat, followed by the rupture and leakage of the pipeline connected to the largest pressure vessel in the process plant as a result of the naval damage (slope)." Evidently, this scenario that could only be defined as a design basis accident after a thorough evaluation of all the other accidental scenarios postulated in the project and of all the response safeguards designed to avoid them.

6.14.2 External origin accidents

There are some types of accidents whose origin may seem totally intangible and uncontrollable by the designers. But even in these cases it is possible to contribute to the reduction of the risks of major consequences, through the adoption of the concept of "external origin accident." However, accidents can have other classifications regarding their origins, as we will see below.

6.14.2.1 Accident classification

Accidents always involve some component of unpredictability and surprise. More often than not it is not one, but there are several causes that contribute to the occurrence of an accident. Aiming at avoiding accidents investigators seek to identify in their reports the root cause, which was the most important in the event to produce undesirable consequences such as victims and environmental and material losses. In this context, an important strategy is the classification of accidents to facilitate the study of the protection measures applicable to each category.

There are different ways to classify accidents. We can classify them based on their sources, as follows:

1. Operational origin: when the origin is a wrong operational action.
2. Project origin: when the origin is a design or conceptual error.
3. Construction and assembly origin: when the origin is in an error that occurred during construction or assembly.
4. Maintenance origin: the accident is caused by noncompliance with the original design during regular maintenance of equipment and installations.
5. External origin: when an influence completely outside the scope of the project is the cause of the accident. External origin can be, for example, a natural catastrophe (earthquake, flood, hurricane, lightning, meteorites, etc.) or a hostile act (sabotage, intentional aircraft crash, military attack, terrorist attack, etc.), or an accident (unintentional aircraft crash, etc.).

A classic example of an accident of external origin is lightning strike. Brazil has the highest incidence of lightning strikes in the world. Even in countries where lightning does not happen so often, nonetheless, its damage is considered so high that it is mandatory to use protective equipment such as lightning arresters.

There are also protections against earthquakes, tsunamis, and hurricanes, but these are included in the projects when justified by the frequency of occurrences of these events. In fact, the decision on whether or not to include a protection system against a specific type of accident of external origin will depend on the combination of two main factors: the frequency of occurrence of the event and its possible consequences. In order to identify the most significant accidents of external origin, experts first assess the potential scenarios and hazards that may occur over the life cycle of the project. Through a risk classification matrix, experts identify the high frequency hazards that occur and/or those with severe and catastrophic consequences. Once these hazards and scenarios are identified, they are studied in detail, and then the engineers will design means of protection to reduce their consequences. The hazard itself is often impossible to avoid because of its external origin that is beyond the control of engineers and designers. Safety systems can be designed and built to reduce the catastrophic consequences caused by these unavoidable hazards in the event of an external origin accident.

Phenomena beyond one's control, such as lightning strikes, earthquakes, tsunamis, and floods, receive more rigorous treatment during engineering design, due to their significant frequency of occurrence. Some industries have more rigorous protection against accidents of external origin. Protection against earthquakes in nuclear power plants is often considered in projects that have as a premise the nonacceptance of the consequences of this type of event, including locations without significant frequency of this type of phenomenon. We can also mention the case of the numerous antiterrorism safety measures included in the projects after the destruction of the World Trade Center in New York. Terrorist act is a typical external origin accident, although it is not a natural catastrophe.

Likewise, acts of sabotage. Another example is the structural design of oil platforms that operate on the high seas. They are designed to withstand the so-called 100-year waves (the largest possible in a 100-year period) produced during tsunamis mostly caused by large earthquakes on the sea floor. Nuclear plants are also protected against intentional or unintentional aircraft crashes. Nuclear plant projects dated before September 11, 2001 had already adopted this level of protection, and for that reason some nuclear plants have an external special reinforced concrete containment about 70 cm thick, and another steel containment vessel for protection against radioactive material leak in case of accidents.

Obviously, each additional protection represents an extra cost and it is often unfeasible. Regarding the risks due to meteors and meteorites, risk management engineering provides the framework for analysis of all these factors related to the frequency and severity of these natural phenomena, thus to deal with the matter within the context of realistic engineering solutions, economic viability, and historical data. Due to the long time intervals between occurrences of these phenomena, the trend is to consider the statistics, weigh the costs, and benefits and, above all, take in consideration the undeniable fact that risks will never be completely eliminated from any human activity, despite the best treatment of their consequences.

6.14.2.2 Example of protection against external origin accidents

Nuclear power plants, including those in Angra dos Reis, Rio de Janeiro, Brazil, are designed for the so-called external origin accidents, which are technically identified by the German acronym EVA (Einwirkungen von Außen). In Angra dos Reis, for example, these are the following postulated accidents for protection against earthquakes:

1. Design earthquake protection: earthquake of maximum intensity that occurred in the past, in the surrounding area within a maximum radius of approximately 50 km.
2. Safety earthquake protection: earthquake of maximum intensity that may occur considering in the surrounding area within a maximum radius of approximately 200 km.
3. Combined effect: safety earthquake plus a shock wave caused by the explosion of a conventional pressure vessel component of the plant as a result of the safety earthquake.

In addition, the Angra dos Reis nuclear power plant has redundancies for cooling water sources and protection against tsunamis, terrorist acts and a commercial aircraft crash on the reinforced concrete containment structure. Despite being an older project, Fukushima was also designed for EVA-type accidents, but it did not withstand the March 2011 earthquake as expected. Three reasons may have contributed to the accident:

1. calculations on natural phenomena may have been underestimated;
2. the construction of the facilities may not have adhered to the structural design; and
3. management and/or maintenance were not adequate causing degradation of the defense layers/barriers.

Regardless of the root cause, any of them raises the suspicion that the theoretically accurate results of the quantitative risk analyses resulted in much lower values than the actual risks, which suggests that the calculation techniques and design are worth being questioned. It is not about calculation errors related to the use of such techniques, but probably there is a lacking analysis beyond the numbers, that is more strongly centered on operational experience in order to identify and correct distortions in the theoretical results.

6.14.3 Beyond design accident

There are accidental scenarios so difficult to be diagnosed that operators could spend days unsuccessfully trying to understand them. This is compounded by the fact that the scenario diagnosis during crisis management needs to be done with precision and extremely rapidly. On top of that, some accidental scenarios can be so severe that all safety systems become unavailable, right from their inception, leaving the facility unable to respond in a way that is compatible with the serious crisis that is established with the scenario in question. High-level risk management projects can adopt the concept of "beyond design accident," which consists of recognizing the extreme situation, without conditions for diagnosis and accurate response as mitigation means. This concept allows the creation of a simplified procedure to be adopted in this type of extreme scenario, based on the monitoring of only six critical safety functions chosen as the most important for the safety of the facility, regardless of the type of developing accident. With this strategy, operators and risk and crisis management experts can take actions without the need for a diagnosis of the current accident and even with damaged safety systems.

Choosing the six critical safety functions adequately is essential. The reason for limiting to only six functions is due to the human innate capacity to manage well up to seven information channels simultaneously. Human being's cognitive capacity to perform such management starts deteriorating with eight simultaneous information channels. So, each of the six critical safety functions represents an information channel, and the seventh channel should remain free for communication between the experts managing the crisis. The choice of the six critical safety functions is a task that requires deep knowledge about the installation that needs to be protected. For didactic purposes, we present below an example of the selection of six critical safety functions for a FPSO offshore rig:

1. Confirmed fire (Is there such a condition?)
2. Confirmed gas (Is there such a condition?)
3. Naval buoyancy (Is the rig sinking?)
4. Naval stability (Is the rig tilting?)
5. Containment of the hydrocarbon inventory (Is there a leak in the rig?)
6. Conditions for launching lifeboats for abandonment (Are all systems available to start abandonment?).

In the event of the recognition of an accident beyond design scenario, operators, and experts from the crisis management leadership team start to focus strictly on the monitoring of critical safety functions. The monitoring continues for as long they are unable to reach an accurate diagnosis of the developing accidental scenarios. In case of a threat or degradation of one of the six critical safety functions, it shall end with an escape and abandonment operation.

6.15 Special safety strategies applied to automation

High-tech risk management projects can adopt special automation and control strategies. An example of a special strategy is the subdivision of all interlocks and automation and control systems in two subsystems. One subsystem may be called a "protection system" and includes all automations and interlocks that can result in the ESD of the unit for safety reasons. The other subsystem may be called the "limitation system" and includes all automations and interlocks that are designed to limit operational activities to avoid ESD.

In other words, we can say that this architecture subdivides automation into two parts, one that monitors data and submits it to safety interlocks in an attempt to shut down the facility for safety's sake, while the other part also monitors data but with the reverse objective of preventing the operating parameters from reaching the shutdown values that put the facility out of operation. A "logical competition" is thus created between the protection systems (which attempt to shut down the installation) and the limitation systems (which try to avoid shutdown by changing the operating conditions).

For the understanding on this architecture at work in practice, suppose that a high liquid level signal from an FPSO oil tank results in the facility ESD. The logic that justifies this type of interlock is that if the tank is completely full, there will be no way to store the produced oil and therefore production needs to be stopped to avoid the tank overflow accident. Considering the concept described above, the interlock would be part of the protection system, responsible for shutting down the installation in the name of safety. But in response to that, an independent system could collect data on the filling level of the same tank and generate a signal when the tank level reaches 70% of its maximum volume, so that the production is reduced to 50%.

This second interlock, within the same concept, would be part of the limitation system and responsible for limiting the facility operation so that its parameters do not reach the limits that generate the ESD through the other system, namely, the protection system.

This strategy can be applied to all facility interlocks. Based on this subdivision concept, each interlock, each alarm and each automatic actuation is labeled in one of the two concepts: protection system or limitation system. The complete automation set associated with the protection system performs the role of continuous monitoring the facility for a threat confirmation signal that justifies an ESD. Competing with the protection system, the other automation set associated with the limitation system also continuously monitors the facility in search of a confirmation signal related to the operational parameters approaching the set points that generate ESD through its counterpart, the protection system. However, when the proximity of the parameters to unacceptable values is confirmed, the limitation system generates signals to limit the facility operations in progress to avoid the risk of ESD. In essence, the protection system tries to shut down the installation in the name of safety, while the limitation system tries to avoid this shutdown in the name of operational availability in a "healthy tug of war" in search of the combined maximum safety and availability.

6.16 Conception of redundancies and ways to start up safety systems

The reliability of safety-related systems can be increased by redundant systems design. For example, FWP systems to be available immediately after the confirmed fire signal. But if the equipment is need in maintenance or if it fails during startup, the facility will be totally unprotected if there are no pump redundancies.

Reliability is the likelihood that a part, equipment or system will perform its function failure free within a certain period of time and under normal operating conditions. One of the reasons for the adoption of redundancies is the increased reliability of the systems. There are typical configurations for the design of systems with more than one device performing the same function, however, even with multiple equivalent devices, not all configurations of its type can be considered technically redundant. Following are some types of system configurations designed for more than one device performing the same function.

1. *Series system:* the components are considered in series when the failure of any one of components causes the failure of the complete system, making it unavailable. For example: two FWP, each with a 50% capacity of the demand required for the system cannot be considered redundant equipment.
2. *Parallel system:* components are considered to be in parallel when the failure of one of the components does not cause the failure the complete system. For example: two independent FWP, each with a 100% capacity of the water demand for fire fighting can be considered a redundancy of one another.
3. *Series-parallel system:* in systems with components in series, if one of the components fails, the complete system becomes unavailable. The series parallel system allows the failure of one of the components of a configuration in series to not cause the failure of the complete parallel series system, since it includes another component in parallel, in addition to the series equipment. This additional component replaces the function of the equipment that fails. For example: three FWP, each with a 50% capacity of the required demand, can be considered equipment assembled in a parallel series system with one redundancy. That is, it is necessary that at least two 50% FWP are available in series, but in the event of failure of one of the FWP, the third pump, in parallel, can avoid the unavailability of the system.

6.16.1 Types of redundant configurations

If reliability is defined as a probability, for calculation's sake the probability is always expressed by a number between 0 and 1. This means that, as we add equipment in series for a system to operate, the reliability of the system is decreased. For each equipment added, a numerical factor less than 1 is concatenated to the product of factors for calculating the total reliability of the systems in series, thus reducing the final value of reliability.

As an example, we can say that three FWP with 50% of the firefighting water demand capacity form a less reliable system than a configuration with two FWP with 100% of water demand (considering that all pumps have the same theoretical reliability value). In practical terms, in the first configuration we will always need to have two 50% devices in series, that is, a less reliable system, whereas in the second configuration, we will only need 100% equipment to keep the system available, which therefore gives this configuration greater reliability. There are many mathematical models and databases developed by reliability experts that allow calculations of the reliability of safety systems. Some of these calculations can become quite complex if the systems studied are composed of multiple equipment and components.

What should guide risk management experts to design safety systems with efficient and reliable redundant configurations? To help answer this question, we will introduce some essential concepts for choosing redundant safety system configurations.

6.16.2 Classical failures

A very useful concept for experts to consider is that a safety system needs to be protected from two main classic redundancy failures, namely:

1. *Unavailability failure:* at the time the use of the equipment is required, it is unavailable due to maintenance, testing, etc.
2. *Startup failure:* at the time the use of the equipment is required; it is operational but does not work as planned in the project.

To overcome the two classical failures, a system should have a configuration of 3 × 100% (three-redundancy parallel system configuration). In the event that both classical faults occur during the accident, a third redundancy will still be available. But this is not the only possible configuration to achieve such a purpose. A 4 × 50% configuration also offers protection against classical failures, although the reliability of the system is reduced due to the characteristics of the series-parallel system. In a 4 × 50% configuration, two FWP in series will always be needed to maintain system availability, with the remaining two redundancies to overcome the two classical failures. The choice among the numerous possible configuration options to overcome classical failures depends on a cost–benefit assessment by the designers.

A lower-cost possible alternative is to concentrate the protection against the two classical failures in a single redundancy. This is the case, for example, with a configuration of type 3 × 50% or type 4 × 33%. For the adoption of this type of configuration, it is necessary to adopt the premise that the two classical failures will not happen during the same accidental event. It is considered that, for each scenario postulated in the design safety studies, either the failure due to unavailability or the startup failure may happen, but never both failures simultaneously.

In summary, the configuration recommendation that best balances costs and benefits is the 4 × 50% configuration, as it considers the simultaneous occurrence of the two classical failures and uses smaller, possibly lower-cost equipment. But these conclusions depend heavily on the characteristics of each safety system, each facility and the safety culture characteristics of each operating company. Only a specific cost—benefit analysis can determine the best configuration option for each project.

It is also important to point out that the characterization of a redundancy for safety systems requires independence and autonomy for each equipment. If two FWP (2 × 100%) depend on a single power supply system, these devices cannot be considered to be completely independent. The system with this configuration can be placed in a condition of unavailability due to a common failure of the power supply system that compromises both FWP.

The independence of physical location is also an important aspect, and the installation of each redundancy as far as possible from the facility also contributes to reduce the risks of a single accidental event compromising more than one safety system equipment being designed. It is also very important to note that the aspects related to the logical interlocks for the safety equipment start up directly influence the success and efficiency of these systems during operation. For example, the classical failure at startup is a justification for a designer to include redundancy as protective measure. However, the effectiveness of the inclusion of this redundancy can be compromised if the logical startup signal does not determine the concurrent startup of all redundancies when the emergency is confirmed. For example, consider a fire-fighting water system with a 3 × 50% configuration that was designed so that at the time of the emergency only two redundancies of FWP will be put in service. FWP take a few seconds and, in some cases, a few minutes to reach their nominal capacity. If the startup of one of FWP fails, precious time will have been lost at a crucial point in the fire-fighting operation. Additional time will be lost until the fault is detected, the third FWP is in operation and its maximum capacity is reached.

The appropriate logic interlock should generate the startup signal for all redundancies. The system should be able to divert or recirculate the excess water volume through a pressure control valve or overboarding valve. After the startup of all redundancies, if the operators in charge of crisis management are able to verify that a redundancy can be safely shutdown, then it should be done manually via specific operation for the specific redundancy. During emergency management, the system's response should always pursue the maximum failure reduction capacity that the system has to offer. Thus time loss can be avoided related to the recognition of delayed startup and reaching the operating condition of a redundancy that is required should classical failures with the main equipment happen. The bottom line is that each redundancy is also a financial investment, with associated investment and maintenance costs throughout the life cycle of the facility. This investment could be wasted if the automation system contained conceptual errors that prevented classical failures from being effectively protected.

6.17 Understanding explosion phenomena

Knowledge of explosion phenomenon can be used for both protection and destruction purposes. Oil and gas facilities are permanently at risk of explosions due to the presence of hydrocarbons in their processes. Therefore some preventive care is adopted in the projects to reduce the risks of occurrence of this type of accident. But in some cases the possibility of including structural reinforcements and other protections is also considered as part of another strategy centered on the protection against the consequences of a possible explosion. In this case, specific studies are conducted to determine the characteristics of probable shock waves that may be generated by the main accidental explosion scenarios considered in the project. In this type of strategy, the objective is to reduce the damage and the consequences of the explosion which can reach areas outside the boundaries.

Another application of technical knowledge on the explosion phenomenon is the use of explosives for construction and demolition works. Implosion techniques allow significant savings in construction schedules that would require much more time without such a knowledge.

However, there is a particular aspect that makes knowledge about explosion a reason of concern for society. Knowledge on the explosion phenomenon also has military applications and regrettably can be used for illicit purposes and for terrorist attacks. Therefore the use of potentially explosive materials as well as training related to the explosion phenomenon deserve special attention and control regarding the dissemination, access and traceability of explosive materials and technical information associated with the subject matter. There are several internationally recognized institutions that specialize in technology related to the explosion phenomenon. However, they are most often associated with military or governmental institutions that have the role of controlling the dissemination of this type of information. Those interested in this area of specialization should understand the context of the topic and seek technical training recognized by local and international authorities. In this book, we will only introduce some general concepts for illustration purpose, since the subject matter is mentioned several times throughout the book. In Section 3.3.2, some concepts related to explosion were presented, which will be complemented below.

In a nutshell, combustion is a chemical reaction between a fuel and an oxidant that produces heat and possible release of hot gases. Explosion differs from combustion by the sudden release of energy into the environment, generating pressure waves and possibly heat. As a review, we can say that flash point is the lowest temperature at which a flammable mixture of liquid emanates vapors with the potential to form an ignitable mixture with air. The flammability limits are the extreme concentration limits of a combustible material in a homogeneous mixture with gaseous oxidizer within which there is flame propagation. Flammability limits establish a concentration range in which a flame can occur in a mixture. Another term also widely used is "stoichiometric ratio." It establishes the proportion of gaseous oxidant and flammable material capable of cause combustion without the lack or excess of oxidant or combustible material.

6.17.1 Types of explosion involving flammable products

The main types of explosion involving flammable products are:

1. explosion in an open space,
2. explosion in a confined space, and
3. vessel explosion with boiling liquid (BLEVE).

6.17.2 Formation of explosive atmospheres in open space

The process of forming an explosive atmosphere in flammable liquids is somewhat slow because it depends on the evaporation and diffusion of vapors into the environment. In open areas, even a breeze can reduce the vapors, thus preventing the formation of extensive explosive clouds. This is the scenario that is established, for example, when a pool of liquid hydrocarbons is formed by a spill. Conversely, the formation of an explosive atmosphere from the loss of containment (leaks) involving flammable gases can be a very fast process due to the dispersion aided by the expansion of the gases. In this case, the breeze in an open place can facilitate the mixture of gases with air. Given the usually large cloud formation from gas leaks, the result is the establishment of conditions of mixture concentration within the limits of flammability in some regions of the cloud, therefore susceptible to ignition.

Flammable clouds that are formed in open spaces do not have a well-defined shape, and their composition can also vary, from region to region in the cloud. In the case of ignition, the burning speed can also vary within the cloud. Some regions of the cloud may have concentrations within the limits of flammability, while others may not. Combustion can be subdivided into phases, the first being the cold flame phase, in which the hydrocarbons are initially preoxidized, forming formaldehyde, ketones, water vapor and aldehydes. The greater the amount of formaldehyde, the greater the combustion speed. Sequentially, the second phase, the hot flame phase is marked by the end of oxidation. At this stage, the carbon dioxide formation releases a large amount of heat. Hydrocarbon molecules larger than ethane oxidize more slowly due to the formation of metastable compounds.

When the burning reaction speed is lower than the speed of sound (subsonic), the term used for the phenomenon is "deflagration." If the reaction occurs at a rate of speed higher than the speed of sound (supersonic), then the applicable term is "detonation." The transition from deflagration to detonation generally occurs in gases in open environment conditions, if formed by highly reactive compounds, with a wide flammability range such as hydrogen, acetylene, ethylene, acrylonitrile, ethylene oxide, propylene oxide, and butadiene.

6.17.3 Formation of explosive atmospheres in closed space

In a confined space, such as within a pipeline, the transition process from deflagration to detonation is associated with the phenomenon of acceleration of the burning front. Assuming that a deflagration was initiated inside a closed pipeline at one of its ends, the deflagration will have the beginning of the flame front generating the expansion of gases and the acceleration of the combustion front itself.

Deflagration occurs in a laminar flow with parabolic profile with an increase in the combustion front area. In a turbulent flow, the increase in the combustion area will be even greater due to turbulence. The speed will gradually increase during the thermodynamic transient that produces adiabatic compression and preheating as the beginning of transition from deflagration to detonation. With the evolution of the adiabatic compression, small explosions will start, further intensifying the phenomenon and increasing the acceleration of the burning front. The phenomenon continues to evolve until the coalescence of the explosions occurs and, finally, the pressure wave (shock wave) formation marking the ending the transition process from deflagration to detonation condition (Fig. 6.21).

The transition process from deflagration to detonation is not only applicable to pipelines. Several areas of the oil and gas facility may combine conditioning factors similar to those described in the previous example. Another example is in the turret area of a FPSO. As a large pipeline, the turret may also generate a reaction comparable to the pipeline closed on one end. Fig. 6.22 is a didactic scheme that shows a possible geometry of the shock wave propagation inside a turret. After ignition, the generated shock wave will expand into an initially spherical volume until the wave encounters the interferences and the inner walls of the turret. When it happens, a burning front is formed (as in the example of the closed pipe on one end), as the geometry of the turret resembles a cylindrical shape in addition to having an opening at the top. The sequence in the figure below shows the progression of the shock wave until it reaches the upper turret opening. From there, the shock wave starts to exert pressure on the external

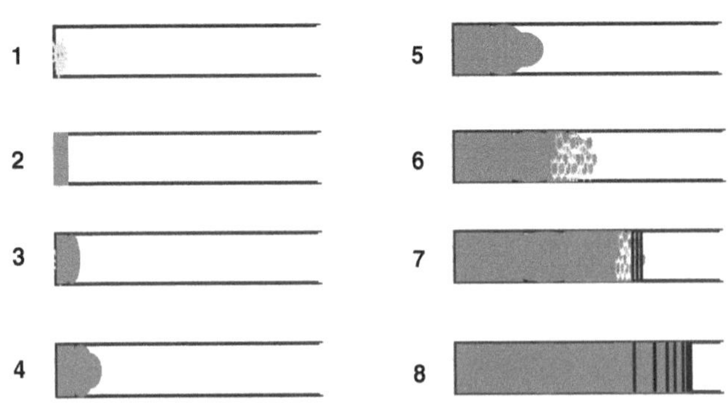

Figure 6.21 Sequence of the transition from deflagration to detonation within a tube closed on one end. Following the sequence, starting from step 6, small explosions start and increase the acceleration of the burning front movement until the complete transition from the deflagration condition to the detonation condition in steps 7 and 8. All steps occur in a split second.

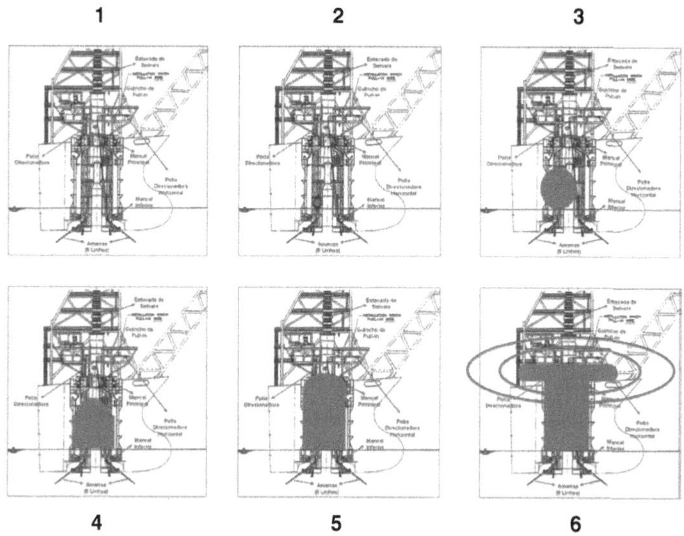

Figure 6.22 Didactic sequence of the staggering of an ignition, deflagration, and detonation of explosive atmosphere inside a FPSO turret. Illustration 2 represents the beginning of the explosion inside the turret and illustration 6 shows the event evolution until the shock wave reaches the outer area above the turret.

regions until the complete dissipation of the energy originated by the phenomenon.

Another example of an explosion in a confined space is one that can occur inside a tank with a significant hydrocarbon inventory (Fig. 6.23). This type of accident can happen in a tank with flammable liquid under conditions that release vapors in the presence of residual air inside the tank. The volume occupied by the mixture of vapors and air can get ignited by an unexpected source, resulting in an explosion. The shock wave produced can rupture the tank releasing the explosive gas mixture to the external environment. These combustion gases undergo a supersonic expansion with the sudden pressure drop. At the same time, a short-lasting vacuum (less than 0.5 seconds) is formed inside the tank due to the supersonic expansion in progress in the region above the ruptured region of the tank. The vacuum intensifies the evaporation of the flammable liquid remaining in the tank, which fuels the combustion. The intensified burning in turn generates more heat, which further fuels the evaporation process of the residual flammable liquid, and for a brief period of time (at least 1 second) the evaporation becomes more intense, forming a volume of vapors that expands until the cloud ignition establishes the sustained

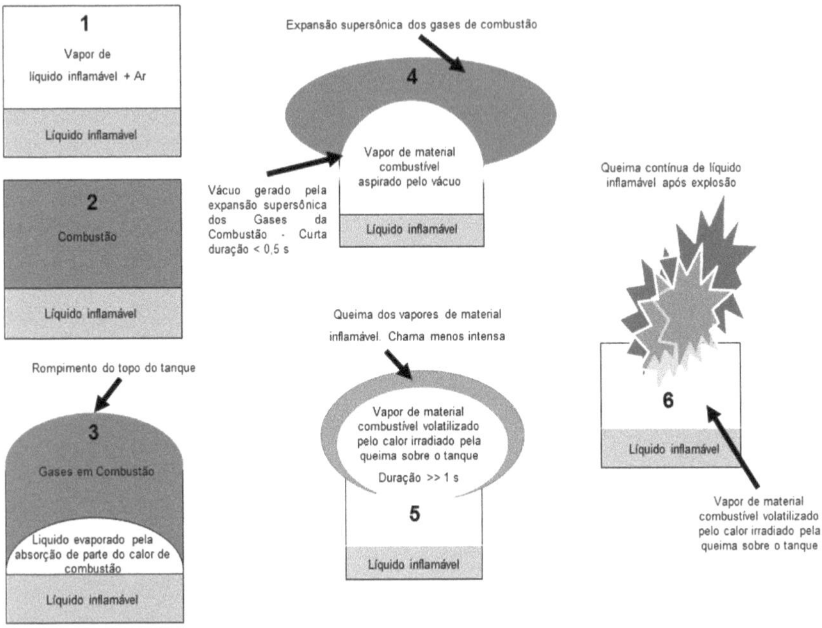

Figure 6.23 Didactic sequence of stages of a tank explosion with significant hydrocarbon inventory. Illustration 3 represents the instant when the top of the tank ruptures, generating the sequence: vacuum caused by supersonic expansion, and in steps 5 and 6 of the explosion evolution sequence, the rapid burning of the remaining vapors, and the sustained burning of the liquid. This burning continues even after the explosion.

burning of the remaining of the flammable liquid, which will burn until the complete inventory is depleted.

6.17.4 Formation of explosive atmospheres by BLEVE

The BLEVE phenomenon (Fig. 6.24) is presented in Section 3.3.2. The BLEVE phenomenon is characterized by a vessel with a significant inventory of liquefied gas that must be exposed to an undesirable external heat source, such as fire flames. The heat establishes the conditions for boiling the stored liquid, generating vapors, and increasing pressure beyond the mechanical strength of the vessel, causing the loss of vessel integrity and the release of an expanding cloud of flammable vapors. Unlike other types of explosions associated with tanks and vessels, in the particular case of BLEVE the vessel is ruptured by the excessive pressure of the vapors and not by the action of a shock wave resulting from an internal explosion.

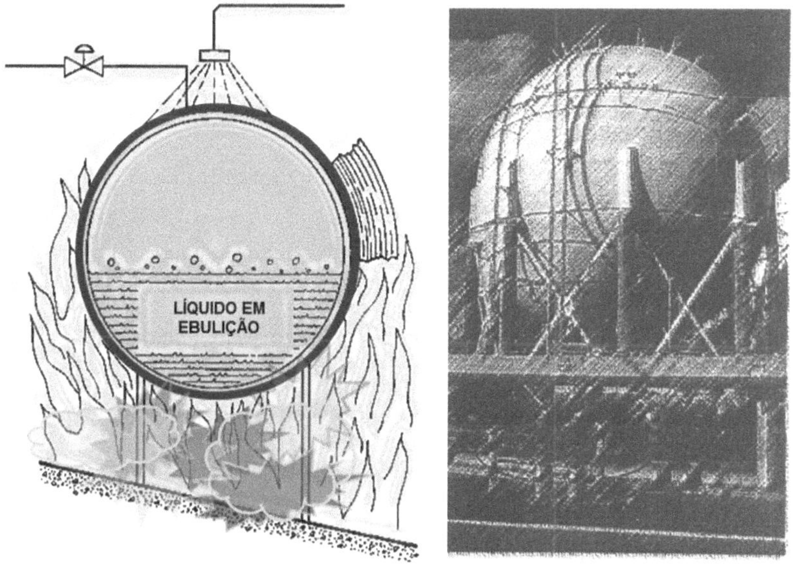

Figure 6.24 BLEVE is the explosion by the expansion of vapor produced by boiling liquid in a pressurized vessel. The loss of structural integrity of the vessel occurs due to the effect of increased pressure. The explosion occurs after the vessel failure and the expansion of the vapor cloud released. On the left, the interior of the pressurized vessel under the effect of external fire and the cooling through the deluge system is presented. On the right, the illustration shows the exterior of the vessel. *BLEVE*, Boiling liquid expanding vapor explosion.

6.17.5 Shock waves and factors that influence explosions

Pressure waves (Fig. 6.25) are composed of a positive phase where compression forces are generated and a negative phase with a suction effect.

The positive phase is caused by the expansion following the ignition of the explosive atmosphere. The secondary effect of this expansion is the formation of a low pressure zone and, consequently, a second wave of effects.

There are three factors that contribute to increase the intensity of an explosion:

1. *Obstruction:* It is associated with the obstacles found in the explosive atmosphere area and their effects in the turbulence and the generation of pressure waves. The obstruction is considered high when the distance between obstacles in the explosive atmosphere region is less than 3 m and the ratio between the volume of space occupied by obstacles and the total cloud volume is greater than 30%. The obstruction is

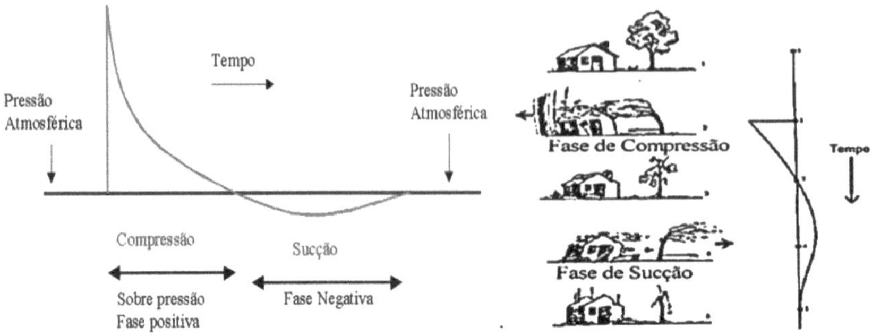

Figure 6.25 Pressure waves and the compression and suction phases.

considered low for distances greater than 3 m and ratios less than 30%. The obstruction is considered null if there are no obstacles in the area of the explosive atmosphere.

2. *Confinement:* the gas cloud is considered to be confined by parallel planes if at least one region of the cloud is delimited by planes on two or three sides, in addition to the ground plane. When the ground plane is the only element acting to limit the cloud expansion, it is considered that there is no confinement.
3. *Ignition energy:* an example of high energy ignition sources are those partially confined in some kind of enclosure so as to launch a spurt of incandescent material into the flammable cloud. Unshielded engines and control panels can generate spurt of incandescent material from internal malfunction. A movable ignition source capable reaching a flammable cloud in the first instants of the ignition is a high energy ignition source. However, ignition sources without such movements, such as sparks, a simple flame, and hot surfaces can be considered low energy ignition sources.

Protection against the undesirable explosion consequences can be achieved by establishing safety distances. Through explosion studies it is possible to identify the likely extent of the damage effects of shock waves that may be generated in the event of an explosion accident at the facility. The protective distance is defined based on the likelihood physical harm to people and damage to the facility. At a region very close to the explosion epicenter will be under thermal effect, with a high likelihood damage. Moving away from the explosion epicenter the likelihood of damage decreases. A region defined by compression effects begins on the boundary of the thermal effect region, also with a high likelihood of damage due to the effects of the pressure waves generated by the explosion. From the boundary of the region of compression effect, a region of debris is formed that the further away it is located from the epicenter the larger it grows, which carries a significant likelihood of damage due to the materials and debris launched by the explosion.

6.18 Lessons learned

6.18.1 Correction of conceptual error results in 50 million dollar savings

After facing difficulties related to competition, a major operating company in the oil and gas industry identified an opportunity to simplify its new offshore rig projects, through a new strategy for the design of the FWP System. The potential for the reduction of investment costs and construction times with the new strategy was significant, since each set of such pumps has an approximate cost of US$ 5 million, without taking into account the costs associated with the pipeline interconnection to the system and the increase in man-hours required for the construction, assembly and commissioning. Each unnecessary FWP added to the project means an extra cost of at least 5 million dollars of investment, without being factored in maintenance and operating costs for the entire operational life of the rig, which can be over 30 years. The designers' idea of demand optimization was to reduce the number of pumps from three to two, increasing the reliability of the system and the installed reserve from 50% to 100%. As a result of this simplification there would also be a reduction in operating and maintenance costs.

If successful, the new strategy could also be adopted in another ongoing project and in eight additional future projects. The potential gains would be even more impressive considering that the new strategy could be adopted in all projects moving forward.

The operator assembled a team for the design of a FPSO offshore rig to be built as a converted very large crude carrier tanker. The rig would require the installed capacity to process and treat 180 thousand bpd of oil, 6 million m^3/day of gas and inject 42 thousand m^3/day of desulfated water. The operation of the FPSO would be at a water depth of 1600 m. The oil to be produced by the FPSO should be exported through an

Table 6.10 Reference data for new FPSO.

New FPSO	
Oil (API grade)	18 degrees
Processing plant (bopd)	180,000
Liquid process (blpd)	250,000
Storage (MM bbl)	1.6
Gas compression (MM m^3/d)	6.0
Water injection (m^3/day)	42,000
CO_2 removal (MM m^3/day)	6.0
Production wells	14
Injection wells	8
Accommodations	110
Anchoring system	Spread mooring
Water depth (m)	1600

API, American Petroleum Institute.

offloading system to relief vessels and the gas through a gas pipeline. Table 6.10 presents the main reference data to be considered in the new FPSO project.

Shortly before the decision for the FPSO project, the operating company had gone through a difficult situation regarding the credibility of its designers. Two FPSO platforms with identical production capacity of 180,000 bpd of oil each had their projects canceled due to high construction prices, which made the projects not economically viable.

After this failure, the operating company decided to invest heavily in the simplification of its projects in order to reduce investment costs and construction times. Because of the operational efficiency and simplicity of competitors' projects, solutions adopted by other operators have become the references for the new FPSO platforms.

The company designers analyzed the results obtained in other projects on the market, which served as the reference for the review of strategies to be adopted in the new FPSO. Internal rules were simplified for better agreement with the international rules.

With this new mindset, the operating company started the implementation of two new FPSO projects with different designers teams within the concept of simplified design and under the following directives:

1. minimum domestic content of 65%;
2. simplified project aimed at reducing investments;
3. use of equipment standardized by the market, with minimal customization possible;

4. standardization and simplification of operational routines;
5. limitation of strategic parameters to some maximum admissible values considered in some systems, such as the water pumps flow rate for FWP; and
6. adoption of successful projects in the market as a reference model.

The new FPSO project would need to incorporate functions to be performed by well-known methods in the oil industry, such as:

1. oil dehydration by gravity and electrostatic separators;
2. oil stabilization by heating and depressurization;
3. removal of carbon dioxide by adsorption of amine;
4. dehydration of the gas through the adsorption of glycol; and
5. water treatment by hydrocyclones and gas flotation.

Due to their independence, self-reliance, remote location, quantities, and access limitations, as much as possible FPSO offshore platforms need to be sturdy, robust, flexible, and simple for construction and operation. The basic engineering project has a fundamental role in the practical achievement of these constructive characteristics related to an offshore unit. The initial systems configuration is one of the key elements that, for the most part, will determine the feasibility of the envisioned simplicity. Moreover, as in all activities of the oil and gas industry, the project should be safe, operational, and optimized, that is, without unnecessary features and redundancies with no added value in terms of cost and benefit.

Within the scope of international standards of the offshore industry, specifications provide a technical freedom space for choosing the various solutions available on the market, while the internal specifications highly customized by the operating company itself limit opportunities for consideration of additional suppliers and, possibly, may hinder the monitoring and keeping up with the industry evolution as the result of constraints imposed by the enforcement of technically outdated requirements.

The adoption of specifications based on international standards allows a better overall assessment of the facility. Rather than specifying the components independently (separately for each equipment) or the functionality and their performance, such specifications allows the standardization and suitability between different components, as well as the possibility of solutions that improve the overall performance of the project.

In general, changes throughout the project affect all disciplines, and as those responsible for each part of the project are not always working in the same office where the project is being developed. Cross-team communication and ongoing evaluation of project changes can be limited, and there is always the risk that team members are not properly updated on the consequences of changes, which greatly increases the effort required to control the project.

After its conception, the concept of project simplification was disseminated in order to exert its influence on the following aspects:

1. project safety strategy;
2. general arrangement (layout) of the FPSO;
3. construction process analysis (ease of construction);
4. cargo handling;
5. need for redundancies, operational flexibility and maintenance; and
6. operational philosophy (working methodology in the new FPSO).

Due to the complexity of interdisciplinary relationships in a project of the magnitude of an offshore rig, it is important to get the commitment of the members of the project team for their proactive and receptive attitude toward important changes. If this commitment does not materialize it is unlikely that a conceptual change and change in the project culture can be achieved.

Team-wide discussions of the impacts of changes on all the disciplines involved are fundamental not only to get the team's commitment but also in the search of solutions that can be incorporated into the project. Aiming to increase reliability and maximize the cost–benefit ratio, the project should use traditional equipment suppliers and whenever possible avoid specification of equipment that is not readily available on the market.

In other words, designers should not oblige suppliers to adapt their products to suit the project, but instead the project should accommodate off-the-shelf solutions available on the market, where it is possible.

6.18.2 Strategy

The calculation of the water demand is the most important input for defining the configuration for the fire-fighting water pump system of the FPSO exploration and production units. In FPSO, the main consumers of this system are:

1. deluge system,
2. fire hydrants,
3. monitor cannons, and
4. foam system.

Among these consumers, the deluge system is the one with the greatest impact for the calculation of water demand. Through standards criteria and corporate guidelines, the minimum flow rates of water sprays on equipment are established. The purpose of water sprays during fire inception is to minimize and delay the escalation of the event to more critical scenarios and to provide sufficient time to abandon the unit as the last resort.

As a strategic driver of investment reduction, the optimization of the demand for fire-fighting water has become the subject of permanent attention by the safety teams of new projects of FPSO exploration and production units.

Designers have identified the following aspects that were considered conservative (overly specified) in previous projects:

1. excessive margins of safety for the hydraulic balance;
2. water sprays located over the entire length of the pipeline racks;
3. water sprays over the gas compression modules without automatic depressurization been taken into consideration;
4. physical arrangement of process plant modules, grouping equipment with large hydrocarbon inventories without strategies for reducing the demand for fire-fighting water been taken into consideration; and
5. submission of documentation based on conservative calculations and assumptions for safety studies simulations (studies of fire propagation, gas dispersion, and explosion).

Studies for the verification of the project's safety level validated previous conservative assumptions and calculations, resulting in overspecified projects.

Due to the high demand for fire-fighting water values obtained in the calculations of the operating company's projects compared to those adopted by its competitors, the designers sought references in internationally recognized projects and with reduced values for fire-fighting water demand as part of their project simplification effort. At the beginning of the project, the traditional procedure for the system design was applied. The calculations led to configuration with three pumps of 2000 m^3/hour at 50% capacity. Based on the configurations of the reference projects with systems composed of two FWP at 100% capacity each, an opportunity to simplify the project was identified resulting in the reduction in water demand for the system.

6.18.3 Interactivity, arrangement, and risk management

The interaction between the risk and safety management discipline and the project's arrangement team became a decisive factor for the effective reduction in the demand for fire-fighting water. The engineers of the safety discipline, while seeking to reduce the water demand, observed in their attempts to subdivide the process plant into deluge zones that there could be a demand reduction if the modules were repositioned to be located further apart. The team's technical training on the techniques for preparing studies and analyses of fire propagation, gas dispersion and explosion enabled a practical and comprehensive understanding of the phenomena associated with the fire and explosion scenarios involving physical, chemical, and engineering aspects.

Based on the stronger interaction between the safety and arrangement disciplines in the elaboration of the general layout, an attempt was made to adopt 15-m distance between the process plant modules recognized as major consumers of deluge water. In cases where this distance is not achievable the longest possible distance was adopted instead, taking into account the limited areas available on the main deck and FPSO operation restrictions.

The target value for spacing between modules, approximately 15 m that was adopted as a reference, was obtained based on the analysis of previous projects and applicable standards. The National Fire Protection Association, through the NFPA 15(5) Standard, recommends a preventive distance of 15.2 m when, for example, there is no computer simulation study that allow a more accurate definition of safety distance. In the initial design phase, when the demand for fire-fighting water and the layout are designed, it is too early for such a study yet be to be conducted due to the lack of data resulting from the early stages of the project.

NFPA 15 Section A.6.3.6 (1) states:

> *(...)* To limit the potential for explosion damage, the following guidelines should be used:
> System actuation valves should be remotely located [at least 50 ft (15.2 m) from the area to be protected, housed within a blast resistant valve house or behind a blast wall designed for at least a gauge pressure of 3 psi (20.7kPa) static overpressure *(...)*

In this item, there is reference to the 15-m distance of related to valves with explosion-resistant body or protected by bulkheads without the need for corroboration by special studies. The longer equipment spacing also brings positive results for the analysis of the explosion scenarios, because in addition to the shock wave and debris, there can also be thermal radiation and escalation of the fire scenario. Distance serves as an additional protection against these undesirable consequences of possible accident.

The innovation inserted by the proposed strategy was the extrapolation of the criterion for the separation between modules and equipment. Thus the adoption of this reference parameter influenced the layout design, relocating the modules as far from each other as possible, targeting the 15-m distance. Classified fire-resistant bulkheads were also added where justified by safety studies (conducted the final design stages). This strategy was successful and the results were reflected in the calculation report. Iterative layout optimization was performed to accomplish the reduction from three to two FWP.

Among the several cases considered for the spacing protection, the main ones are:

1. *Electrical panels room:* action by team members related to the electrical engineering, layout and safety disciplines to improve the interface between the electrical equipment room and the process plant—risk area—in order to preserve the safety of the operation and eliminate the need for passive protection (with associated cost increase).
2. *Elimination of the exclusive module for the treatment of produced water:* the equipment for treating water produced by a water injection module was incorporated in the layout. The produced water treatment module was initially located on the FPSO's port side. As the pumps for water intake were located on the starboard side, large diameter pipelines would interconnect these pumps (crossing the pipe rack) to the exchangers located in the produced water treatment module. These pipelines would be routed back to starboard for connection to the main injection pumps, as shown in Fig. 6.26.

   Based on multiple arrangement studies, a proposal was made to combine the module for the treatment of produced water and the module closest to the intake (water injection). The elimination of the module for the treatment of produced water made it possible to increase the distance among other modules in the area, manifolds and separators, as shown in Fig. 6.27. The resulting distancing favored the protection and efficiency aspects related to the fire fighting and propagation.

   The main advantages of the optimization were:

   1. reduction of large-diameter pipelines length and construction and assembly time,
   2. less cluttered pipe rack,
   3. elimination of passive protection in one of the bulkheads of the module, and
   4. reduced demand for fire-fighting water.
3. *Subdivision of spill containment basins:* the subdivision was done to reduce the cascading effect of fire scenarios that could occur if the spill containment basins were interconnected and shared. The use of shared spill containment basins spreads the spilled hydrocarbon, mainly due to

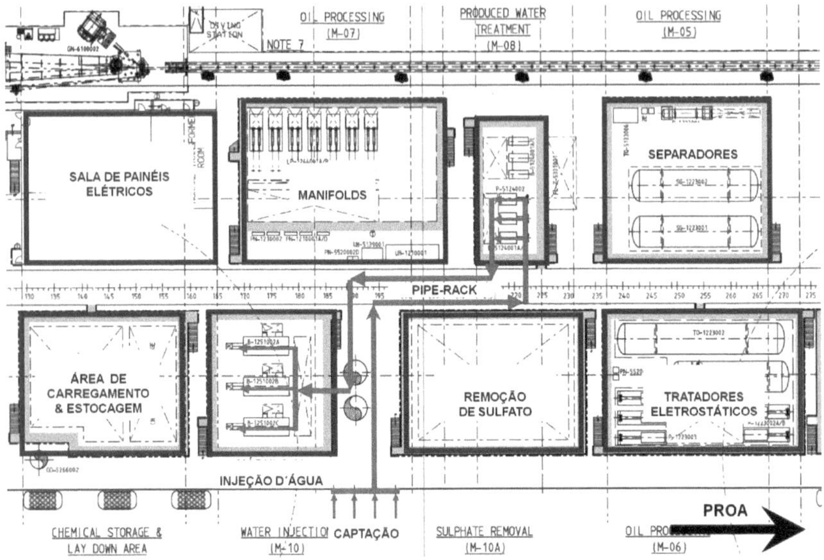

Figure 6.26 Original layout prior to the optimization work. The intake pipes for the water injection system run through four modules of the primary processing plant.

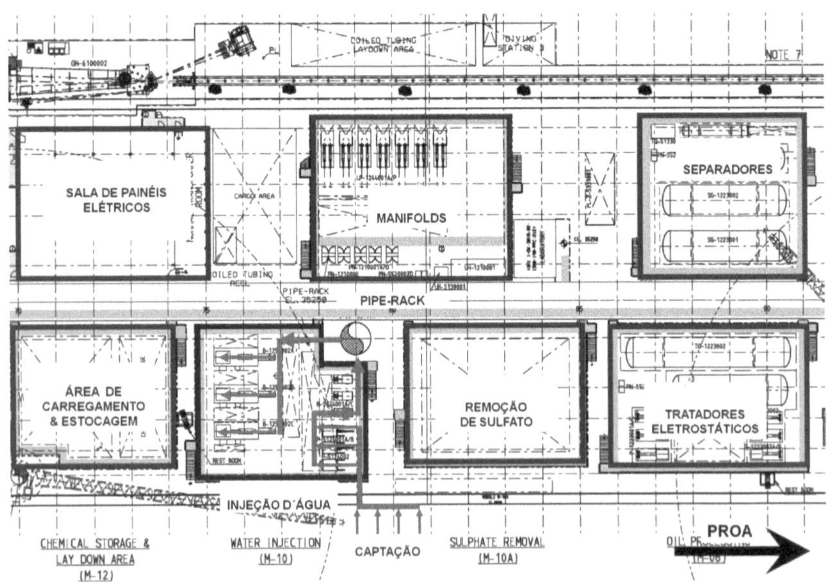

Figure 6.27 New layout after the joint optimization work. The intake pipes for the water injection system pass through only the region of one of the modules of the primary processing plant. Note that the produced water treatment module has been eliminated and its equipment was incorporated in the water injection module.

the FPSO motion at sea that amplifies the displacement of the oil pool subjected to ignition in the event of accidents.

The layout changes also contributed to a significant improvement in the ventilation conditions of the process plant due to additional spacing obtained between modules, positively influencing the results of gas dispersion and explosion studies, reducing the number of gas detectors in the plant and also reducing overpressure in explosion scenarios. For the implementation of this technique be possible, it is necessary commitment by the project team to strategically established directives. The team should also be receptive to changes, since each FPSO layout change causes a ripple effect with consequences for the other project disciplines.

6.18.4 Criteria and results

The following criteria were adopted to simplify the system:

1. 15% margin of safety for hydraulic balancing;
2. automatic depressurization of equipment that operates with gas, reducing the consumers of the deluge system;
3. elimination of deluge on pipe racks and pipeline areas;
4. deluge in the module where the fire event starts and on the adjacent sides, based on international regulations, and skipped in the modules located along the diagonals with respect to the reference module; and
5. orientation of the general FPSO layout for the purpose of optimization of the fire-fighting water supply.

The concept of simplified projects made it possible to design a general plant layout with more spaced modules, consequently a more ventilated unit, which minimizes the escalation of the severity of fires and significantly reduces confinement, with positive effect on the results of explosion studies.

Table 6.11 presents a comparison of the results obtained for fire-fighting systems—water demand values and number of power pumps—between the strategies adopted in previous projects and the new simplification strategy.

The assumptions adopted were validated by safety analyses carried out by two internationally recognized societies for offshore project classification. The study of the fire spread following the project found no damage or loss of safety functions as a result of the new technical approach adopted in the calculation of the fire-fighting water demand.

Safety studies (Fig. 6.28) based on computer simulations, such as fire propagation, gas dispersion, and explosion, are recognized as important resources as aid in designer's decision making, since they represent an

Table 6.11 Comparison of results for fire-fighting systems.

Project		**Type**	**Process plant (bopd)**	**Demand[a] (m³/h)**	**Pump flow rate (m³/h)**	**Number of pumps**	**Excess capacity (%)**
Previous concept	1	SS	180,000	2986	1800	3	20.6
	2	SS	180,000	3196	1800	3	12.6
	3	FPU	180,000	3336	1800	3	7.9
	4	FPSO	180,000	4097	2200	3	7.5
	5	SS[b]	180,000	4283	2350	3	9.7
	6	FPSO[b]	180,000	6015	2200	4	9.7
New concept[c]	7	SS	180,000	1916	2000	2	4.4
	8	FPSO	180,000	NA[d]	2400	2	–
	9	FPSO	180,000	2306	2400	2	4.1
	10	FPSO	180,000	2330	2400	2	3.0

[a]Contingency to balance the system during executive project is considered. In the previous concept, it was 30%.
[b]*Discontinued projects; SS, Semisubmersible.*
[c]Projects executed based on the simplification principle.
[d]Data not available.

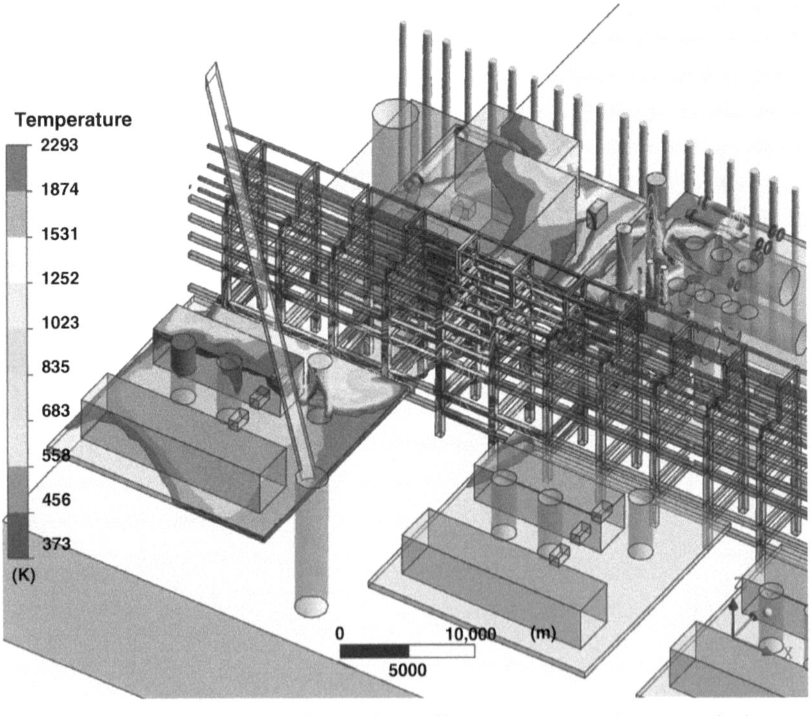

Figure 6.28 Temperatures on the surface off equipments and structural elements were affected during the accident simulated by a computer. The figure represents the effects of a fire event in gas-lift risers as part of a fire propagation study.

indicator of the installation likely behavior in response to the most critical postulated accidental scenario. However, like any computer simulation, safety studies do not consider all decision-making factors.

According to developers of one of the leading simulation software used in this type of study, the "results of safety studies should not be viewed as a decision, but rather as one of the factors that make up a technical decision, which will always be associated the designer's experience and expertise" Conceptually, computer simulation is a tool whose results should be analyzed and used to assist engineering decisions, which additionally require extensive operational experience and technical training in the field.

6.19 Conclusions

The optimization of fire-fighting water demand, obtained through the new strategy presented herein, has reduced the number of pumps in the system from three to two and increased its reliability by providing a 100% reserve (two pumps at 100% each), that is, twice the initial reserve at the beginning of the project (three pumps at 50%). In addition to ensuring greater reliability, this new configuration also complies to:

1. current international standards,
2. requirements of any classification society, and
3. internal rules.

The result obtained is similar to those of competitive and reference projects in the offshore market. The new strategy reduced the initial flow rate from 3040 m^3/hour to the optimized value of 2330 m^3/hour.

Each set of FWP has an approximate investment cost of about 5 million dollars—equipment only. In order to calculate the total investment reduction, the pipeline costs for their interconnection to the system and the increase in construction time, assembly, and commissioning need also to be added to this value. As a result of the simplification achieved in the system, there is an additional reduction in operating and maintenance costs.

These results for the new strategy were obtained, for the most part, by the assembly of a high-performance project team aligned with the strategic project directives, motivated, and open to new ideas, which made it possible to identify the opportunity for this improvement and its implementation.

The new strategy presented herein was also adopted by the operating company in a similar project that was in development, as well as in the eight subsequent FPSO projects, with a total savings of 10 FWP, which is the equivalent to approximately 50 million dollars, in the main equipment alone (FWP with no accessories and pipelines). The potential gains are even more significant when it is taken into consideration that new projects will continue to be developed over the years based on this new strategy. This confirms that a good risk management engineering strategy can achieve better results than those obtained with projects where the safety system equipment capacity is simply increased in the specifications. And the most important is that the simplest projects, with the least level of complexity possible, also contribute to the improvement of the quality and efficiency of risk management from project through the offshore rig operational life.

6.20 Exercises

1. In regard to electrical systems, what are consumers of the safety system and how do they differ from the so-called essential consumers?
2. Why are ventilation, heating and air conditioning systems important for safety?
3. Can an oil tank that is kept completely empty in a facility explode? Explain.
4. How are the locations of gas and flame detectors selected?
5. Why does not the "deluge" system fed by fire-fighting water have as its main objective direct fire fighting?
6. What are coamings and what role do they play in association with fire-fighting foam systems?
7. What is the advantage of using a "fire ring" in the fire-fighting water distribution system?
8. What is the main advantage of the water mist system over the CO_2 flooding systems?
9. What is the strategy for applying PFP in oil and gas facilities?
10. What is a cryogenic product and why is the importance of this type of product growing in the oil and gas industry?
11. What is the design basis accident concept for?

12. Why should the startup of most safety systems with redundancies be done by automated systems with a simultaneous startup signal for all redundancies, even though it is known that at least one of them is a backup?
13. In general terms, why can we say that a shock wave produced by an explosion generates two different "destructive waves"?

6.21 Answers

1. They are those directly associated with the protection of people and the rig integrity. The main difference in relation to essential consumers is that safety consumers cannot be affected by power outage even during the transfer from the main power generation system to the emergency power generation system, besides being subdivided into groups according to autonomy requirements that can vary from 30 minutes to 96 hours of availability.
2. These systems are important for safety because they can contribute to the environment control so that flammable gases and vapors are kept in concentrations below the flammability limit. Besides, they can act to prevent the spread of smoke and heat from regions affected by accidental scenarios through fire dumpers and exterior air intakes.
3. If the tank has been used previously and no flushing, purging or inerting operation is performed, an explosive atmosphere may form within it, if left over vapors mix with air in concentrations within the limits of flammability.
4. The choice of the location of the detectors should be made based on the operational experience and the experience from other previous projects. In addition, studies of fire propagation and gas dispersion conducted using fluid dynamic computational tools should be used for possible corrections.
5. The objective of deluge system is to cool equipment that is subjected to heat radiation from a fire in its vicinity. The main objective is to prevent external heat from causing an increase in pressure and temperature inside the equipment with significant hydrocarbon inventories. The cooling provided by the deluge system reduces the rate of heat increase, allowing precious extra time for attempting accident mitigation.

6. Coamings limit the movement of liquid hydrocarbons pools and foam. They are some type of flat bars strategically distributed to prevent pools of hydrocarbons from spreading quickly, what makes emergency control difficult. They also help to contain the foam released during the control of this type of emergency, keeping the foam layer on the liquid hydrocarbons pool.
7. The "fire ring" strategy allows greater flexibility in the supply of fire-fighting water in the event of ruptures or damage to the distribution lines. The fire ring are equipped with isolation and shutoff valves that allow, through alignment maneuvers, to divert the feed flow from the unavailable branches to the branches in operating conditions, maintaining the conditions of fire-fighting water supply to consumers.
8. The water mist system does not pose risk of suffocation for people in the location subject to this type of protection, whereas the CO_2 flood system requires special care during startup, signaling, and alarm to avoid accidents with suffocation victims.
9. Passive protection should be installed to protect structures that may be affected by heat, delaying the loss of mechanical and structural strength. It also serves as protection against the effects of shock waves resulting from explosions. The choice of locations for application of passive protection should be made based on operational experience, reference projects and results from fire and explosion propagation studies.
10. A product is considered cryogenic when its boiling temperature is below 150°C. The importance of cryogenics has increased in the oil and gas industry due to the growth of the LNG market. This cryogenic product can be transported by vessels, specifically designed for this purpose for large consumers who process the product back to natural gas through regasification. LNG is a product that adds flexibility to the energy market and has very specific safety requirements related to cryogenics.
11. The design basis accident concept defines the worst-case scenario for which the facility ought to respond through the safety systems that are part of the project. The application of this concept allows designers and operators to have a clear view of the facility's response limits, avoiding crisis management failures such as hesitation in making decisions on escape and abandonment and the inappropriate fire brigade dispatch to a scenario in which they have no chance of success.

12. The purpose of the addition of redundancies in safety systems is to cover classical failures due to unavailability and during startup. For most of the redundant equipment, if one of the possible failures occurs, it takes time to detect it and for the backup redundancy to start up. This time is considered lost and, in some systems, as in the case of fire-fighting systems, it can make the difference between success and failure of the fire-fighting operation. Therefore whenever possible, all redundancies in a safety system must have simultaneously startup, and if the timely shutdown of the redundant equipment should preferably performed manually at the discretion of the crisis managers.
13. Shock waves from explosions usually generate two phases. The first is called the positive phase that imposes a compression effect on the environment. The second phase, called the negative phase, imposes a suction effect. Thus during the initial instant, the shock wave produces an impact followed by the suction period resulting from the negative phase. The negative phase creates a region of low pressure that ends up generating a secondary shock wave due to the air masses bouncing back after being displaced by the first phase of the shock wave. The affected environment is perceived to undergo the effects of "two destructive waves."

6.22 Review questions

1. Explain the priority position of the "emergency control" component in the risk management strategic line?
2. What are the fundamental technical requirements that characterize emergency power generation systems?
3. Regarding the power generation system, what are essential consumers and safety consumers?
4. What is the purpose of area classifying in oil and gas facilities?
5. Define ignition temperature, explosive atmosphere, combustion, combustion speed, deflagration, detonation, and explosion.
6. What does the acronym BLEVE mean and what is the mechanism of this phenomenon?
7. What does the term flash point mean?

8. Define vapor pressure?
9. What is the difference between flammable liquid and combustible liquid?
10. Explain the meaning and the applicability of the flammability limits?
11. Describe the main types of ventilation used in oil and gas facilities and provide an example?
12. What is the meaning of zone 0, zone 1, and zone 2 in terms of area classification?
13. What are the special requirements for the ventilation system of rooms protected by a CO_2 fire-fighting system?
14. What is the difference between flushing, purging and inerting?
15. Explain the strategy for quantification and location of flammable gas detectors?
16. What is the difference between open path gas detectors and point detectors?
17. Explain the operation of IR gas detectors and provide examples of its application.
18. What is the purpose of catalytic gas detectors and how do they work?
19. What are the actions resulting from the confirmation of the presence of hydrogen gas?
20. What is the operating principle of H_2S detectors and what are the actions resulting from the confirmation of the presence of this gas?
21. What type of detectors is used for CO_2 monitoring?
22. Explain voting logic for ESD and voting logic for gas detection alarms?
23. What are the main technologies used in flame detection and what is the sensor operating principle for each technology?
24. What are combined UV/IR detectors and what is the application of this type of equipment?
25. What are the main recommendations for the definition of the location of smoke detectors?
26. What is the difference between thermovelocimetric detectors and fixed temperature detectors?
27. What is an ADV valve and what is its function in fixed fire-fighting systems?
28. What is the basic principle of foam for fire fighting and in what situations is this type of resource applicable in oil and gas facilities?
29. Describe the main characteristics of FWP for oil and gas facilities?

30. What are the main types of FWP?
31. Explain the differences in terms of power requirements for the main FWP redundancy configurations?
32. What is fire-fighting water demand and how is the maximum design demand calculated?
33. How is the fire-fighting water supplied by the FWP distributed in oil and gas facilities?
34. What are the types of activation of CO_2 flood systems?
35. What are the alarm requirements to be observed in CO_2 flood system designs?
36. Explain the functioning principle of the water mist system and its advantages over the CO_2 system?
37. In what locations should mobile foam generation equipment be used?
38. What is the strategy for use of passive protection, including examples of the classification of spaces adopted in offshore rigs?
39. What are the characteristics of LNG and what is its basic composition?
40. Describe the main accidental scenarios related to LNG?
41. How does the evaporation control mechanism of LNG spill work during high-expansion foam operation?
42. How does the burn control mechanism of LNG spill work during high-expansion foam operation?
43. Explain the phenomenon of rapid phase transition that occurs with LNG?
44. What factors should influence the decision to send the fire brigade to an emergency operation in an oil and gas facility?
45. What is the meaning of "manophobia" and provide examples of the use of this term?
46. What is a "design basis accident" and how is this concept applied?
47. How are accidents classified according to their origin?
48. Explain the meaning of "external origin accident" and provide examples of scenarios applicable to this concept?
49. What does beyond design accident mean and what are the advantages of a project that adopts this concept as part of its risk management?
50. Explain the term "critical safety functions" and the uses of this strategy?
51. In terms of special automation strategies, what is the meaning of the terms "protection system" and "limitation system"? Where are these concepts applied?
52. What are the main types of redundancy configurations applicable to safety systems?

53. What are classical failures in safety systems?
54. Why is access to knowledge on explosions somewhat restricted?
55. What are the main types of explosion of flammable products?
56. How are explosive atmospheres formed?
57. Explain the explosion sequence in a tank with a significant inventory of liquid and gaseous hydrocarbons?
58. What are shock waves and how are they formed?
59. What are the phases of a shock wave and what are its effects?

CHAPTER 7

Reducing unpredictability

Every real accidental scenario includes components associated with unexpected factors (factors that we had not been prepared for). Even if an operation team is prepared to respond to a specific fire scenario or it is postulated in the project, the fire event indicates some unexpected factor that was missed in the design safeguards or the operators were unprepared for. The better the prior knowledge about what may happen in a real accidental scenario the less likely we will be caught off guard both in terms of the lack of design safeguards and in terms of inadequate technical operational skill. Previous well-developed technical knowledge on the largest possible number of accidental scenarios reduces the number of unexpected factors in a possible accident, because they are incorporated in the theoretical scenarios, postulated by the designers and risk management experts. These elements are no longer unpredictable factors and instead become part of the scenarios studied in the design phase or are considered in safety analyses to be conducted throughout the operational life of the enterprise. Therefore the reduction of unpredictability is achieved through studies and safety analyses on accidental scenarios which identify vulnerabilities and issue recommendations for the creation of project safeguards and preventive operational practices. The more representative the accidental scenarios postulated in the studies and analyses can become, the less the unpredictability component affecting future risk management of the enterprise will be.

But there is an important finding that needs to be pointed out with respect to the concept of unpredictability reduction. The most important and central point that really makes a difference towards unpredictability reduction is not the techniques or risk analysis tools and safety studies. The most important point is technical knowledge.

Understanding the physical and chemical phenomena related to the rig activities and benefitting from the experience accumulated by those who are already familiar with such activities, these are the factors that can make the difference between being prepared or not to respond to a particular accidental scenario. The risk analysis and safety studies tools serve only to organize technical operational knowledge. It is not sufficient to have solid knowledge on computational fluid dynamic (CFD), statistical analysis

Risk Management in the Oil and Gas Industry.
DOI: https://doi.org/10.1016/B978-0-12-823533-1.00003-4

techniques, complex 3D computer simulations, among other tools, if these are not supported by technical engineering knowledge and operational experience (the most important factor of all leading to successful risk management). Being able to master the tools and techniques of risk analysis does not ensure the correctness and relevance of the results from the analyses and studies. It is possible for absolutely perfect studies and analyses to be conducted, in terms of CFD application and statistical analysis techniques, and even so, the results obtained might be totally wrong, or else completely irrelevant. Such results, although obtained via the use of good techniques related to study and analysis tools, may not add any improvement to the risk and safety management of the enterprise due to its lack of adequacy and representativeness to the project in real terms as well as the operation of the enterprise.

For this reason, the risk management strategic line (Fig. 7.1) ranks the reduction of unpredictability in the fifth priority level. It does not mean that reducing unpredictability and the analyses and studies that promote it are not important. On the contrary, they are among the top five priorities that define the risk management strategic line. However, the reduction of unpredictability

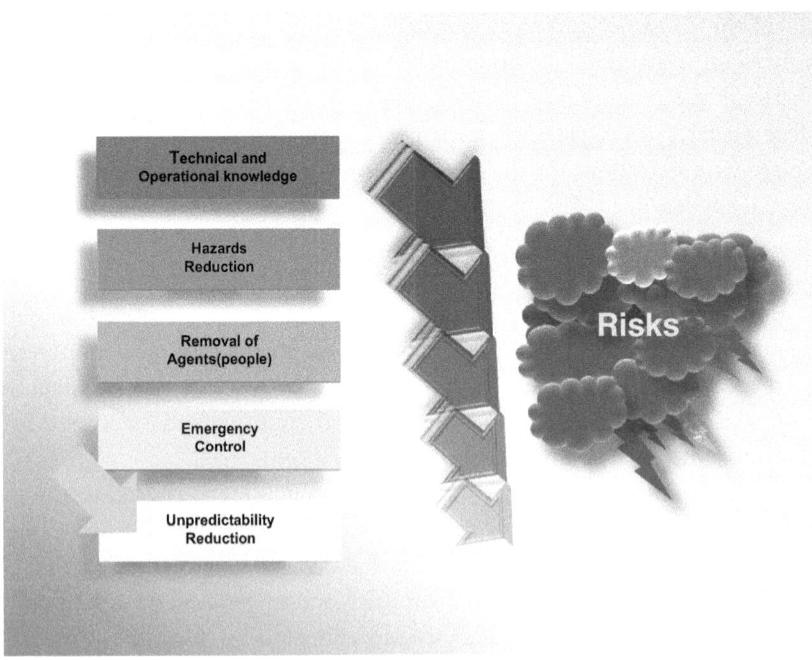

Figure 7.1 Risk management strategy line, with emphasis on the fifth component: "unpredictability reduction."

can only be achieved through technical and operational knowledge that is five levels above it in terms of strategic priority related to risk management.

Unfortunately, there are a significant number of professionals working in the area of risk analysis and safety studies without the technical and operational background required for the support of the analyses and studies they perform. In some cases, professionals' skills are solely concentrated on the tools for the preparation of analyses and studies, which compromises the results and recommendations that often are inconsistent with the operational experience accumulated over years by those who have worked not only on the design but also on the systems operation on a daily basis. There is still a false sense of technical competence by laymen or engineers who are not experts in risk management. Some professionals confuse the ability to use tools with the competency to make technical and operational assessment of accidental scenarios. Especially with respect to computer simulations, the ability to use tools can generate 3D scenarios and illustrations with apparent realism, but which in fact can lead to distorted conclusions and recommendations due to the lacking technical and operational knowledge.

Safety studies and analyses need to be considered as complementary to the technical and operational knowledge already accumulated. If, for instance, an offshore company has been in the market operating more than 100 units for over 40 years, this accumulated experience should be the main guide source of recommendations for projects and decision-making in the area of risk management. Computer simulations, studies and analyses should not be discarded vis-a-vis this vast accumulated experience, but instead, they can be useful if used along with the technical and operational foundation, for the identification of points of interest for adjustment and correction of the means of risk management in new projects. In summary, studies, analyses, and computer simulations are perfect for specific issues where previous operational experience can only provide limited guidance. These matters are always part of technological ventures, because of the nonstop advancements, either in terms of design techniques or operational practice, which justifies carrying out studies to complement the operational experience. Studies, safety analyses, and computer simulations are valuable tools for risk management experts, but they only have value when backed by a solid operational technical background. This is currently an area of focus for risk management experts.

Experts should not be misguided by the appearance of very well-justified results in terms of statistical arguments and poor in terms of technical knowledge and operational experience based on the professional profile of the authors. Specialists should focus their attention on risk management to avoid that market pressures and commercial interests prevail over the technical aspects and cause safety studies and analyses to be misused, including to justify decisions that could be unfavorable to the safety of technological enterprises.

Quite often, there is a lack of basis for the argument that computational studies, analyses, and simulations are objective, quantitative, and more accurate tools than the simple consideration of operational experience records. On one hand, this argument is a cover up for the fact that behind all the accuracy of the analyses, studies and computer simulations are previously established premises where decisions just as subjective are present in a less visible way. On the other hand, the operational experience has the advantage of not being based on premises or the accuracy of the numbers generated from these subjective premises. Good technical knowledge and operational experience are exclusively fact based, and therefore they should always occupy the highest priority in the risk management strategic line, without sacrificing the correct use of safety studies and analyzes and simulations computational tools that are very important instruments, but at lower priority in the pursuit of genuine technical, scientific, and operational knowledge.

7.1 Risk analysis techniques

Risk analysis techniques are employed to organize information about risks in order to reduce the unpredictability of accidents. One of the main methods for the classification of these techniques is applicable to two complementary lines of study:

- Qualitative techniques: based on experience, perception and technical operational knowledge. Their foundation is the deep knowledge on the phenomena related to the operation and the technological enterprise project form. Qualitative techniques consider historical data, recorded experiments, previous accidents, besides promoting broad multidisciplinary discussions, and being open to the influence of a wide range of diverse opinions. Qualitative risk analyses are marked by subjectivity, very complex, and of difficult assessment regarding the right way to be conducted. But they are more robust technically when developed by professionals with the appropriate technical training.
- Quantitative techniques: aiming to promote greater objectivity in the treatment of issues related to accidents, these techniques "try" to

quantify the risks through mathematical modeling and association with numbers. As a result, a large extent of the decisions that depend on discussion and consensus that are usually difficult to be reached, can be guided by analyses that can be related to numerical values, thus reducing the difficulties associated with handling discussions with divergent opinions. It is not uncommon for exact sciences experts to repeat the maxim "numbers don't lie." Although we can really agree, we would rephrase this maxim as follows: "the numbers do not lie, but they help hiding some parts of the truth." That is, quantitative risk analyses, when well conducted, can achieve real results. But it only occurs under very specific, very precise, very limited conditions where such results can indeed be considered in fact truthful.

Qualitative and quantitative techniques complement each other and whenever possible technological enterprises should use qualitative and quantitative tools for risk management.

Risk analysis techniques deal mainly with the concepts of severity (related to the consequences of accidents) and frequency (related to the number of events), whether in a qualitative or quantitative way. They define each accidental scenario to be studied, which is the set of specific conditions that generate a particular accident with undesirable consequences. The purpose of the analyses and studies is to assess the RISK, that is, the probability that a specific accidental scenario will occur.

Every risk assessment always includes, to a greater or lesser extent, some subjective component, which means that the unpredictability of the scenarios can never be completely eliminated, even if all possible safeguards are in place. In other words, even a quantitative risk analysis has premises and criteria that include some subjectivity. Despite all the tools and technical resources, not even the most accurate calculation can guarantee that an accident will not happen with 100% certainty.

Therefore the unpredictability related to accidents, although it can be reduced, it can never be totally eliminated regardless of how the risk is studied and analyzed, whether by qualitative or quantitative technique. Usually this concept is associated with the term "residual risk" (that risk that remains due to the difficulty in predicting it). When this book uses the term "reducing unpredictability," the author's intention is that it is understood as the reduction of "residual risk."

Risk management experts, especially those who work in the preparation and evaluation of safety studies, need to have a clear understanding of the main limitations of risk analysis techniques (Table 7.1).

Table 7.1 Questions × limitations of the risk analysis techniques.

Question	Limitation
Scope and coverage	It is not possible to ensure that all risks, causes, and effects will be identified
Reproducibility of results	Different analyses using the same information produce different results
Applicability	The method of application of the technique makes the result difficult to be applied and used by other technicians
Relevance and experience	The team's experience is extremely relevant for the identification of the most significant risks
Subjectivity	Some events end up being "deliberate" due to lack of knowledge or information, causing results to be questionable

Risk analyses and safety studies are not flawless methods for identifying all possible risks related to a technological enterprise. Some techniques fail to identify spatial problems (layout, arrangement, escape routes, etc.). Since all systems are somehow "interrelated," then the consequence of a deviation identified at one location in the facility might apparently have no undesirable consequences in a limited context of an analysis, but it can generate many undesirable consequences elsewhere in the facility. Some techniques do not have mechanisms that allow the identification of hazards due to acts of sabotage or vandalism, and also some are hardly capable of identifying accidents of external origin, especially those caused by natural phenomena. The risk analysis and safety studies consider as a premise that the project will be built perfectly as designed and in compliance to all applicable standards, which is not always a reality.

7.1.1 Quantitative and qualitative risk analyses

Some experts characterize quantitative risk analyses as controversial because in many cases, when they require numerical data on frequencies related to a specific scenario, the analyses seek this data in international databases, collected using scientifically recognized methods. But to obtain apparently accurate results, these analyses need to consider premises that contain some component of disguised subjectivity. For example, although data on failures of a particular type of valve are available in international databases, the technological enterprise under study will always have unique characteristics. These local characteristics almost always have a direct influence on the values of failure frequencies, which creates disparities with respect to international databases.

The criteria and means of correcting these disparities is itself subjective in nature. For many facilities the end result is that in practice the international databases provide data that are often unconfirmed when compared with the data collected as part the local operational routine. Qualitative analyses, although apparently can be considered less accurate, they are more realistic because of their dependence more on quality and experience operational than on uncertainties of representativeness and agreement related to international databases.

Quantitative analyses should be used to complement the operational experience and qualitative analyses as well. They are more useful in the lack of operational experience. For this reason, some operating companies establish criteria for applying risk analysis techniques and safety studies, identifying which types of analysis tools should be used in each phase of the technological enterprise. Table 7.2 shows an example of this type of criterion.

There is a wide variation in the qualitative and quantitative risk analysis techniques, being differentiated in some cases by details or subtleties that justify their names. But basically, they consist of strategies capable of organizing the multidisciplinary and complex information that are associated with the discussion of accidental scenarios of technological ventures. There are so many acronyms and terms created to identify different techniques that often can confuse even risk management professionals because, in addition to the technique itself, their application method can also be defined based on practices adopted by each company and by each expert. In other words, a well-known technique can be applied in so many different ways that might cause it not to be recognized as a single technique.

Amid such overabundance of techniques, the risk management expert should identify the technical objective to be achieved by the technological enterprise in each situation and then select and adapt the most appropriate

Table 7.2 Example of criterion for applying risk analysis techniques.

Enterprise phase	PRA (preliminary risk analysis)	HAZOP (hazards and operability analysis)	Consequence analysis (fire propagation, gas dispersion, explosion, etc.)
Economic feasibility assessment	X		
Conceptual project	X		
Basic project	X	X	X
Executive project	X	X	X
Decommissioning	X		

technique to the conditions required to fulfill the objective. In strict terms, there is no well-defined criterion for determining which specific techniques should be employed when the managers of a technological enterprise identify the need for safety analysis. Even when the recommendations by standards and procedures indicate the adoption of particular techniques, it is up to the experts to verify that the risk management objectives will be effectively achieved through these techniques. If it cannot be confirmed, then complementary techniques should be used. Even techniques recommended by standards and procedures may require adaptations both in terms of their methodologies and how they are applied, according to the particularities of each technological enterprise. The purpose of risk analysis techniques is the structuring of multidisciplinary knowledge to facilitate team work. Risk analysis techniques do not provide technical, scientific, and operational knowledge on the activity to be analyzed, that is, it is only a tool and as such it should be adjusted to be the right fit to the work to which it is applied.

Based on the criterion of the previous table, the three types of studies most used in the daily risk management activities in oil and gas installations were recommended. In the following sections, we will outline these main techniques and some other complementary techniques and variations of those. But one needs to remember that there are big differences both in terms of formulation and the way they are applied, that is, depending on the experts involved, the safety culture of each enterprise, other forms of application, and methodological variations may be adopted as adjustment.

7.1.2 Preliminary risk analysis

It is a technique internationally known also by the name preliminary hazard analysis (PHA; when not including risk classification) whose objective is to only identify the accidental scenarios to be postulated in subsequent studies and analyses of the technological enterprise. The preliminary risk analysis (PRA; when including risk rating) identifies these scenarios, their causes and consequences and adds the risk classification based on a risk matrix. The application of the technique consists of gathering a multidisciplinary group of experts involved in the design and operation of the rig or other kind of facility. A leader, usually independent of the enterprise under study, uses a guiding spreadsheet, and the technique results in a final report of recommendations that can contain suggestions for adjustments, corrections, and the inclusion of safeguards for each identified risk. This approach makes it possible the verification of the safeguards all the way to the application of the technique and, through the recommendations, additional measures can be proposed to improve the risk and safety management of the enterprise under study.

The classification of the identified risks is done through a risk classification matrix (criteria) that was previously established among the participants along with all the assumptions related to the analysis work. PRA is a qualitative, deductive, and very subjective technique with basis on the knowledge of the team of experts and the independent leader who applies the technique. The flowchart below shows the steps of a PRA.

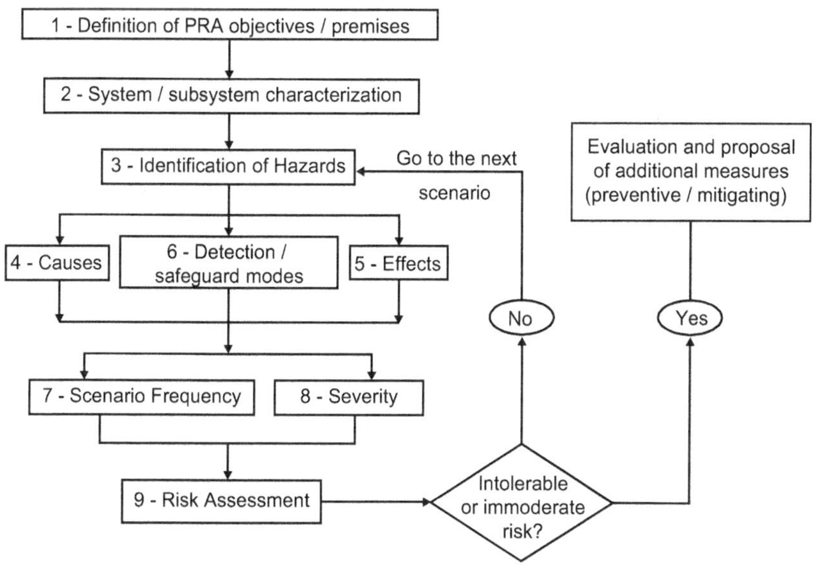

The spreadsheets (Table 7.3) used in the application of the PRA can vary widely, but basically they need to include a row for each hazard to be assessed and columns for recording causes, effects, safeguards, and risks. They can be subdivided further into categories such as risks for people, rig/installation, environment, and the image of the management organization.

The frequency and severity categories of each risk are recorded in the spreadsheet based on the risk classification matrix (Table 7.4). This matrix serves not only for applying the PRA but also for other risk analysis and safety studies techniques.

This matrix is also applicable to other risk analysis techniques and safety studies.

In combination with the risk classification matrix, Table 7.5 describing the level of control required indicates which control measures need to be

Table 7.3 Example of spreadsheet used in the application of preliminary risk analysis (PRA).

PRA			Client																						Date:
			Project: LPG Caustic Treatment Unit																						Rev: 0
			PRA—Preliminary Risk Analysis																						
Unit: Gas Treatment Unit				System: LPG Caustic Treatment Unit																					
Subsystem: general				Description																		Drawings:			
Item	Hazard	Causes	Effects	Safeguards detections	Freq.	People				Rig/installation				Environment				Image				Recommen-dations observations	Resp.	Deadline	
						s	Risk			s	Risck			s	Risk			s	Risk						
							T	M	NT		T	M	NT		T	M	NT		T	M	NT				
1	Presence of explosive gas in the system during unit startup	Inertization failures	Confined explosion, damage to equipment, possibilities of casualties	Use of oximeter (D)	B	IV		M		IV		M		II	T			III	T			O1			
																						R1		February 14	
																						R2		March 15	
2	Small LPG leak	Failures at joints, flanges, connections/improper opening of drain valve	Flammable cloud formation, fire and explosion	Gas detector (D)	C	II	T			II	T			II	T			II	T			R3			

3	Large LPG leak	Line rupture/ catastrophic failure of 232101A			B	IV		M		IV		M		III	T			III	T			R3		
4	Caustic soda leak 50%	Unplugging/ rupture of tank supply hose	Personal injuries		D	II		M		I	T			I	T			I	T			R4		
		Failures at joints, flanges, and connections			D	II		M		I	T			I	T			I	T			R5		

Table 7.4 Example of a risk classification matrix used in the preliminary risk analysis (PRA) application.

Categories of Consequence Severity		Description / Characteristics				Frequency Caetgories				
						A Extremely remote	B Remote	C Unlikely	D Possible	E Frequent
		People	Property/Availability	Environment (See Note 1)	Image	Conceptually possible, but without references in the industry	Not expected to occur, despite available references facilities in the industry	Low likelihood of occurrence during the operational life of a set of similar units	Possible to occur once during the operatinal life of the facility	Possible to occur multiple times during the operational life of the facility
V	Catastrophic	Multiple casualties within the facility or causalties outside the facility (See note 2)	Catastrophic damages potentially to the loss of the industrial facility	Severe damages to sensitive areas or extended to other locations	international impact	M	M	NT	NT	NT
IV	Critical	Casualties within the facility or serious injuries outside the facility (See Note 3)	Severe damage to systems (slow recovery)	Severe damage with localized effect	National impact	T	M	M	NT	NT
III	Medium	Serious injuries within the facility or light injuries outside the facility	Moderate damage to systems	Moderate damage	Regional impact	T	T	M	M	NT
II	Marginal	Light injuries	Light damage to systems/ equipment	Light damage	Local impact	T	T	T	M	M
I	Negligible	No injuries or first aid cases at most	Light damage to equipment without affecting the operational availability	Insignificant damage	Insignificant impact	T	T	T	T	M

Table 7.5 Example of a description table of the level of control required used in the application of preliminary risk analysis (PRA), also applicable to other techniques of risk analysis and safety studies.

Risk category	Description of the level of control required
Tolerable (T)	There is no need for additional measures. Monitoring is required to ensure that controls are maintained.
Moderate (M)	Additional controls should be evaluated with the objective of obtaining a reduction in risks and implementation of those considered feasible ["As Low As Reasonably Practicable" (ALARP) region].
Intolerable (NT)	Existing controls are insufficient. Alternative methods need to be considered to reduce the probability of occurrence or the severity of the consequences, in order to position the risks within regions of lower risk magnitudes (ALARP or tolerable regions).

formalized through recording in the PRA results report for each risk category, also applicable to other types of risk analysis and safety studies. Risks should be reduced to an acceptable risk region as much as possible –[as low as reasonably practicable (ALARP)].

7.1.3 Preliminary hazard analysis

The PHA (when not including risk classification) or hazard identification (HAZID) study are techniques similar to the PRA (when including risk rating), but even more general, also taking into account the hazards of accidents originating outside the technological enterprise (such as natural disasters and aircraft crashing on the rig). The main difference is that HAZID aims to identify HAZARDS (potential threats) and not necessarily completely defined accidental scenarios. Hazards can be identified without the need for an assessment of the frequency and severity of the associated risks. But PHA can also use the risk classification matrix, as long as this is established in the premises. The PHA is also applied through a guiding spreadsheet and a final report is produced with the conclusions and recommendations.

HAZID analysis is a qualitative technique for the early identification of potential dangers and threats to people, property, environment, and image. The greatest benefit of HAZID analysis is to support decision making in risk management activities. This means in practice the prior identification of hazards, threats, their descriptions, and the respective mitigating actions required to reduce human, property, environmental, and image losses.

The methodology of the HAZID technique combines the attitude of identification with the capacity of analysis of the participating specialists through a "brainstorming" conducted by a specialist leader in risk and safety management who coordinates the application of the technique. "Keywords" and technical concepts are used to organize the participants' ideas in order to facilitate the identification of possible potential hazards, their consequences, and threats.

The application of the HAZID technique requires that the analyzes take place in sessions with approaches that group the related processes that characterize a given system or operation. Participants are selected under the requirement to be experts in these processes because the success of the technique depends on the level of experience and technical operational knowledge of the participants. Participating specialists identify and classify the dangers, in order to evaluate their consequences and formulate actions to mitigate risks and threats. These actions are recorded in the form of recommendations.

It is also important that the identified hazards are postulated on the basis of reliable documentation. In each HAZID session it is necessary to register the documents used as a reference and as a basis for the identification of hazards.

Regarding the participants of the risk analysis carried out by the HAZID technique, in addition to the specialist risk management leader ("coordinator") and the rapporteur ("secretary"), specialist participants in the processes related to the system under analysis are required. As an example of areas of specialty that can be represented we have:

- senior management and decision-making representatives,
- specialist in the design of the physical installation,
- specialist in production, installation operation,
- specialist in the field of knowledge of the main activity,
- maintenance specialist (electrical, mechanical, information technology, etc.),
- specialist in providing basic utilities such as energy, water, and Internet,
- specialist in internal and external communication,
- specialist in statistical and operational data if necessary,
- specialist in the operational routine,
- specialist in logistics and supply if necessary,
- infrastructure and transport specialist if necessary,
- financial resources specialist if necessary.

The smaller the number of participants, the more agile and quick the application of the technique will be. But without experienced experts and with credibility in the various areas of knowledge, HAZID's mitigating actions and recommendations may become irrelevant. That is why it is necessary to have technical representation of all important areas related to the system under analysis and at the same time the smallest number of participants possible.

So that the size of the group does not make the application of the technique impractical, a number of between 3 and 8 specialists is recommended, in addition to the coordinator/editor, paying particular attention to the maximum limit of participants if the application session is not in person (by videoconference). A strategy for reducing the number of participants is to include specialists who accumulate knowledge on related areas and/or keep on-call, outside the HAZID session, technical support team to be consulted if necessary.

As for the specialist leader in risk and safety management (coordinator of the application of the technique), he must meet the following requirements:

- Lead the group as a facilitator in the application of the HAZID technique, highlighting in a systematic and structured way the identification of hazards, their consequences, the existing safeguards and recommendations with risk mitigation actions.
- Use concepts and "keywords" to level the understanding between specialists in different areas.
- Control discussions between experts so that the focus, objective, agility, and fluidity of the analysis are preserved.
- Record the relevant points of the discussion on the electronic form of HAZID which will constitute the report of the risk analysis.

Some of the desired results in the risk analysis carried out with the HAZID technique, the following stand out:

- Identification of opportunities for effective improvement in process safety.
- Identification of dangerous scenarios and the respective prevention and mitigation actions.
- Opportunity for advance planning and special preparations to respond to identified dangerous scenarios.
- Identification of required physical changes, corrections in processes, and improvement in methodologies that can be evaluated and established at stages well before the possible occurrence of the dangerous scenario.
- Increase the training of specialists participating in HAZID to promptly respond to the dangerous scenarios analyzed, improving operational reliability and availability and reducing contingencies.
- Mapping, still in the initial phase, of the dangers and vulnerabilities associated with operational activities and their respective equipment, facilitating the demonstration of evidence to prove to the inspectors and auditors the effective search for excellence in risk and safety management.
- Early identification of hazards, threats, safeguards, and recommendations for mitigating actions to support decision making at all levels.

The technique requires meeting the participants in a face-to-face session in a room with a notebook for records and a projector. In the absence of a face-to-face meeting, the technique can be applied with limitations by videoconference, requiring great discipline from the participants to compensate as much as possible for the losses resulting from interaction difficulties. It is necessary to use an electronic spreadsheet containing the risk matrix for decision making and HAZID form to be used for the records of the application of the technique.

The risk analysis sessions bring together experts from various fields. Even under the guidance and coordination of a risk management specialist, these discussions can become unproductive requiring the careful registration of the technical content applicable to the objectives of the technique.

All discussions between experts are important for the application of the HAZID technique because even those that are not registered, in one way or another, contribute to the broad discussion of the topics and the achievement of the tool's objective. But it is necessary to discard marginal information, even if important, and to accurately record the dangers, consequences, safeguards, and recommendations for mitigation actions. To assist in the selection and accurate recording of this information, the technique requires a registration form.

When HAZID is used also to classify hazards and risks, a risk acceptance matrix for decision making must be included. The configuration of this matrix can vary according to the safety culture of each organization, institution, or company. The important thing is to establish as a premise a risk acceptance matrix for decision making before the analysis begins. There are many general examples of a risk acceptance matrix aligned with the most accepted methodologies of risk and safety management.

7.1.4 Hazards and operability analysis

Hazards and operability (HAZOP) analysis is one of the most used risk analysis techniques in the oil and gas industry whose objective is to identify accidental scenarios generated by operational failures, risks, and hazards arising from process operations, as well as the possibility of deviations of process variables and their consequences. From the hazard and risk scenarios preliminarily identified through the PHA and PRA techniques, the HAZOP technique assesses, with a greater level of refinement and accuracy, which improper operations (alignments, maneuvers, start-ups, shutdowns, etc.) may result in the establishment of additional accidental scenarios of significant importance. HAZOP allows the verification of existing safeguards or that will need to be created to prevent such accidental scenarios and their undesirable consequences.

HAZOP requires a certain level of completeness of the project documentation because of the drawings required, such as, pipelines and instrumentation flowcharts, so that the alignments and maneuvers are evaluated. This technique requires a multidisciplinary group of experts from the various project disciplines and in the operation of the facility, and a trained leader to coordinate each step of the activity. In some cases where the facility is more complex and with a wealth of documents, it is also necessary to have a technical secretary to handle the records that will be necessary for the preparation of HAZOP's final report. During the application of the technique, a guiding spreadsheet is used, based on which the HAZOP report is written, including the conclusions and recommendations.

HAZOP analyses the accidental operating scenarios, their causes and consequences. The technique focus on the verification of the safeguards envisioned to prevent and control the events related to these accidental scenarios. When a vulnerability is discovered by the experts team by the application of the existing safeguards, mitigating measures are proposed, which may include the addition of equipment, changes in interlocks, design modifications, recommendations in the operation manuals, among others. Risks are classified by a previously defined risk classification matrix (Table 7.4). HAZOP is a qualitative, deductive, highly subjective technique, supported basically by the team's knowledge (Tables 7.6 and 7.7).

HAZOP technique uses a structured methodology to identify operational deviations that consists of the combination of words and the filling out the worksheet associated with the study of each "NODE": process segment compiled from the engineering flowcharts where process deviations are analyzed.

HAZOP evaluates the design intent, that is, the operational values (set points) defined for the process variables during the project. Deviations correspond to the departure from the design intent. Deviations are obtained by combining process parameters with guide words: words applied to the process parameters to qualify deviations from design intent or operating conditions.

Based the practical use of the technique, some guide words were defined by the HAZOP leaders to express the observations and conclusions related to possible deviations:

The application routine of the HAZOP technique can be summarized in the flowchart below, but one needs to remember that variations of it can be adopted based on the HAZOP leaders criteria, the experts team, the culture of the organizations involved.

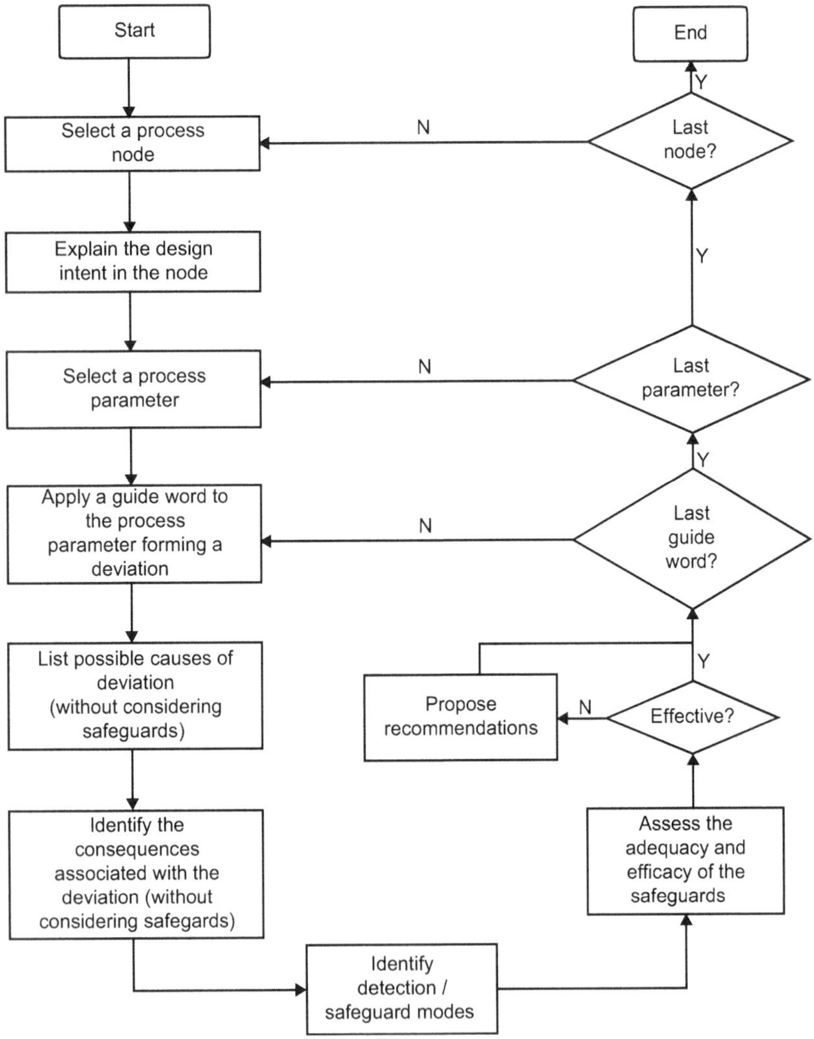

Identifying failures may seem confusing to those new to the technique.

Accidents and the management of its risks are multidisciplinary activities, therefore complex. For those beginners of the practice of the HAZOP technique may have difficulties, including the use of words with related meanings, synonyms, in addition to technical terms from different disciplines. Sometimes the same technical term can have one meaning for

Table 7.6 Example of a spreadsheet used in the application of Hazard Identification Analysis (HAZOP).

HAZOP			**Client**												**Date:**	
			Project: LPG Caustic Treatment Unit												**Rev: 0**	
			HAZOP—Hazards and Operability Study													
Area: Gas Treatment Unit								**Unit: LPG Caustic Treatment Unit**								
System: soda tank supply			**Description: # 1—from soda CT 50% up to TQ-232101 A/B/C**										**Drawings: DE 3A02.01.-CFM-006 rev. A**			
Item	**Deviation**	**Possible causes**	**Possible effects**	**Safeguards detections**	**Freq.**	**People**		**Rig/ installation**		**Environment**		**Image**		**Recommen-dations observations**	**Resp.**	**Deadline**
						S	**R**	**S**	**R**	**S**	**R**	**S**	**R**			
1.1	Lower or no caustic soda flow from CT to the soda tanks	Failure of B232101 A/B	Delay in the unloading of CT. pump captivation.	Backup pump (S) PIT-024(D)		C	I	T	I	I						
		Improper blocking of valves		PIT-024(D)												
		Unplugging/ rupture of hose		Containment basin 1.6 m^3 close to access ramp (S)												
		Obstruction of line filter of B-232101 A/B		PIT-024(D)												
		Improper opening of line drain		Containment wall (S)												

(*Continued*)

Table 7.6 (Continued)

HAZOP	Client															Date:
	Project: LPG Caustic Treatment Unit															Rev: 0
	HAZOP—Hazards and Operability Study															
Area: Gas Treatment Unit							Unit: LPG Caustic Treatment Unit									
System: soda tank supply		Description: # 1—from soda CT 50% up to TQ-232101 A/B/C												Drawings: DE 3A02.01.-CFM-006 rev. A		
Item	Deviation	Possible causes	Possible effects	Safeguards detections	Freq.	People		Rig/ installation		Environment		Image		Recommen-dations observations	Resp.	Deadline
						S	R	S	R	S	R	S	R			
		Alignment failure	Loss of product (alignment for the TQ of spent soda or for the loading hoses of the CT of spent soda)	LIT-008/009/010 (D)												
1.2	Higher caustic soda from CT to the soda tanks	Not applicable (limited by the flow of B-232101 A/B)														
1.3	Reverse flow	Not applicable														
1.4	Lower pressure in the soda tank supply line	Ditto, lower or no flow of soda														
1.6	Higher pressure in the soda tank supply line	Improper pump blocking	Pump motor heating	PIT-024(D), pump thermal relay (S)												

Table 7.7 Process parameters × guide words.

Process parameters	Guide words for qualifying deviations from project intentions	
Flow rate	No	Negation of design intent
Pressure	Less	Qualitative decrease of the design intent
Temperature	More	Qualitative increase of the design intent
Level	Reverse	Logical reversal of the design intent
Velocity	Part of	Qualitative reduction of the design intent
Composition	Other	Complete replacement of the design intent

Table 7.8 Categories × failure examples

Category	Failure examples
Failures in instruments and controls (logical interlocks)	Control valves with wrong indicators, problems with automation equipment, poorly designed control grids, masking of measurement
Operational failures	Alignment error, dosage error, improper opening/blocking of valves,
Equipment failures	Pumps, scales, conveyors, tanks, compressors, holes in heat exchangers
Others	Equipment obstruction (filters, foot valves, ducts), disconnection of hoses, lack of water, power outage

one discipline and another meaning for other disciplines. The proper use of words requires leveling up the meaning of the most common terms, to be recommended right from the initial application of the technique. Another frequent problem is confusion related to the characterization of failures. Many failures in different equipment might appear to have similarities, leading to confusion. There are indeed similarities, even though there may be major differences between equipment, that is, the chance of some confusion is real. This frequent discussion at HAZOP meetings can be facilitated by the leader with some initial guidance on the types of failures through examples. This will significantly promote discussions throughout the application of the technique. Therefore in addition to the correct usage of the terms, failures in general can be classified into certain categories, as exemplified in Table 7.8:

The experience in the application of the HAZOP technique shows that failures in instruments and controls (logical interlocks) are the ones that generate the most discussions and consequently the most loss of time,

Table 7.9 Defining control failures.

Control failures	Definition
Hardware at random	It occurs as a result of a multitude of degradation processes acting on the internal components of a device.
Common cause	Failures in more than one device, component or system caused by the same direct cause.
Demand	Failure of a safety function—SIF when it is subjected to real demand (see note).
Hidden	It is only noticed when a safety function—SIF is in demand or being tested.
Dangerous	It can potentially prevent a SIF safety function from operating when there is a real demand.
Safe	It can potentially cause a safety function—SIF to act when it is not required.
Systematic	Deterministically related to some cause.

Notes: SIF, safety function of an instrumented system. SIS (Section 7.1.5), whose objective is to achieve or maintain a safe state of a process or equipment through automatic action when facing a specific operational deviation. *SIF*, Safety instrumented function; *SIS*, safety instrumented system.

which reduces the efficiency in applying the technique. To reduce this problem, some additional clarifications can be provided by the HAZOP leader, right at the beginning, as suggested in Table 7.9.

7.1.4.1 Establishing premises for a HAZOP

The HAZOP analysis needs to establish the premises that will be considered during the application of the technique, starting from its preparation phase. Some premises characterize the HAZOP technique itself and are adopted in all applications of the tool. One of them is that the risk is always classified in its "mitigated" form, considering the detection and control devices planned for in the project, therefore already implemented in the enterprise. For example, if a pump sends a fluid to a tank with a risk of overflowing, the assessment of this risk should not only consider the pump and the tank, but all existing protections to prevent overflow. For example, if there is a high level limit meter that generates a pump shutdown signal, the risk needs to be assessed considering this protection.

Another usual premise establishes that if the risk is classified as "Intolerable," and protection measures are included as a recommendation, then that risk needs to be reclassified taking into consideration the implementation of additional protection measures.

The premise that simultaneous failures generated by independent events should not be considered in the analyses also needs to be adopted in every HAZOP.

The probability of occurrence of two simultaneous independent failures is so low that this scenario can be neglected for risk analysis purposes. At most, the escalation of events should be considered, that is, one failure causing another. Even so, these cases need to be clearly defined with the delimitation of the boundaries of the accidental scenario to be analyzed.

An important premise is to establish a frequency level above which an accidental scenario deserves to be analyzed. As a result, a prior assessment of the frequencies of the scenarios under discussion should be made, albeit subjective.

A major source of problems during the application of the HAZOP technique is the attempt unseasoned participants to try to take advantage of the technique aimed at improving safety for other purposes, such as operational comfort or simply in the design in nonsafety aspects. This behavior completely distorts the technique's objective, significantly increasing the application time, generating confusion, ambiguous records, and a compromised and confused final report where it is not possible to clearly tell apart the important conclusions and recommendations for safety from those conclusions and recommendations with interest only for the other technical disciplines of the project. Therefore the use of risk analysis techniques exclusively to provide conclusions and recommendations for improvements to management risk and safety should be adopted as HAZOP's premise, with focus on its objective, avoiding that other themes cause distraction to the problems related directly with the safety of the premises. In some cases, managers can hold an event prior to HAZOP just to analyze questions from other disciplines considered irrelevant to the management of risks and safety. This strategy avoids the misuse of HAZOP, and the compromise of the credibility of its results and recommendations.

Another premise is that the HAZOP leader needs to undergo formal trained performing the role and to have operational experience related to the type of facility that is being analyzed. It is also an essential premise that all technical documentation needs to be consistent and accessible to participants in advance.

Oftentimes, due to tight deadlines and schedules, experts perform a HAZOP without the minimum technical documentation. The direct consequence is a wasted effort for the application of the technique and a final report that will be a vulnerable and unreliable document due to the numerous voids that can be identified due to poor or insufficient documentation.

An essential premise is that the experts team in charge of the application of the HAZOP technique must necessarily count on the presence of operators. The name of the technique itself refers to operators' activities and without them the HAZOP technique loses its technical meaning and objective.

The event needs to have a previously established agenda of activities that includes an opening session with a general presentation on the technological enterprise to be analyzed, on its hazards, on the premises to be considered, as well as on the definition of the "nodes" to be analyzed.

7.1.4.2 Examples of division by nodes

As we have already said, a node is the process segment defined based on the engineering flowcharts where the process deviations are analyzed. The facility to be analyzed needs to be strategically subdivided into nodes. In the example in Fig. 7.2, the flowchart fragment details the alignment for receiving water from a truck that is to be transferred to a tank for use in a given process plan. This alignment would be a node, but other alignments are also possible for the same plant, such as, receiving concentrated soda illustrated in Fig. 7.3, which would represent another node to be analyzed through HAZOP.

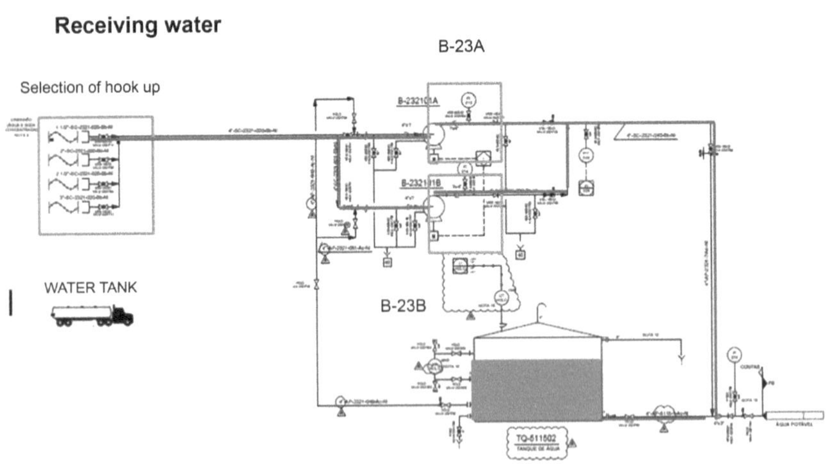

Figure 7.2 Example of a flowchart fragment detailing a node to be analyzed in a HAZOP, representing the alignment for receiving water in a given process plant. *HAZOP*, Hazard operability analysis.

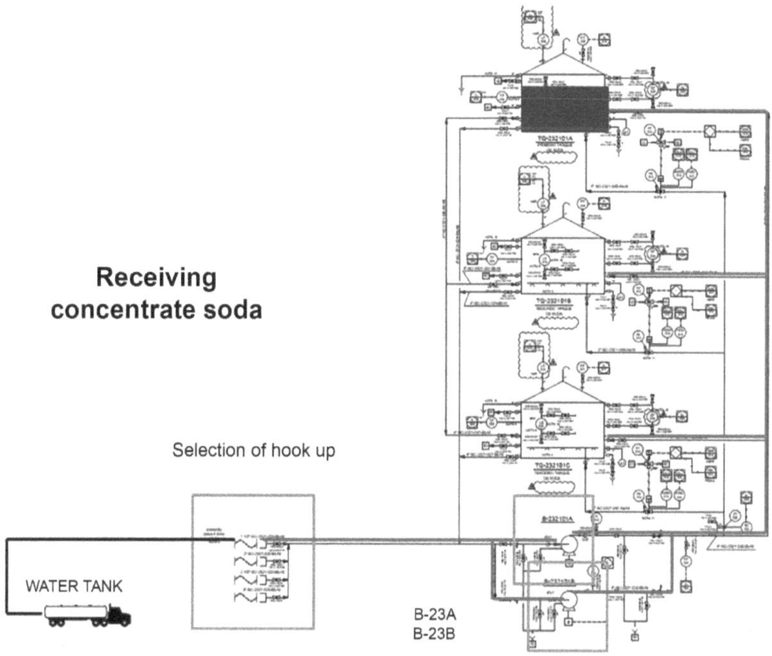

Figure 7.3 Example of a flowchart fragment highlighting a node to be analyzed in a HAZOP, representative of the alignment for receiving concentrated soda in a given processing plant. *HAZOP*, Hazard operability analysis.

7.1.5 Other risk analysis techniques

7.1.5.1 Brainstorming

Brainstorming is a free discussion among members of an experts team. A facilitator prepares some points of interest or important safety matters to guide the discussion and prevent participants from getting side tracked regarding the purpose of the application of the technique. The brainstorming success depends directly on the facilitator's capacity and technical knowledge both on the technique and on the problem being analyzed. A good facilitator should be able to promote the participants' creativity by encouraging them to think about various possibilities, sometimes confusing and conflicting. Brainstorming should not have established rules about the scope of the discussion. This freedom ensures the tool's objective of seeking hidden hazards and risks and bringing them up for discussion. All contributions should be accepted and recorded and no views expressed by the participants should be discouraged, challenged, or criticized. The skillful application of the tool aims to provide a comfortable environment for the experts. So they can "think in parallel" to the usual lines of approach adopted in conventional discussions, where the fear of making mistakes or being reprimanded inhibits the imaginative power of experts seeking the reduction of accidents unpredictability.

As an advantage, brainstorming is very useful for identifying new hazards in systems, new situations, confusing multidisciplinary scenarios, and lacking a guideline to direct the initial approach to the problem. Another advantage is that the various areas that may be passively awaiting each other's initiative are placed at the same level of responsibility to try to tackle the problem. Brainstorming is a rather easy and quick tool to be applied and performed. It can be applied in a wide range of situations, including in other areas of technological management.

As disadvantages, brainstorming is a somewhat unstructured tool and, therefore, not necessarily conclusive and complete. It depends to a great extent on the training and experience of the facilitator and the profile of the participating experts. The results are also very susceptible to the influences of the mood and work dynamics of the team. The brainstorming success relies on the facilitator's skills and operational technical knowledge.

7.1.5.2 Checklist

Checklist is a list of important items related to the known hazards and causes of accidents based on previous technical experience. Technical experience builds on previous risk analyses conducted on similar systems or on the direct operation of similar equipment or past accidents. The technique is characterized by the systematic and frequent use of a checklist that is adapted, item by item, to the enterprise to be analyzed. Generic checklists are not recommended because its efficiency is based on the ability of its items to indicate precisely the most important aspects related to the safety of the specific technological enterprise being analyzed. Even in the case of identical operations on identical equipment, the checklist needs to be adapted to reflect possible differences in each situation or, at a minimum, the checklist needs to be submitted to a revalidation process to corroborate its applicability.

The checklist validation is part of the technique's methodology, that is, regardless of whether the checklist has been applied previously, or how often, prior to every use of a checklist its validation needs to be confirmed as a first step. It consists of verifying not only whether the checklist is still valid but also to check if any new or specific factor requires a new validation at the time of the ongoing application.

Checklist is a technique recommended for analysis and simple activities that do not require much mental effort. Complex systems, plants, facilities, and equipment also contain simpler operational steps, but that nonetheless can be subjected to catastrophic failures and accidents. In these cases, the checklist can be evaluated as an applicable tool for a specific analysis of these simple steps. The efficiency of checklists depends as much as possible on the ability to reduce the answers by the applicators to "yes or no" regarding the hazards and risks that are being analyzed. Oftentimes, the conditions for applying a checklist cannot afford time for elaborate analyses that require concentration. Checklists can be applied in the industrial environment, under noise, heat, etc., and without additional sources for queries and clarifications. That is the main reason checklists need to be easy to read, cannot be too long or leave room for doubts regarding what needs to be verified in an objective and clear way.

As part of its advantages, the checklist containing "yes or no" answers can be used by inexperienced technicians with limited knowledge about the system. The checklist allows to capture conformity in a wide range of items important to safety, chosen based on technical knowledge and previous operational experience. The systematic repetition of the checklist ensures that common items related to obvious problems can be formally verified, avoiding that they can be overlooked due to routine. The tool becomes more efficient when the checklist is applied by a pair of professionals allowing double check.

As disadvantages, checklists have many limitations when applied to new systems or equipment. As previously mentioned, a good checklist depends on the technical knowledge and operational experience of its authors. In new systems, equipment and facilities, there is limited accumulated experience that consequently limits the quality of the checklist being prepared. Another disadvantage is that a side effect can occur with less experienced operators during the application of the checklist.

Systematic repetition of its application may convey the false sense of completeness that all possible hazards and risks are described in the checklist items. This is not true and every emergency situation or accidental scenario includes an unpredictable factor, not previously thought of and often not anticipated in the routine procedures. The continuous, systematic checklist use among inexperienced professionals and within a culture of poor safety may inhibit the imagination necessary to anticipate accidents that may occur from subtle elements. The identification of these important conditioning factors that form accidental scenarios is of great value for safety increase, but in many cases they may not have been incorporated in the checklist in use. Unfortunately, if the applicators are unaware of these facts, during the application of the checklist they may fail to identify hazards, risks, and accidental scenarios not yet covered by it.

7.1.5.3 Failure modes and effects analysis

Failure modes and effects analysis (FMEA) is a technique to be applied "from the bottom up" (regarding the operational sequence to be studied, from the most basic tasks to those at the highest level) in order to investigate possible situations where the basic components of a system may fail to fulfill its design intent. This failure can be at the equipment/ component level or at the functional level.

The technique evaluates the detailed description of the systems and considers the failure modes in which each component. The methodology is employed with the objective of verifying the fulfillment of the original design intent, the consequences of failures when this it is not met and the extent of these consequences that extend beyond the boundaries of each subsystem.

For each subsystem to be evaluated using the FMEA technique the following aspects should be considered:

- All possible modes of failure of the component or function should be included.
- All effects of each of these failure modes that can have an impact on the system environment as a whole should be considered.
- All possible causes of the failure modes identified should be considered.
- The possible ways in which failure modes can be mitigated internally in the subsystem or its environment of influence should be investigated.

The influence environment at the level of a complete and specific system is formed from the failures of its components, mainly those with some degree of importance and consequences associated with identified hazards. The level of detail required by the application of the technique can vary widely and is determined by the level of detail of the description of each system available during the application of the FMEA. Depending on the nature and complexity of the system to be assessed, the analysis may require an expert on the specific system or even a team of experts. The advantages of FMEA include the systematic and rigorous characteristic this technique, which is capable of generating detailed and auditing-ready records regarding hazards identified in the processes. FMEA can be applied in wide range of system types. As disadvantages the FMEA is only capable of identifying current hazards associated with specific single-source failures in the process, instead of combinations of failures. Another limiting factor is that the efficiency of the tool is dependent on the participation of highly trained people in the knowledge of each system internals. The systematic, rigorous, and detailed nature of the technique ends up generating a great consumption of time, which represents costs.

7.1.5.4 "What-if" (Swift—structured what-if technique)

The Swift technique was originally developed as a more simplified, efficient alternative to HAZOP technique. Like HAZOP, Swift involves multidisciplinary expert teams and a facilitator to lead in the application of the technique. One of the differences in relation to HAZOP is that it explores elements of a previous brainstorming activity, thus conducted at a higher level of detail of the systems under evaluation. When compared to HAZOP, the number of subelements analyzed is smaller and, in general, fewer discussion topics are considered. Another important difference is that the Swift technique is not only focused on the systems but also their association with the relevant procedures. Other techniques such as HAZOP and FMEA are focused only on the process flowcharts and on the detailing of the systems and their equipment.

The Swift technique considers possible deviations from normal operation identified by a brainstorm, and uses typical expressions like "What if . . .?" "How could . . .?".

Brainstorms associated with the application of the Swift technique should be performed under the guidance of a checklist to avoid that hazards can be ignored due to failure during the application of the technique.

The leader should prepare a list of topics in advance for discussions about the systems. These topics are based on questions such as:

- What would happen if . . . (What if. . .)
- Could someone . . .
- Has anyone ever. . .

The leader should use this type of topic to start the group's internal discussion and stimulate the imaginative capacity of the experts involved with the application of the technique underway.

As advantages, the Swift technique allows the creation of a set of rather detailed and auditing-ready records regarding the hazards identified in the processes. The Swift technique is less time-consuming in comparison to other systematic techniques, such as HAZOP.

Among the disadvantages is the amount of work required during the preparation for the application of the technique regarding the organization of the topics. The technique also requires a team composed of experts who are highly qualified both technically (on the systems under evaluation) and for the application of the Swift technique itself. As could be expected, the requirements for the profile of the leader responsible for applying the technique are yet more stringent, which contributes to additional difficulty for the application of the tool due to the need to gather high-level experts in large numbers.

A possible strategy for applying the technique is

1. Definition of the systems and processes to be analyzed.
2. Analysis of each system/process one at a time.
3. To conduct brainstorming to identify and list hazards without discussions.
4. Structuring of hazards in a logical sequence for discussion. Start the discussion from the highest to the lowest hazard following the logical sequence.
5. Analysis of each individual hazard one at a time, considering the possible causes of accidental events; the possible consequences of these events; the safeguards to be included to prevent events; the frequency level and consequences.
6. To conduct review for identification of any omitted hazards.

The discussion records should be prepared using a spreadsheet of the Swift technique. Additional checklists and previous experience (when available) should also be used to complement the results obtained.

The main advantage of Swift is the possibility of using a simpler methodology that requires less time and consequently lower costs. The main disadvantage is that the tool is less well known and the reason why experts are not very familiar with the technique, requiring prior training.

7.1.5.5 Layer of protection analysis

Layer of protection analysis (LOPA) is a quantitative risk analysis technique that utilizes information about hazards, severity, initiating causes, data on the probability of the occurrence of events as well as information obtained from the results of other previously applied techniques, such as example, HAZOP. The application of the LOPA technique begins with data collection from previous studies such as HAZOP, PRA, etc., related to the identified hazards, their safeguards and protective layers. Based on criteria established as the premises for the application of the LOPA technique, it is possible to obtain a theoretical risk reduction value required for a facility or system to be adjusted to an acceptable risk level. This risk reduction is achieved by the insertion of multiple layers of protection through design changes such as the inclusion of interlocks, changes in alarms and operational actions, inclusion of relief valves and safety valves, etc. The total reduced risk with the safeguards and layers of protection in place can be recalculated and if there is a need for further risk reductions, it can be investigated and analyzed. If even after the inclusion of these additional layers of protection the assessment of the risk level still exceeds the acceptable risk value, then the inclusion of a safety instrumented function (SIF) may be required. SIF is a function of a safety instrumented system (SIS) with the objective of reaching or maintaining the safe state of a process or equipment through specific automatic action against a specific operational deviation. When risk reduction requires the inclusion of a SIF, the LOPA methodology allows the determination of the safety instrumented level (SIL) required for this SIF. SIL is the reliability indicator required for instrumented SIF security functions.

The SIL value is dependent on the level of risk protected by the SIF. The higher the SIL the greater the reliability required.

As an advantage, the LOPA technique allows a presentation of risks apparently less affected by subjective influences, since the LOPA technique is typically a quantitative technique, that is, it translates risk assessment into numbers. However, quantitative risk analysis tools are not always subjectivity free as it seems, because although the risks are presented quantitatively, the initial information base for the application of the LOPA technique is made up of many premises based on subjective analyses. As a disadvantage, the LOPA technique relies on information from databases related to values of the frequency of failure occurrences. These databases, although developed according to scientific standards, do not always offer data that are the proper fit to the specific problem being studied, which also contributes to additional hidden source of subjectivity in the results obtained with the methodology. The LOPA technique is considered difficult to be applied and very complicated for most professionals without specialization in quantitative risk analysis. In some cases, the methodology may require many hours of analysis and calculation, and yet, the results may appear to be unable to reflect the operational reality adequately. The improper use of quantitative tools by professionals with no experience in this type of technique can cause problems such as purely mathematical results, without connection to operational experience. Rushed conclusions based on misuse of the technique can lead to overly-conservative design of systems or to failures in protection and safety. The studies and results obtained by the LOPA technique applied without the proper technical quality often require further adjustment by experts with operational experience to avoid the total loss of time spent in the application of the technique. The concept of protective layers used by the LOPA technique is correct, but its quantification is not as accurate as it may appear.

7.2 Studies and consequence analyses

Studies and consequence analyses consist of the evaluation of the most severe accidental scenarios, disregarding the safeguards implemented or established in the project to avoid them. The objective is to study the consequences of catastrophic accidental scenarios that, in theory, the technological enterprise is already protected against by project safeguards.

Therefore the scenarios studied in the consequences analysis are improbable, since risk management experts have already included safeguards to prevent them. The purpose of these studies and analysis is to complement the risk management of the technological enterprise through the investigation of possible consequences of a catastrophe, both for the facility itself (internally) and for the community and the environment under the influence of the technological enterprise under study (external consequences). Basically, studies are developed from premises and definitions of specific accidental scenarios to be considered. In general, the premises assume that certain safeguards are not effective, which allows catastrophic accidents scenarios to become possible. Therefore studies and analysis of consequences are a theoretical exercise in the investigation of "what would happen if" the enterprise did not have or lost a certain safeguard and the accidental scenario "X" or "Y" were established.

Engineers and technicians without specialization in risk management often add inadequate value to studies and analysis of consequences. This inadequacy can be better understood by comparison with medical imaging diagnostics generated by sophisticated, state-of-the-art equipment. When a patient undergoes a CT scan or a MRI scan, high-tech tools accurately identify an anomaly, locating it perfectly in the patient's body. Unlike the physical exam, though, when a consequence study uses sophisticated computational tools to generate an image of an oil and gas facility, the risk analysis experts who apply the tool need to create an anomaly artificially, based on assumptions, and also need to create possibilities for the evolution of this anomaly, and besides need to simulate the facility to be studied through a virtual 3D model. In other words, in medical examinations an anomaly is detected that is confirmed in a physical body, whereas in the analysis of consequences, experts carry out an experiment on an anomaly that may or may not exist in a facility that in most cases is yet be built. It is a mistake to draw definitive and unquestionable conclusions to studies and analysis of consequences such as those that can be obtained, for example, in medical imaging diagnostics or in X-ray and gamma-radiography nondestructive tests with used to investigate problems in parts and machine components.

It is not unusual for less experienced professionals to be surprised when they learn about reports of analysis of consequences such as fire propagation, gas dispersion, and explosion. Unaware of the philosophy behind these studies and analyses, some professionals without specialization in risk management may mistakenly conclude during their analyses that large areas of a processing plant may be at unacceptable risk of being affected by temperature increase caused by fires, or by a plume of flammable gases. This is because these studies assume, for example, the rupture of 100% of the cross section of a gas pipeline or the explosion of a large pressure vessel as a basis for defining the scenario to be studied.

Obviously the project and/or facility is already protected by several safeguards that prevent the rupture of explosions, but the risk management exercise requires the assumption as an initial premise that such catastrophic events will happen, regardless of the safeguards in place. The most appropriate term to describe this technique is "exercise", that is, studies and analysis of consequences allow risk management experts to assess the consequences of unlikely scenarios, as some kind of exercise whose results can lead to recommendations and increased safeguards when a particular unacceptable risk is identified.

However, studies and analysis of consequences need to be conducted and assessed by professionals specialized in risk management. In addition to the technical knowledge of employment of the analysis tools, sound operational experience is required to avoid distortions and the lack of agreement of the results with the operational reality. Any study or risk analysis is subjected to influences unrelated to the best engineering practices. Some technological enterprises can be questioned by the managers involved in the studies and, depending on the interest or lack thereof in making them viable, as well as economic and political pressures can influence the application of the risk management technique. This happens both in qualitative and quantitative techniques. However, qualitative techniques are more transparent regarding this type of interference, since the professionals and experts involved in the application of the tool express their opinions openly. But the quantitative risk analysis techniques, especially those that involve computer simulations, require a large amount of data and premises to be performed. Moreover, small changes to the data or just the exclusion or inclusion of certain data can completely affect the final results of studies and quantitative risk analyses.

For this reason, the experts involved in the application of these techniques should redouble their attention to the possibility of manipulation imposed by external, economic, political, and strategic pressures. Independence and technical impartiality are paramount for the results of these studies to be scientifically based and unbiased.

Unfortunately, in poor safety culture environments, experts can use safety studies and analyses as a power achieve strategic objectives, circumventing techniques based on interests extraneous to risk and safety management. Professionals specialized in risk management should have, among all other technical skills, the ability to maintain a neutral and independent position, always prioritizing the scientific application of risk analysis techniques, reconciling them with the interests of the managers involved with the technological enterprise, without however compromising the scientific quality of the work, responsible for safeguarding lives, property, and the environment.

7.2.1 Fire propagation study

The fire propagation study assumes the establishment of an accidental scenario with a major fire at the facility. The choice of the accidental scenario to be studied can be made based on data from previous risk analyses. In general, a classic scenario to be studied use in the oil and gas industry it is the hydrocarbon spill with the formation of a puddle and its ignition, or the jet fire formed from the gas leakage from a pressurized pipeline.

Evidently the heat generated and the duration of the fire will have a direct relationship with the volume of hydrocarbons that form the puddle, or the gas flow that is released in the leak. For example, the volume of a puddle depends on the quantities of hydrocarbons stored and the type of loss of containment (leakage) postulated. Experts depend on the precise definition of the accidental scenario, including materials involved, volumes, wind conditions, characteristics of the hydrocarbons involved. There are several computational tools capable of performing an investigation of fire propagation, but the most common tools are those based on CFD. Due to the large amounts of data to be processed as part of these tools, and the increase in the data and analysis' complexity for each scenario, there are major limitations on the results of studies and analysis of consequences.

These limitations can be easily understood with the realization that CFD simulations are generated from a large amount of data related to a very small number of scenarios. In fact, each of the scenarios requires a large amount of data, which makes its processing costly in terms of time and economically. The simulation results are only valid for each specific scenario, which greatly reduces the applicability of the results of simulations of this type. Note that the results are valid for a given volume of oil spills, under a given wind condition, in a specific area of the facility being studied, due to a specific failure of a specific equipment. It is a fact that in an oil and gas facility spills can occur with countless volume possibilities, under countless wind conditions, in countless areas of the facility under study, due to countless possible failures. Therefore the results of the consequences studies are like a "snapshot" of a single instant in time and serve to give an idea about the scale of the consequences of unexpected catastrophic events for a facility with the project safeguards in place. The purpose of these studies is to complement risk management with the refinement of existing safeguards should any unacceptable risks be identified and not yet adequately protected against. It is possible, for a given fire scenario, to identify the affected areas at different temperature levels and the consequences for the structural integrity of equipment and structural components (Fig. 7.4). The results aid in the specification of passive protection, assessment of the need for structural stiffening, and the location of flame detectors.

7.2.2 Study of dispersion of gases and smoke

Gas dispersion studies are performed using computational tools similar to those used in fire propagation studies (CFD). One of the main

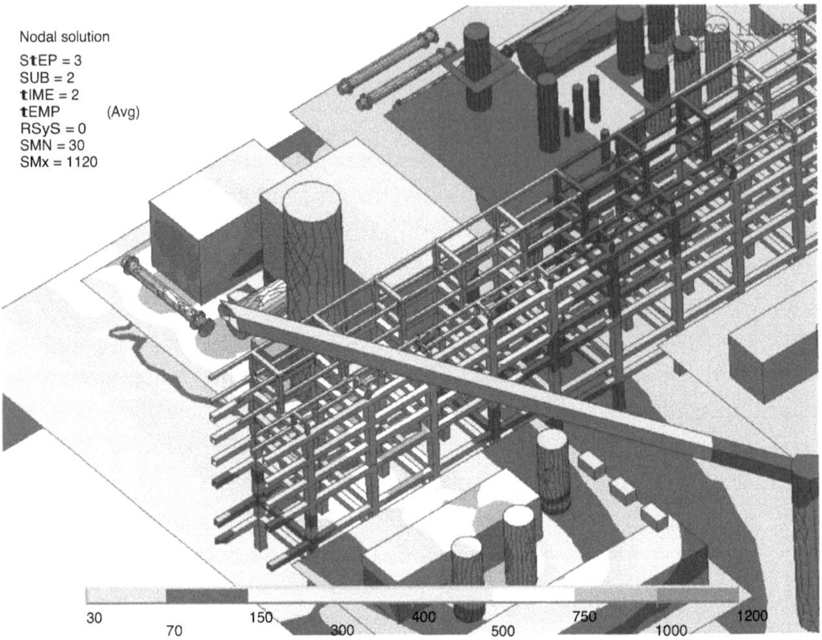

Figure 7.4 Temperature distribution due to a jet fire in an offshore rig, from failure in the third stage of a compressor. The flow rate at the leak point is 50 kg/s and the tone indicators represent the temperatures after 900 s since the start of the accidental event. The 3D view was generated by a computational fluid dynamic (CFD) tool.

applications of gas dispersion studies is to provide complementary information to assist experts who study the best gas detectors locations. Although the location of detectors is best defined based on the operational experience and the experience obtained in previous projects, for novel situations or even more ordinary projects, a complementary verification of the detectors location drawings based on the results of gas dispersion studies (Fig. 7.5).

For buildings and closed facilities, a smoke dispersion analysis should be carried out to verify the conditions of survival and moving ability in the case of emergencies. This applies, for example, to closed offices, accommodations, restaurants, and leisure areas.

7.2.3 Explosion study

The computational tools for conducting explosion studies are a little more sophisticated, although they also use computation fluid dynamics, as in the case of fire propagation and gas dispersion studies. This sophistication

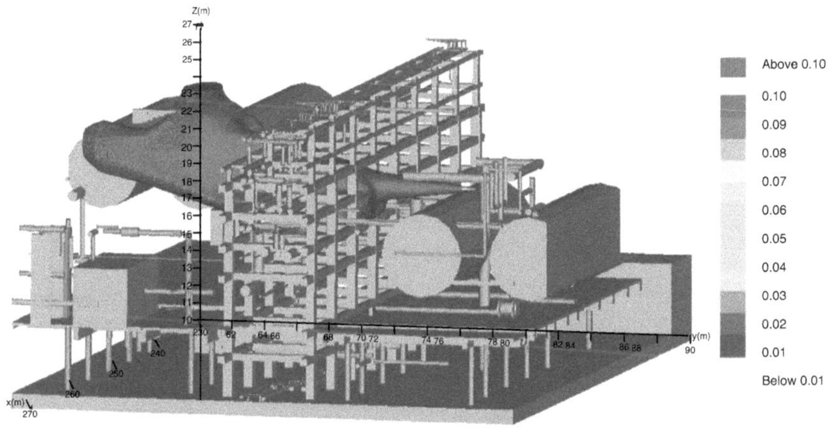

Figure 7.5 Leak in an oil and gas separator subject to wind blowing in the north direction with a speed of 0.5 m/s. The plume and the concentrations can be observed considering the tone mapping relative to the gas concentrations. The 3D view was generated by a computational fluid dynamic (CFD) tool.

is justified by the higher complexity of the explosion phenomena. Based on explosion studies, structural stiffening and changes in arrangement and layout can be recommended as additional safeguards to the project. Fig. 7.6 shows an example of a 3D view produced by a CFD tool for an offshore rig.

Once again, we highlight the importance of technical training and, most importantly, operational experience for deep understanding and better use of the results of studies and analysis of consequences. It might not be clear to inexperienced professionals that the scenarios studied do not take into account the existing safeguards in the project so they are excessively severe and unrealistic. From the operators and field technicians perspective, the misinterpretation or misuse of 3D views, such as in Fig. 7.6, can generate unsubstantiated worries and fears. Lack of preparedness during presentation of this type of results may lead to mistaken perception that the facility is subjected to unacceptable risks, when in fact the safeguards existing in the project were intentionally suppressed in the simulated scenario due to the premises established at the beginning of the study.

7.2.4 Escape, abandonment, and rescue study

The escape and abandonment system is the most important safety system for saving lives. Totally aimed at protecting people present in accidental

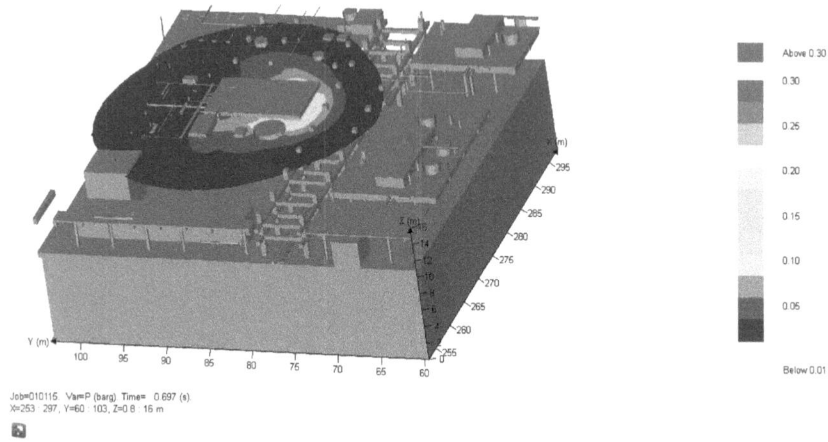

Figure 7.6 Tone mapping of the overpressures produced by a shock wave generated by an explosion in an offshore rig module. The 3D view was generated by a computational fluid dynamic (CFD) tool.

scenarios and throughout the technological enterprise, the removal of agents (people) occupies the top position among the components of the risk management strategic line (Fig. 7.1). It is a requirement by international standards to conduct specific studies on the conditions of escape and abandonment of facilities for some technological enterprises. Depending on the development level of the safety culture associated with the technological enterprise, the approach adopted to the design of escape and abandonment systems can be more technical and scientific or more legal oriented, being restricted to the compliance to basic rules without taking into consideration the specific characteristics of the enterprise or behavioral and safety culture aspects related to the agents.

There is a technological tendency by specialists in risk management to combine the results of fire propagation studies, gas dispersion, and explosion in a single study (Section 7.3), which may also include human factors aspects and safety culture to partially address the behavior of agents both individually and in groups. For high-level risk management, it is strongly recommended the use of the full safety analysis (FSA) technique presented in Section 7.3. For other risk management levels, less sophisticated techniques remain applicable, with coarser results, but nonetheless useful for the improvement of escape and abandonment systems. One of the techniques for the study of escape and abandonment systems is known for its acronym EERA (Escape, Evacuation and Rescue Assessments) report.

The objective of the study of escape, abandonment and rescue strategies is to analyze the conditions of the resources and equipment available to perform a safe escape and abandonment operation, as well as the rescue of agents at sea or in locations of difficult access.

The scope of the studies is defined based on the boundary of influence of the facility being evaluated. Broader emergency plans involving communities in the vicinity of the facility should be dealt with separately in other types of analyses, taking into account also additional information on external resources. In some cases, the escape and abandonment study includes a checklist to be applied from the basic design phase of the technological enterprise, in compliance with the standards and recommendations adopted by the project. Checklists performed throughout the project development should be revised for more details.

Some initial premises for the realization of escape and abandonment studies need to be established as well as the definition of reference values for some parameters. Some typical values are: the maximum travel time for people to reach the meeting point should be 10 minutes; the displacement speeds to be considered are 1.0 m/s for horizontal displacement, 0.8 m/s for displacement by stairs and 0.3 m/s for displacement by vertical ladders; the reaction time from the alarm announcement is 2 minutes, taking into consideration the minimum required operational activities and the selection of the escape route to be used.

Also related to premises, the studies should establish strategies based on procedures adopted by the managers of the technological enterprise being evaluated. Some managers prefer agents to pick up safety equipment at a specific location before traveling to the meeting point. As an example, some offshore operating companies require that agents travel to their respective cabins to get their life jackets before heading to the meeting point. Studies should consider the worst route to the cabin and from the cabin to the meeting point. Regarding the checklist, it should contain:

- Safety studies and the implementation of the recommendations resulting from these studies.
- The conditions of the escape routes to be considered should be at least 1.20 m wide and 2.10 m high for the primary routes and 1 m wide and 2.10 m high for secondary routes.
- The conditions for signals and markings indicating the direction of the meeting point and the abandonment posts adjacent to the lifeboats.
- Floor and lighting conditions to consider the availability of electricity including in an emergency.

- No long corridors with more than 7 m in length without exit.
- Opening doors that may interfere with escape routes.
- Existence of two escape route options from any location in the facility.
- Investigation of possible obstructions and blockages in the escape routes resulting from the escalation of accidental scenarios through safety studies and analyses, as well as the mitigation measures adopted.
- Choice and location of meeting points (muster stations) considering the space adequacy with respect to the number of agents.
- Adequacy of the locations chosen as meeting points (muster stations) to maintain the survival conditions for 60 minutes considering the presence of fire, smoke and gas.
- Availability of all personal protective equipment necessary for the operation of escape, abandonment and rescue.
- Capacity of boarding areas near rescue boats in offshore facilities for the preparation of the abandonment.
- Availability and adequacy of additional equipment necessary for the effective abandonment of facilities such as vehicles, cars and rescue boats.
- Protection provided by fire fighting systems for escape routes meeting points and drop-off points.
- Adequate inclusion of passive protection at the locations required for the protection of escape routes.
- Dimensions of manways and passageways that can be used as escape routes, considering ellipses of 0.6 m × 0.8 m.
- Adequacy of means of access to escape routes in confined locations or of difficult access.

The study of escape, abandonment, and rescue should also include tables for calculating travel times, including from all points of the facility to all meeting points. EERA studies continue to be widely used, but for high-level risk management systems there is a growing trend of the application of the FSA technique (Section 7.3) that uses escape and abandonment computer simulation tools, considering human factors and aspects of the safety culture.

7.2.5 Analysis of loss of liquid containment and environmental control

Emergencies may occur in oil and gas facilities that result in loss of containment or disposal of oil into the environment. Losses of containment that can lead to damage to the environment needs safeguards provided for in the project, but unfortunately such a scenario can be established, despite all the prevention efforts by the designers and operators.

Considering the accident beyond design concept mentioned in Section 6.14.2, safety systems should be designed to ensure that complementary mitigation measures can be adopted even in scenarios as degraded as those causing oil leakage to the environment. Safety studies and analyses may contribute to the investigation of possible leak scenarios and their consequences. A typical problem is oil leaks in rivers and bays, contaminating water, beaches, fauna, and flora. Hydrocarbons from some facilities in the oil and gas industry such as marine terminals very often are transported by vessels. Special attention is paid to the treatment of risks involved in the movement of large vessels with potentially polluting hydrocarbon inventories.

Accidents resulting from the collisions of these large vessels with the terminals or other vessels are subjected to analysis during the terminal's design phase. These analyses also consider the environmental impact from the various accidental scenarios postulated. The use of simulations based on environmental computer modeling in risk and safety analyses allows a more detailed and accurate investigation of the effects of accidental scenarios on the environment. Analysis using computational tools can demonstrate that scenarios previously considered catastrophic in fact have a negligible impact or restricted to a specific area.

Conversely, such analyses can be used proactively to identify problems and risks of impacts on the environment in scenarios that can only be technically identified through such simulations.

Computational tools have been used in studies and projects involving modeling of natural bodies of water. These studies investigate the environmental impact of hydrocarbon leaks in rivers and bays using data such as:

- polluting loads,
- bathymetric information,
- tidal information,
- wind information, and
- flow rate of affluent rivers.

By means of the hydrodynamic characterization of the area where the accident is postulated, it is possible to obtain guidance on contaminating plumes and their consequences for the environment through computer simulations.

This type of simulations can improve risk analyses regarding the environmental impact of the following accidental scenarios:

- Collision of hydrocarbon cargo vessels with terminals.
- Collision between hydrocarbon cargo vessels in the vicinity of the terminals.
- Collision between hydrocarbon cargo vessels and other ships in the vicinity of the terminals.
- Oil spills/leaks from collisions or other accidents.

Some computational tools allow simulations using the Lagrangian transport model (in Lagrangian mechanics, the trajectory of a particle system is obtained by solving the Lagrange's equations), which is particularly useful to solve problems related to emissions at effluent release points along from the coast, in oil spills, besides supporting analyses of the movement of floating debris and the calculation of pollutants residence times in natural bodies of water (Fig. 7.7). The results can be analyzed using a stochastic model, providing direct partial answers to the most common questions in the environmental area, regarding the transport of contaminants. The probabilistic analysis can be performed both in terms of the number of events and their duration.

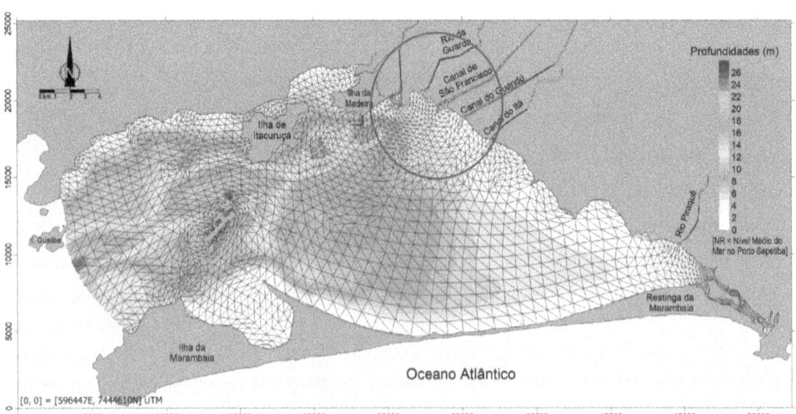

Figure 7.7 Example of the modeling of a domain for environmental studies of a coastal region, in this case, Baía de Sepetiba, Rio de Janeiro Brazil, illustrating the finite element mesh discretization and the current bathymetry referring to the average sea level in the Sepetiba region. The axes represent distances in meters. The ellipse highlights the region of specific interest. In the case of the oil and gas industry usage, a similar mesh can be generated for any of the regions under the influence of industrial facilities, for example, for Baía de Guanabara—RJ, Todos os Santos—BA. *Courtesy www.sisbahia.coppe.ufrj.br.*

In the analysis related to the number of events (*N*), the term in the denominator is *N*. Each event and each contamination source are independent events. The duration of each event is similar and the variable is the conditions that trigger the event. This situation is typical of spills: the duration is considered to be the same, but a range of hydrodynamic and meteorological initiating scenarios can occur, such as: starting at different tidal stages, that is, from low to ebb tide, during the day or at night, with different wind conditions, etc. The total number of events (simulation scenarios) is of the order of hundreds. The possibilities of results within the stochastic model are:

- % *N* of spill patch passage (isolines): when selecting this option, the simulation will perform internal calculation to determine the number of times a spill patch is detected at one or more points in the domain, recording each occurrence. With this data, a map of isolines passing through the spill patches is produced.
- % *N* of passage with concentration > limit: similar to the previous one, except that in this case the passage of particles with concentration above the limit will be recorded, generating a concentration isolines map.
- % *N* of passage with lifetime < limit.
- % *N* of times a spill patch reaches the coast.

Another type of approach is the probabilistic analysis with respect to time. In this case, the denominator is the simulation time (*T*). The user can select the sources to represent the time variability of the same event, such as, flow rate, concentration, or decay characteristics that are variable over time. The % *T* stochastic model is often used to estimate frequencies of detections of emitted plumes.

The possibilities of results within the stochastic model are:

- % *T* of passage of the spill patch (isolines).
- % *T* of passage with concentration > limit.
- % *T* passing with lifetime < limit.
- % *T* times a spill patch reaches the coast (table).

Conventional risk analysis studies establish some accidental scenarios. In general, the most critical scenarios are chosen to be analyzed and it is assumed that the other scenarios are covered, since they are less critical in theory. The number of scenarios is limited. However, a probabilistic analysis based on the Lagrangian transport model allows events (simulation scenarios) to be repeated hundreds of times. If, for example, the impact of the spill resulting from the collision between two vessels in the area of a terminal is studied, via the conventional analyses, the location is chosen where the most critical environmental impact is expected. However, a probabilistic analysis makes it possible for this this type of accidental scenario to be multiplied hundreds of times considering events each with its specific spill location, at different times, and also take into consideration the effects of tides, winds, etc. for each instant when each of the various spills are studied.

The results obtained in this case allow for a much richer and more accurate analysis where many catastrophic hypotheses can be discarded by side-by-side comparison with the results obtained by the simulations. It also becomes possible to identify important scenarios that are not perceived to be critical via conventional analyses. The use of computer simulations based on the Lagrangian transport model in probabilistic analyses improves the quality of the results of risk and environmental safety analyses related to the movement of vessels in the area of the maritime terminals. This type of tool can also be used in accidental scenarios of leaks from pipelines, tanks, and losses of containment in other equipment and industrial facilities that can cause contamination of rivers and bays.

7.2.5.1 Practical results

The following types of results can be obtained via probabilistic analysis associated to number of events:

- The frequency with which a spill patch resulting passes through certain locations in the area of the terminal where the accident occurred.
- Once it has been established that a particular spill patch is important in terms of environmental impact when it maintains a minimum concentration "*x*," it is possible to obtain, through simulations, the frequency of passage of the spill patch through specific points in the terminal area while that patch still keep the minimum concentration "*x*."
- In the case of a spill patch that has an established, predefined life span, for example, a patch resulting from a liquefied natural gas (LNG) spill (in natural evaporation)—it is possible to obtain the frequency of passage of that spill patch by specific points during its life span.
- The frequency and locations where oil or LNG spill patch reaches the coastline while maintaining minimum concentration or during its life span.

Mutatis-mutandis, probabilistic analyses in the time domain support the following results:

- The time that a spill patch due to an accidental oil spill passes through certain locations in the area of the regasification terminal where the accident occurred.

- Once it has been established that a specific oil or LNG spill patch is important in terms of environmental impact, if a minimum concentration "x" is maintained, through the simulations it is possible to determine the time of passage of the spill patch through specific locations in the area of the regasification terminal, while its minimum concentration "x" is maintained.
- In the case of a spill patch with a predefined life span, for example, a patch resulting from a LNG spill (in natural evaporation)—it is possible to determine the passage time of that spill patches through specific locations during its life span.
- The time and locations where an oil or LNG spill patch reaches the shore while maintaining minimal concentration or during its life span.

There is an increasing need for risk management experts to be informed and technically prepared to associate their typical activities related to the safety of people and property of technological enterprises with environmental matters. Risk management activity has a multidisciplinary characteristic, and the technical sensitivity of problems related to risks and threats to the environment has become an indispensable requirement for specialists who wish to keep up-to-date with the field.

7.2.6 Studies of stability and naval damage condition

A suite of naval engineering studies may be required for offshore rig projects. These studies are important to define the most severe naval damage condition for which safety systems should be designed for. In general, the naval damage condition establishes a maximum heeling angle as well as other relative positions in which the offshore rig may require the availability of safety systems, even when damaged. Also considered are the atypical draft and stability conditions during the transport from the shipyard to its final location. Some safety systems need to be operational to respond to specific accidental scenarios that can happen during the transport.

The set of FWP pumps needs to be designed to meet the required demand, when the rig is listed due to a naval damage. This means that if the damage scenario identified by the stability and naval damage studies indicates 16.5 degrees heel angle then, if the configuration is 2 × 100% FWP, at least one 100% FWP must be available. That is, the rig slope might such that the suction of one of the FWPs is in an unfavorable position in terms of its manometric height. However, the other FWP necessarily needs to be in operational condition for the same scenario. The same reasoning is used to check the lifeboats availability. Up until the accident evolves to the point where the rig reaches the heeling angle defined by the naval damage condition, the number of lifeboats required to meet 100% of people on board (POB) capacity should always be available ready for launching. Some of the lifeboats can be unavailable due to the large heeling angle (which prevents launching of the lifeboats), but a minimum number of lifeboats needs to be positioned at another point on the rig (favorable launching conditions) to ensure the escape and abandonment of the entire POB of the rig.

7.3 Full safety analysis

FSA studies are elaborated through efforts to combine the results obtained in different safety and risk analysis studies into a single simulation tool. The objective is to group the data obtained in several scientific studies and analyze them together utilizing a tool that supports the maintenance of the validity of the data while generating more complete simulations, with results closer to the operational reality. Researchers are still working on the development of software and computational tools capable of handling of such large amount of data and simultaneous functions. But there have already been analyses of this kind performed for installations in the oil and gas industry like for state-of-the-art offshore platforms.

The approach to risk and safety management matters adopted via this type of tool is based on the principles of human factors and safety culture. The most comprehensive results are useful in the investigation of problems related to harm to people in various accidental scenarios and in operations to escape from hazards and abandonment of emergency scenarios. There are several software on the market for simulation of escape and abandonment in buildings, hospitals, urban areas, airplanes, and ships. One of the most complete software is Exodus, developed by the School of Computing & Mathematical Sciences (University of Greenwich, UK). Companies specialized in this type of technology offer prototypes of computer simulation tools of escape and abandonment that include human behavior elements. These tools are used to improve the safety projects of technological enterprises and are capable of simulating scenarios that include fire, explosion, smoke, temperature, flooding, human behavior, and the effect of natural disasters such as floods and earthquakes, and also emergency situations arising from terrorist attacks.

Most of the related software involves a considerable level of programming complexity for the definition of scenarios, for building models as well as significant training required for their use by engineers. Some of them also require considerable computational processing time to achieve objective results. Any software for this type of analysis applied to oil and gas facilities requires the adaptation of its functionalities to the specific characteristics of the industry. An applicable example for the offshore oil exploration and production facilities reality is a software used for the simulation of escape and abandonment operations on passenger ships in emergency and regular boarding and disembarkation scenarios.

Some of these software are capable of including in the simulations of fire scenarios the effects of heat and smoke on people, the inclination of the float, the behavioral aspects and the rig motions, in damaged or normal condition. A technically correct adaptation allows the approach to the escape and abandonment issue based on the principles of factors and safety culture within the FSA study concept.

As a reference of a tool applicable to a FSA study, Environment Editor module (EVE)/Evacuability Index module (EVI) was developed by the company Safety at Sea Ltd. (specialized in maritime safety analysis) in collaboration with the Department of Naval Architecture at the University of Strathclyde, Glasgow, UK.

EVI (simulation software module) is a tool used to simulate the movement of people in any type of environment, including onshore. This software has been used to model the circulation and abandonment of people on ships, offshore wind power structures, and buildings. The program works with a 3D interface that allows the user to prepare a realistic presentation of scenarios and make changes in real time.

The agents (people) are modeled individually and interact among themselves and with the rig environment that they are in. There are no limitations on the number of people or the size of the environment to be modeled. Demographic and anthropometric variables that can impact people's behavior, such as age and gender as well as people's speed, can be assigned to agents as part of a probabilistic approach.

EVI programming may include the assignment of tasks to be performed by specific people in the simulated scenarios. This way, it is possible to simulate a very complex circulation of people, with previously defined rules of interaction and the influences of the procedures adopted in the facility (for instance, the operational procedures of a unit in operation, or the construction and assembly procedures of ship still in the shipyard).

A wide array of results can be obtained by playing back the recorded simulation concurrently with utilities for measuring the various technical parameters involved. The 3D visualization allows the analyst to observe the event from any view point, making it possible to record the simulations and play them back in batches for statistical treatment. A set of analysis functions can provide graphs, statistics, and the identification of congestion, in addition to assessing the efficiency of the means of escape and abandonment.

The software is split into two independent modules, namely: EVE—example evacuation analysis of IMO ship, which is a software specifically for editing the 3D model of the environment to be simulated; and the main module called EVI—example evacuation analysis of IMO ship, which is the software responsible for importing the EVE (software modeling module) 3D model that allows setting the variables related to population and all characteristic parameters of the scenario under study, thus preparing the simulation data input and environment. In Figs. 7.8 and 7.9 screenshots of the main software graphical user interface are shown.

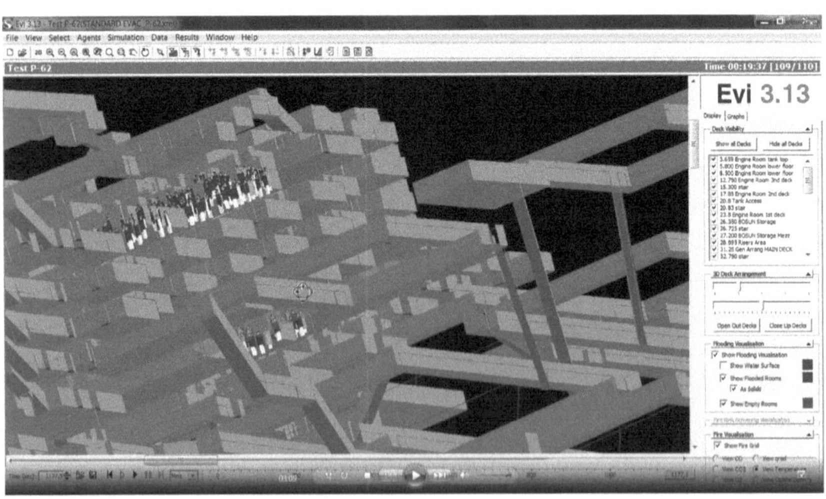

Figure 7.8 Graphical user interface of the EVI software (simulator). *Courtesy Safety-at-sea, Glasgow UK, http://www.brookesbell.com/service/software.*

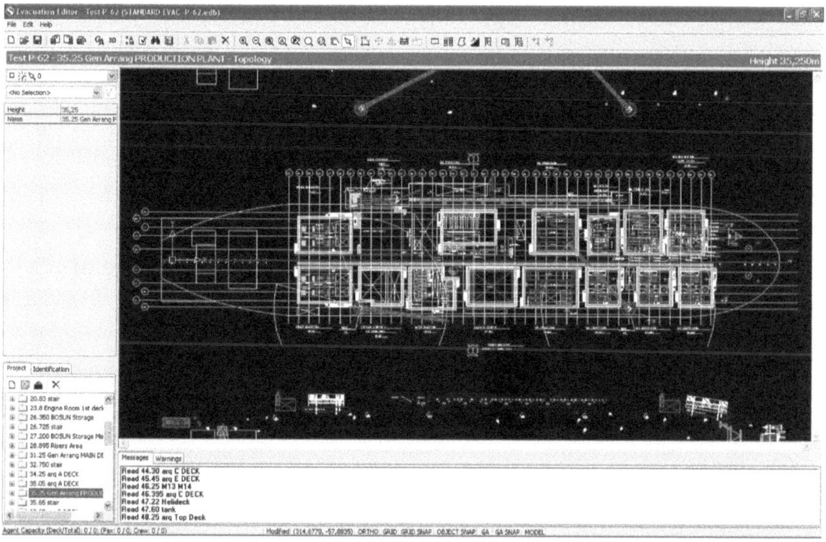

Figure 7.9 Graphical user interface of the EVE software EVE (3D Model Editor). *Courtesy Safety-at-sea, Glasgow UK, http://www.brookesbell.com/service/software.*

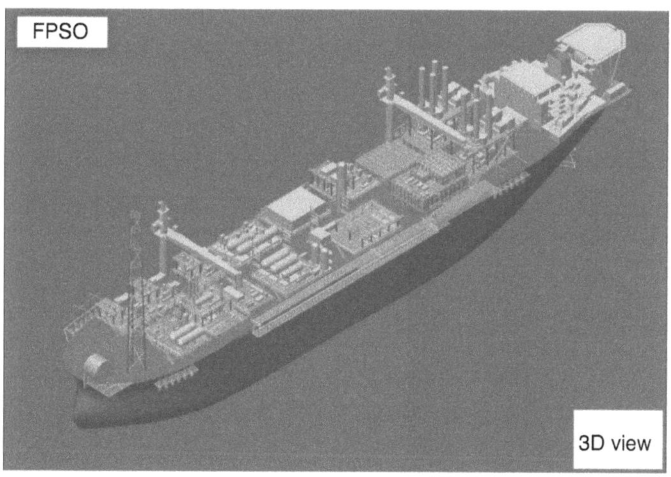

Figure 7.10 3D model of FPSO under study. *FPSO*, Floating production storage and offloading.

7.3.1 Features of the analysis of offshore rig

For this type of analysis, complete information about the rig under study is required. As an example of the application of the FSA technique, let us consider a floating production storage and offloading (FPSO; Fig. 7.10), which is an offshore oil and gas exploration and production unit resulting from the conversion of a very large tanker crude carrier (VLCC). The FPSO chosen for the study has an estimated installed capacity for processing and treatment of 180,000 barrels of crude oil per day (bpd), 6 million m^3/day of gas and to inject 42 m^3/day of desulfated water. The unit is supposedly installed in 1600 m water depth. The unit's POB is planned for 110 people in the project.

Let us consider that the system of hazard escape and abandonment of scenarios of the FPSO was designed in compliance with the following standards:

- ISO 13702—Control and Mitigation of Fires and Explosions on.
- Offshore Production Installations.
- IMO SOLAS: Safety of Life at Sea.
- IMO MODU CODE: Mobile Offshore Drilling Units.
- NORMAN 01 Brazilian Maritime Authority Standard for Vessels Employed in Open Sea Navigation (Norma da Autoridade Marítima Brasileira para Embarcações Empregadas na Navegação de Mar Aberto).
- *Classification society requirements* [American Bureau of Shipping (ABS)], Det Norske Veritas (DNV—GL), Bureau Veritas S.A. (BV), and Lloyd's Register Group (LLOYD'S).
- API RP 14J American Petroleum Institute—design and hazards analysis for offshore production facilities.
- *Operating company:* technical specifications and standards applicable to the offshore safety of each operating company.

7.3.2 Importing documents to build the 3D model

The FPSO project under analysis is recent and the original documents of the basic design were made available on MicroStation format. The 3D model editing software requires that the original documents be converted from MicroStation (.dgn) format to the Data Exchange Format (.dxf), so that they can be imported without major loss of information. Converted drawings from the physical, naval, architecture and safety disciplines resulted into a set of about 90 drawings that made up the input data for editing the FPSO 3D model under study. In addition to the original drawings, technical specifications and safety analysis reports were used to complete the set of information necessary for building the 3D model.

7.3.3 Building the 3D model

After the original design documents being imported into the software, they are automatically organized by level relative to the FPSO keel. The documents are also referenced through the definition of the origin point (0.0,0) in each drawing for the purpose of positioning in the 3D space.

Each document is associated with a FPSO deck level. The level of details required for building the 3D model is greater than what is available in the design documents, which makes it necessary to create intermediate decks, such as the intermediate staircase floors. This level of detail is very important for the 3D model to work perfectly, since all accessible locations on the unit must be properly interconnected. Any failure or technical inconsistency will void the application of the model and consequently impair all simulations.

After the documents were imported, the necessary adjustments and the creation of the intermediate decks were completed, the FPSO 3D model under study was built with a total of 45 decks. When editing the 3D model, the location of the muster stations (meeting points) is established based on the definitions of the original project. It was noticed that the muster stations for the 3D models used for escape simulations and the ones for abandonment simulations were defined distinctively. The operational procedures adopted in the FPSO under study establish that after the escape, people will gather in specific rooms in the living quarters. The gathering of people in the abandonment simulation is planned for the lifeboats (rescue boats) boarding area. Likewise, the decision for the abandonment via port or starboard lifeboats will also require different locations for the muster stations. Therefore three 3D models were needed to satisfy the different categories of escape simulations, port abandonment, and starboard abandonment.

7.3.4 Adaptation of the process plant area

The greatest difficulty encountered during editing of the 3D model for technical adaptation is in the oil and gas processing area of the offshore rig. The software used was originally developed for passenger ships and the process area is totally alien to the 3D model editor. It is not that software lacks the technical resources to represent the process area, but the understanding of routes, equipment, and the movement of people in offshore units is not readily available in the software editing interface. Solid field experience, operational knowledge, and in projects of offshore rigs

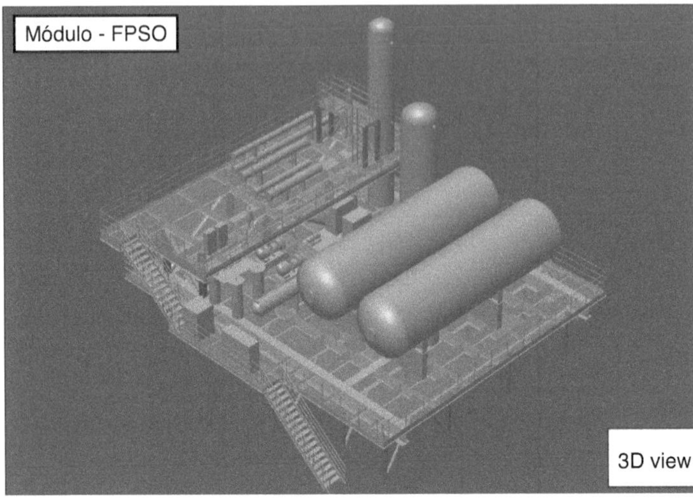

Figure 7.11 Detail of process module and escape routes belonging to a FPSO. *FPSO*, Floating production storage and offloading.

are required for establishing correspondences between modules, vessels, tanks, and numerous equipment from the offshore industry and the functionalities available in the software regarding the construction of the 3D model.

Some modules have levels and mezzanines that can often be occupied by people, while others, for safety reasons, are rarely visited. Ladders are not supported by the software, needed to be adapted since this type of access is typical in the process modules. Circulation routes that require bypassing numerous normal interferences in the field are not ordinarily supported by the software. In some cases, it was necessary to create equivalent routes to maintain the maximum correspondence between the actual rig and the 3D model.

The adaptation of the process area in the 3D model was done to an acceptable level of approximation that incorporates technical features of the actual rig, including all modules and all levels established in the FPSO project under study. Fig. 7.11 shows the schematic example of the internal arrangement of module 10 (water injection), including stairs and accesses.

7.3.5 Adaptation of FPSO hull internal areas

The areas underneath the main deck within the hull of the FPSO under analysis can be split into a set of smaller occupation areas at the bow and a set of densely occupied areas at the stern. This latter set can be subdivided further into several decks and mezzanines, being significantly compartmentalized. Most of the volume within the hull contains areas considered to be uninhabited for the purposes of FPSO operation. This area, contains the unit's large service tanks, which in this work are considered uninhabited.

Although there is some correspondence with the equivalent area from passenger ships, the aft decks within the hull have many differences regarding the distribution of rooms and equipment. For example, the FPSO under study had its main engine removed after the conversion from VLCC to FPSO. Many equipment and operating rooms typical of offshore rigs are located in this area in the unit and required adaptations of the software functionality to be modeled. Specific technical knowledge of the offshore field, hands-on experience in similar units and project knowledge is essential during the editing of the 3D model to preserve its technical association with the actual FPSO. Ladders are also widely used in the interior areas of the hull and several mezzanines and walkways have open grid floors, requiring specific offshore design knowledge for a precise interpretation of the drawings of the area, especially the interlinks that are indispensable for editing the 3D model, since these are not always explicit represented in the original drawings but only indicated for further interpretation by a specialist in offshore projects.

The adaptation of the areas within the hull was carried out successfully, preserving a technically acceptable level of correspondence with the original FPSO project, making it possible the editing the unit's 3D model to a high degree of representativeness.

7.3.6 Adaptation of the superstructure internal area

The superstructure areas (also known as living quarters) are those with closer correspondence with the software functionalities, since they are quite similar when compared with the arrangements of passenger ships. Although they can be edited more easily, they are also the most labor intensive due to the number of compartments and amount of details. The recreation, bedroom and work areas can be edited rather easily using the software built-in resources. Special attention is needed in the definition of muster stations, helideck access, and the cafeteria, as these are the most critical compartments in terms of people circulation during the simulations and require accurate adjustments in the connections with corridors, doors, and access stairs.

The adaptation of the superstructure areas was carried out with an acceptable degree of correspondence that was sufficient for the edited 3D model to be considered adequate for the simulations.

7.3.7 Definition of agents on board and their behavioral parameters

After editing the 3D model of the FPSO under study, it is ready to be debugged and exported to the simulation program (EVI), the software directly responsible for the definition of the scenarios and performing the simulation. The definition of agents on board (people) can be made by the software in several different ways, as well as setting agents attributes related to behavioral characteristics. Below, we present the adaptations made in the use of EVI to generate scenarios representative of escape and abandonment operations in emergencies in the FPSO under study.

7.3.8 People on board definition

The POB considered in the design of the unit was set to 110 people. EVI is a tool created for passenger ships with up to 7000 POB. However, the level of complexity of the tasks performed by people in an FPSO is higher than on passenger ships. Therefore it was necessary to adapt the use of the simulation program so that a comparatively small number of agents involved in more complex activities would not affect the quality of the simulations results.

This adaptation was done through normal usage of the software programming commands, driven by decisions based on real-world experience in offshore units and FPSO project experience, which allowed the program to work perfectly for POB of 110 people, taking into consideration the operational tasks and their influence on POB behavior.

7.3.9 Operational experience of agents on board

Basically, the population on board is split by the software into two major groups: crew and passengers. This division is incompatible with offshore oil and gas exploration and production units. As a workaround we replaced these two groups as follows: the first group is composed of the most experienced agents related to offshore activity and the second composed of the remaining agents.

For the purpose of performing the simulation, the following offshore experience evaluation criteria were created:

- On the specific knowledge of the unit under study

- o Agents with less than 3 years of work in the unit—no experience.
- o Agents with more than 3 years of work in the unit and less than 10 years of offshore experience—medium experience.
- o Agents with more than 3 years of work in the unit and more than 10 years of offshore experience—experienced.

  As the FPSO was studied it was still under construction, all agents were classified as inexperienced by the above criteria, since no one yet has specific experience in this unit. For this reason, we move on to the second criterion (applicable in new units) described below:

- *On the offshore experience*
  - o Agents with less than 3 years of offshore experience—no experience.
  - o Agents with 3–10 years of offshore experience—medium experience.
  - o Agents with more than 10 years of offshore experience—experienced.

In the case of the FPSO under study, since it was analyzed when it was still under construction, only the second criterion is valid, and for that reason, only people with over 10 years of work in any offshore oil and gas exploration and production units. For simulation purposes, a total of 25 people with highest offshore experience (23% of POB) were estimated. This value was estimated based on operational and project experience, as previously justified. The impossibility of collecting real data while the study is being conducted does not prevent future adjustments at any time, once the data is available. A similar procedure was also adopted in all other data estimates relating to agents, which will be described below.

7.3.10 Gender of agents on board

The Maritime Safety Committee, through Circular MSC.1/Circ.1238 (2007), provides statistical parameters related to the agents to be included in simulations such as those performed by the software. This is a specific standard for passenger ships and establishes split 50/50 between men and women. This premise is not compatible with offshore installations and, for the purposes of performing the simulations, the general distribution of agents by gender was adapted through an estimate of 86% men and 14% women, totaling 95 men and 15 women a board. This estimate was made based on the observations obtained from 10 offshore rigs embarkation records, and can be adjusted as more accurate data are available over the operational life of the rig. To complement the data preparation, we also consider that the group of agents with the highest level of offshore experience is composed of 96% of men and 4% of women, that is: of the 25 people considered with the highest level offshore experience, according to the criteria in Section 7.3.9, one of them is a woman.

7.3.11 Age of agents on board

The software assigns ages to agents as required by the Maritime Safety Committee Circular MSC.1/Circ.1238 (2007). Therefore the age assignment for passenger ships needs to be adapted for offshore oil and gas exploration and production facilities.

The Committee's Guideline establishes a uniform age distribution for three age groups, with the mean for each group being, respectively, 20, 40, and 60 years of age and standard deviation of 10 years. In addition, POB percentages are assigned to each of group, subdividing them by gender and personal agility, to a total of 10 groups for passengers and 2 groups for the crew.

The adjustment of this age distribution for application in the FPSO under study resulted in 3 age groups, with the mean for each group being, respectively, 30, 40, and 50 years of age, with a standard deviation of 10 years. The subdivision by gender and personal agility has been reduced to just six groups (for less experienced agents) and two lanes (for more experienced agents), due to more homogeneous distribution of the POB of an offshore unit than the POB of a passenger ship. The assignment of percentages concentrated most of the POB in the male groups with average of 30 and 40 years of age, as estimated based on the operational and project experience of FPSO.

7.3.12 Travel speeds of agents onboard

The software also assigns travel speeds to agents as required by the Maritime Safety Committee Circular MSC.1/Circ.1238. These parameters need to be adjusted for compatibility with offshore oil and gas exploration and production units.

For each group of people with the same characteristics of age, gender, and reaction time, the software assigns three speeds, according to the Committee's Guideline: speed of movement on level surfaces, speed when climbing stairs and speed when descending stairs. Specific values for each person are assigned statistically considering a standard deviation of 0.25 m/s. This enables people with the same characteristics of age, gender, and reaction time to assume different speed values considering the calculated mean and standard deviation, which makes the simulation quite realistic.

The adaptation of the speed parameters to values compatible with offshore oil and gas exploration and production units was done by adjusting these three speeds for each of the groups with the same characteristics of age, gender, and reaction time. The speeds defined in the Committee's Guideline were preserved, but only for the age, gender, and reaction time ranges compatible with the FPSO POB under study.

7.3.13 Reaction times for agents on board

The reaction times are assigned by the software based on the Maritime Safety Committee Circular MSC.1/Circ.1238 with different values for day and for night. This reaction time assignment is done through the normal Log function to reproduce the process-triggering effect popularly known by the term stampede (when there is a first initiative by an agent, the others tend to have similar reaction). The software allows the adjustment of the parameters: average, standard deviation and displacement (additional time). The mean and standard deviation values adopted in the Committee's Guideline were maintained.

However, the displacements were adjusted in order to account for the reaction time at night, which despite being in general slower than during the day, in the case of offshore units, they are not significantly different due to the state of preparation including during night time, as the result of professional offshore training.

Another adjustment vis-a-vis the passenger ship parameters defined by the Committee's Guideline was the inclusion of the displacement parameter (additional time) greater than zero including in daytime tasks, since offshore operators cannot always interrupt their job immediately, as in some cases they have to perform some minimal tasks to enable the start of the escape and abandon operation. Fig. 7.12 shows the software's user interface used for setting values and the statistical functions associated with agents that influence their behavior during simulations. Each agent can be considered an encapsulated set of mathematical functions in an FSA study.

7.3.14 Physical positioning of agents in the rig

EVI allows people both to be located individually (at the user's discretion) and using random distribution. In the first case, the user can position each person in specific locations, considering age, gender, work experience,

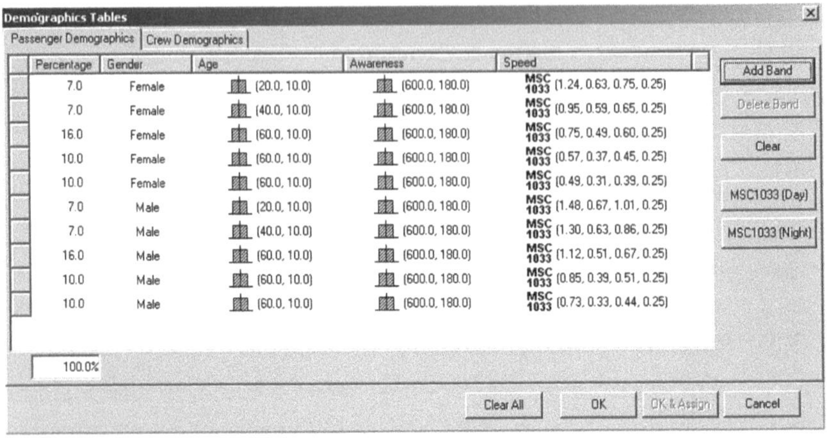

Figure 7.12 User Interface for setting human factor values for the POB. *POB*, People on board. *Courtesy Safety-at-Sea, Glasgow UK, http://www.brookesbell.com/service/software.*

reaction time, and individual specific speed characteristics. In the latter case, the software distributes the POB randomly in all areas of the unit.

As part of the analysis of the tool results, we identified vulnerabilities in the random distribution of agents, due to the software original application for passenger ships, normally with POB dozens of times greater than for FPSO offshore units, which creates some distortions for comparably small POBs. For this reason, we have adopted a mixed approach for the distribution of people in the unit.

Based on the operational experience and from FPSO projects, as well as the experience on board offshore units, we have created several possible configurations for agents distribution. Some are done totally manually, where each person is placed at a specific FPSO in the 3D model. Other random configurations, limited by areas, that is, for a given FPSO area, a number of people and their respective characteristics were defined, however, such agents were distributed randomly within that area, for example, on a deck. Mixed configurations were also defined, where certain agents with specific characteristics and tasks are previously positioned at strategic FPSO locations, while others are randomly distributed by area. Fig. 7.13 shows the positioning of the agents in the superstructure area prior to the beginning of a simulation. For example, agents wearing yellow vests are the most experienced and prepared, those with blue vests are for male agents and green vests represent female agents.

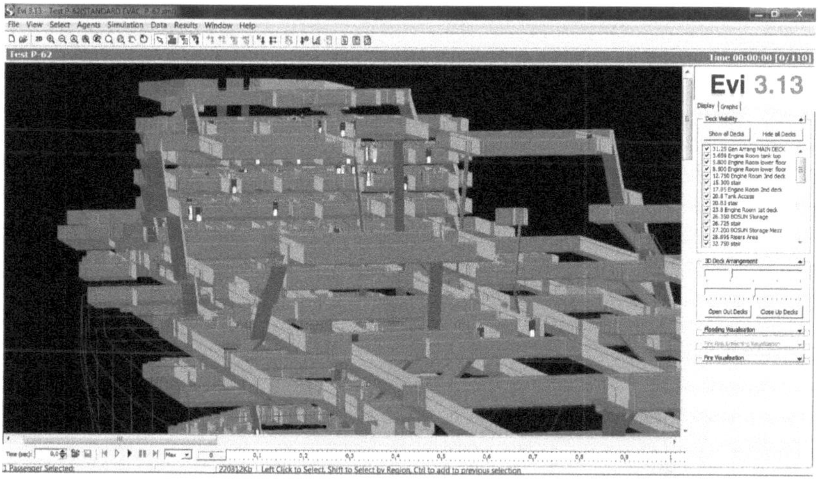

Figure 7.13 Example of agent positioning in the FPSO. *FPSO*, Floating production storage and offloading. *Courtesy Safety-at-Sea, Glasgow UK, http://www.brookesbell.com/service/software.*

7.3.15 Special tasks for specific agents during the emergency

The simulation software allows the creation of tasks to be performed in the middle of the escape and abandon operation. This means that, although the POB is performing escape and abandonment procedures, some agents may travel in the opposite direction for different purposes, unlike the rest of the POB who are performing tasks associated with emergency mitigation.

The software allows the programming of virtually any task, and its functionality can also be used for the purpose of improving operational efficiency outside emergencies, in normal operation related to tasks that involve movement of people.

This software functionality was explored, for example, to program the performance of two offshore rig operators. During a fire scenario, they were programmed to move against the flow, from the safe area of the FPSO to the process area, with the objective of manually identifying and opening a deluge valve that, due to some failure, has not been automatically opened by FPSO safety systems. Extra time can also be allocated to perform the task after the agent arrives at the area where the valves are located.

7.3.16 Measuring the effects of emergency on people's integrity

Each agent (person) under study through the emergency simulation is under the influence of the effects of temperature, visibility, and smoke intoxication recorded during the simulation. The parameters related to these influences are recorded by the software in real time. So, it is possible to identify the people affected at the end of the simulation through the analysis of the results, and also during the simulation when, for example, an agent is fatally injured, which is indicated via the color change associated with the physical condition of the agent.

For this functionality to be available, it is necessary to import data in CFD related to accident scenarios under study in a format compatible with the simulation software that includes the measurement of the effects of the accidental scenario on people. An optional configuration allows the insertion of a point cloud to represent the spread of fire and smoke in the area of the living quarters (which is the area with the closest similarity between FPSO and passenger ships) for the purpose of demonstrating this simulation feature. Fig. 7.14 shows the temperature grid in the zoomed in region of a fire (postulated to happen in the kitchen). The grid colors represent the temperature range, and when an agent is selected using the mouse its color changes to white, while a window shows the

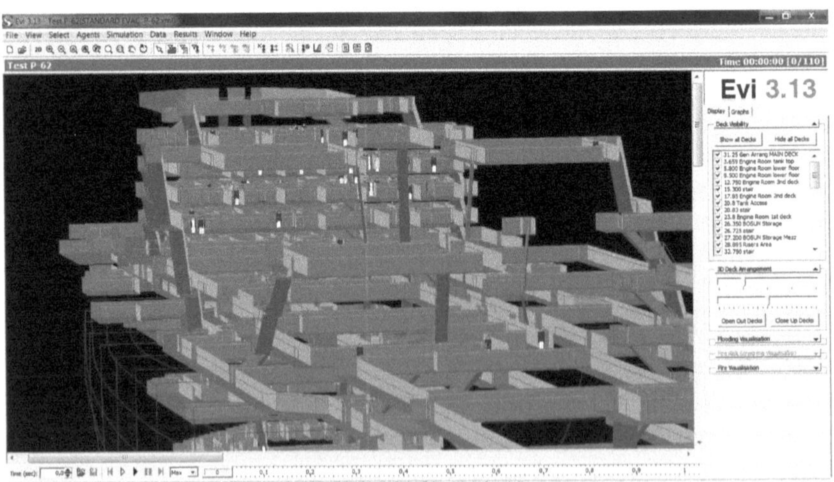

Figure 7.14 Presentation of properties and effects on the agent. *Courtesy Safety-at-Sea, Glasgow UK, http://www.brookesbell.com/service/software.*

characteristics and percentage of harm on the agent as a result of the effects of temperature, smoke, and loss of visibility.

7.3.17 Conceptual definition of accidental scenarios

The conceptual definition of scenarios is the most important part of risk analysis using computer simulations. The software and tools only process the information, premises, and parameters defined in the conceptual scenario of the emergency to be simulated.

Every conceptual definition of a scenario has its limitations. It is impossible to anticipate all the possible accidents, natural influences, human behavior, and the complex interaction among these factors. When we define a conceptual scenario for study or simulation, in reality we are establishing its limits and scope. In real-world situations there are no such limits, and any scenario not considered in the study can happen, which justifies the use of the term risk management as more appropriate than simply safety.

The greater the number of scenarios evaluated, and the influences considered in the definition of these scenarios, the less limited the results of the analysis of the associated risk management will be. Evidently, the work complexity grows with the increase in parameters, variables, and scenarios evaluated. Therefore there is a threshold beyond which computational tools are the only options for risk management analysis and studies to be viable. Only computational tools allow the processing of a large volume of data, its logical consequences, and the reactions of the interlocks to these data and consequences.

Considering an oil and gas exploration and production unit, with a POB of 110 people, the risks due to the large hydrocarbon inventory, human behavior, the operational complexity of offshore activities, maritime risks, and many others sources of influence, we can conclude that the risk and safety management of the escape and abandonment system without the use of computer simulation tools is extremely limited. But, even with the utilization of computer simulations, it is essential that scenarios be defined based mainly on operational experience, accident history, and prioritization of the most general and probable cases to be considered. The frequency of occurrence of a scenario can never be calculated with absolute accuracy. Despite the great effort towards this direction, the choice of scenarios based on history and operational experience continues to be much more reliable and realistic. The values calculated based on databases and statistical analyses can have a mathematical foundation, but it is not sufficient to ensure that such numbers have correspondence with the operational reality. There is no substitute for operational experience and technical operational knowledge when it comes to the definition of the best scenarios for risk and safety management.

7.3.18 Standard and gas leakage scenarios

These scenarios consider the leakage of gaseous hydrocarbons without ignition of the released volumes. It also considers that the physical damages to the installations are not severe enough to affect the efficiency of the abandonment means planned in the project and also that the concentrations of gases present are not capable of intoxicating the simulated agents.

Basically, a gas leak is supposed to be caused by damage or a human error, which justifies the escape order and if necessary followed by the abandonment. Therefore in this scenario, it was considered that all escape and abandonment routes are available and that there is no gas ignition during the accident or intoxication of people. For this reason, this scenario also corresponds to the order of escape and abandonment during training (standard escape and abandonment scenario) or for any other reason unrelated to significant damage to the FPSO. This scenario considers that all FPSO safety systems are available and operational.

In addition to the gas leak scenarios, the simulator also has the capability to consider toxic gas leak, such as H_2S, and to assess its impact or of any other gas on people during the simulation. For this purpose it is necessary to import a point cloud in the CFD software as described earlier, and whose data are representative of the leak to be simulated.

7.3.19 Fire scenarios

Fire scenarios can be simulated by software that uses CFD. The results of these simulations can be imported by transferring data in the form of a grid (point cloud), as previously mentioned. After the grid (point cloud) is imported, the program simulates the movement of people and the impact on them caused by the influence of temperature, toxicity, and visibility, based on the information contained in the grid (point cloud).

Based on this grid, the agents are monitored either when traveling or stationary in the FPSO for some reason. A monitor attached to each agent calculates the toxic gas absorption rates and accumulates the result, as well as monitors the visibility at their locations in the simulation model. Likewise, for temperature, as the agents' conditions are considered fatal, color changes for these agents are reflected in the simulation, along with an indicator of casualty. At the end of the simulation it is possible to obtain the data of heat, gas, and visibility difficulty for each agent, at each location, at each instant of the simulated event.

A simplification technique can be employed in fire simulation for the purpose of studying escape and abandonment systems. It depends on the conclusions of the consequences analysis. It is necessary that the reports of fire propagation, explosion and gas dispersion studies confirm that even in extreme scenarios, at least one main escape route will always be available and operational. When this conclusion is confirmed, it becomes possible to use a fire propagation and gas dispersion grid (point cloud) standardized by the Kelvin Hydrodynamics Laboratory—University of Strathclyde UK, to simulate the effects of a fire in the living quarters.

Another essential strategy in parallel with this resource is related to the simulation of escape and abandonment in fire emergencies. The FPSO project under study considers that a person shall not need to walk more than 7 m to access an escape route from any location in the FPSO at any time. In general, it also considers that there will always be two alternative options for escape for any agent on board the FPSO. In strategic terms, we consider that if a fire event starts at a location in the FPSO there will always be one available access to an escape route and the second one blocked by the accident itself, whether due to smoke, heat, or any other consequence of the fire.

Thus strategically, the options for access to the escape route for each simulated agent on the side affected by the accident are reduced to 50%. For example, if a fire starts on the port side, 50% of access to port side escape routes is blocked by the simulation. Similarly, if the fire starts on starboard, 50% of the accesses to the starboard escape routes are blocked. This way we simulate the difficulties experienced by the agents during the fire, considering that each agent will find an access to the escape route blocked because of the fire and the other available.

7.3.20 Naval damage condition scenarios

The naval damage scenarios were defined based on the naval damage condition established in the design as 16.5 degrees inclination angle.

All FPSO safety systems under study must necessarily be operational and available up to this angle value as established in the design phase by naval engineers. Moreover, the fire fighting systems and means of escape and abandonment also need to be available and operational to enable accident mitigation and the safe escape and abandonment of the offshore rig. The calculation of this angle allows the conservative conclusion that at 16.5 degrees the studied FPSO shall already be considered in a state of beyond design accident and therefore it should be abandoned.

From the original project documents of the FPSO, reports of movements, accelerations, stability, and naval damage were used to generate the effects of the maximum predicted naval damage condition. Safety systems (including escape and abandonment) must be operational and available in the maximum damage condition.

The Hydrodynamics Laboratory—University of Strathclyde UK prepared the simulation of the FPSO movements under study based on the documentation of its original project. Through this information, the EVI included the effects of the ship motions and its interference in the escape and abandonment operation (Figs. 7.15 and 7.16).

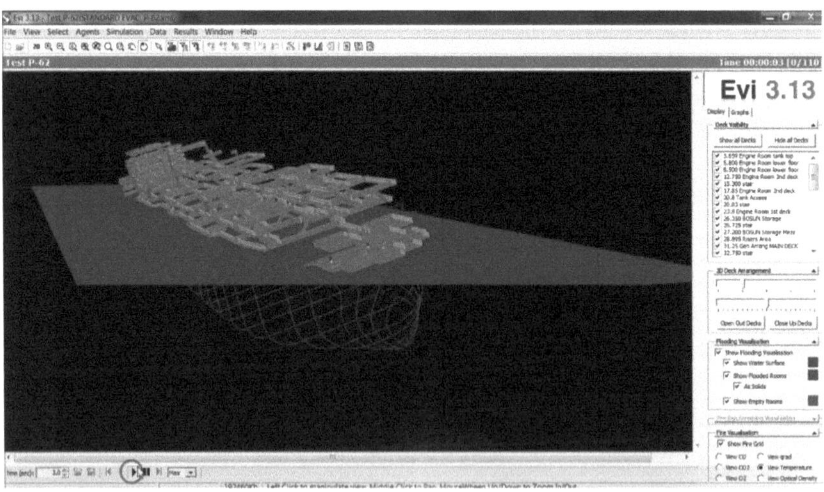

Figure 7.15 Simulated naval damage condition, with heel of 16.5 degrees. The simulation viewing angle allows the visualization of the movement of people in the bow region, the process plant, and the superstructure of the FPSO. *FPSO*, Floating production storage and offloading. *Courtesy safety-at-sea, Glasgow UK, http://www.brookes-bell.com/service/software.*

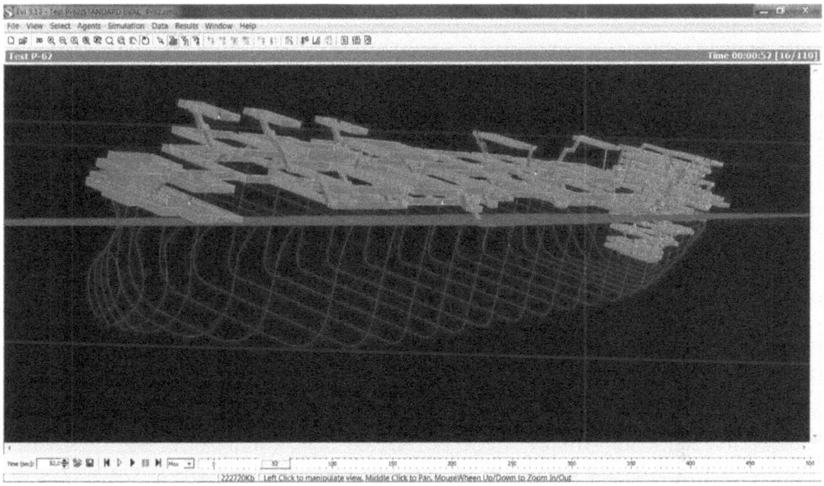

Figure 7.16 A demo video with a complete didactic simulation with a scenario of naval damage condition and fire in the houses in FPSO can be accessed via the link http://www.youtube.com/ (Gerardo Portela YouTube Channel). *FPSO*, Floating production storage and offloading.

7.3.21 Theoretical scenarios for comparative purposes

Special scenarios were created for unlikely situations, whose simulation results are of interest for risk management and safety as a reference for comparisons. Examples were scenarios created respectively based on 100% concentration of POB in the bow and in the stern, situations that are rather unrealistic operationally.

In this type of scenario, people were concentrated in the most difficult areas to access in the FPSO under study. The movements to the meeting points were studied. The objective is to obtain conservative results as a reference. As these are extremely unfavorable and improbable situations, their results can be used in comparisons and to identify possible distortions in other simulations that require corrections.

7.3.22 Representative simulations for offshore rigs

In offshore risk and safety management, it is not postulated that two independent accidental scenarios occur can occur simultaneously. It is possible to consider in offshore safety analyses, though, the cascading effect from one scenario to another, as the first scenario will gradually increase the degradation of the FPSO until the second scenario starts. That is, one scenario can cause degradation and trigger others, but it is not postulated that, from a normal operating standpoint, two independent accidental scenarios can occur simultaneously in the same study.

For this reason, each postulated scenario should be studied separately. The software has a functionality that allows repeated simulations in batch mode. This means that the same simulation or set of simulations can be repeated as many times as intended, where specific results are generated for each of the repetitions. The non-deterministic nature of the results for each repetition of the same simulation are possible due to the use of random variables established in each scenario, such as age, reaction time, speed, etc.

In this study, 1000 simulations of different escape scenarios were carried out, 1000 simulations of different abandonment scenarios, and 2000 comparative simulations between scenarios. For each batch (set of simulations based on the same scenario), the mean and standard deviation of the results were calculated. The maximum duration time of each scenario was considered to be the mean time value plus three standard deviations, which based on the Gaussian distribution, means a maximum error of 0.03%. The number of repetitions of each scenario was defined based on the relevance of its influence in terms of determining the maximum escape and abandonment time for the FPSO under study. During the repetition process in batches of simulations, if the initial estimate of the relevance of a given scenario is not confirmed, the number of repetitions is adjusted to a value compatible with the maximum results obtained in the batch, that is, if the results of a scenario are more relevant than expected, the number of repetitions is increased accordingly.

The complete results, conclusions and recommendations generated by the FSA study for the FPSO used as an example are presented in detail in the book *Risk Management Based on Human Factors and safety culture*, published by Elsevier, by the same author. Figs. 7.17–7.19 show a summary of the results through some illustrative graphs obtained through the analysis.

7.4 Lessons learned

7.4.1 Risk analysis and team work

Qualitative or quantitative risk analysis techniques, most often require the efficiency of team work. Many techniques rely on meetings with several

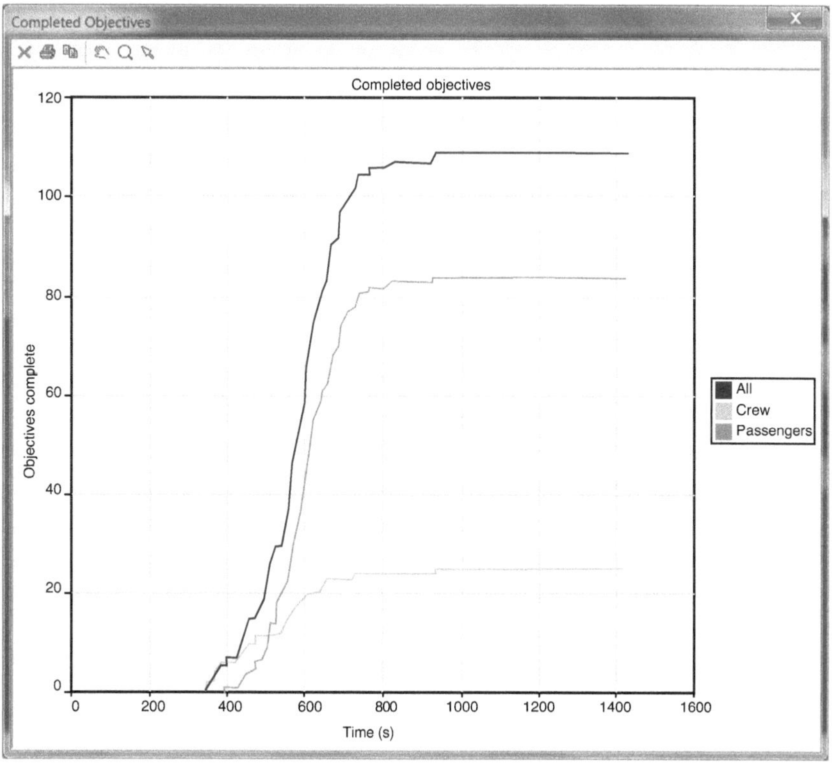

Figure 7.17 The graph shows three curves: one with the least experienced operators (passengers); another with more experienced operators (crew) and a third curve with the sum of all FPSO operators (all). On the abscissa axis, it is possible to identify the time values in seconds until each operator fulfills all pertinent tasks during the escape operation and reaches the meeting point (muster station). On the ordinate axis, we have the number of operators that have already completed all the tasks expected for the escape operation. *Courtesy safety-at-sea, Glasgow UK, http://www.brookesbell.com/service/software.*

experts, led by a leader with adequate training to guide the team to prevent that the discussions lose the focus on the objective of obtaining an effective improvement in the risk management of the technological enterprise. It is also applicable to quantitative techniques as well, for example, based on computer simulations, it is impossible to carry out a high-level risk analysis without meetings with experts from various fields to adjust the simulation model and for discussions on the best strategy for the use of the data to be considered.

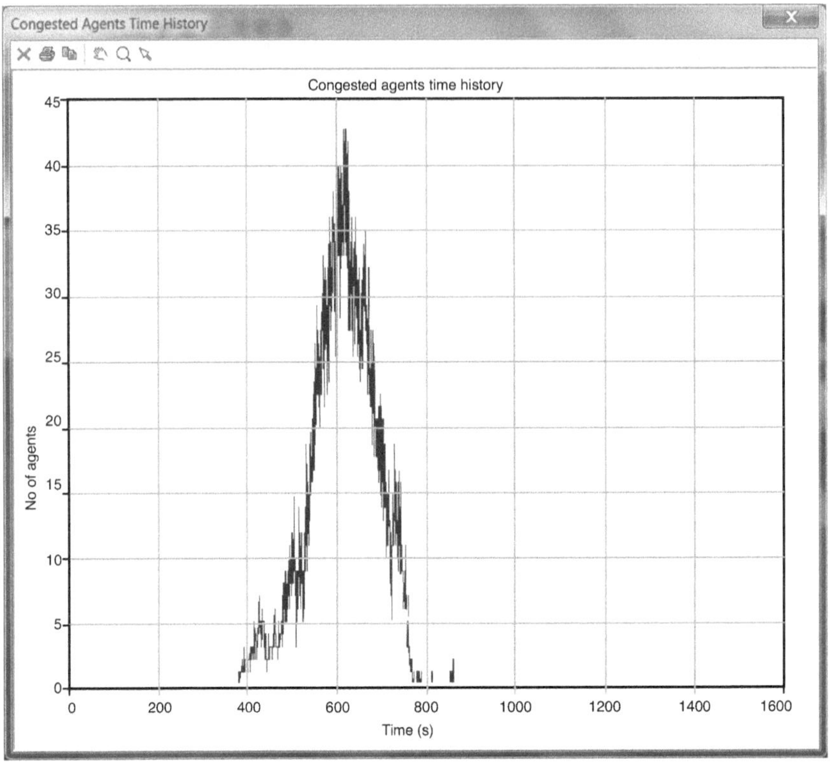

Figure 7.18 The graph shows the number of congested agents (with movement speed less than the minimum acceptable) at each instant of the escape operation and movement to the meeting point (muster station). The abscissa axis shows the time in seconds from the start of the escape and abandon operation. The ordinate axis shows the amount of congested agents. It is possible to observe that the peak of agent congestion occurs between 400 and 600 s after the start of the escape and abandonment operation, which varies between 40 and 45 operators moving at a speed lower than the acceptable minimum. Other functionalities, such as 3D simulation, also allow the identification of specific locations where congestion occurs, such as doors and hallways. This type of information can also be shown graphically. *Courtesy safety-at-sea, Glasgow UK, http://www.brookesbell.com/service/software.*

As we have stated earlier, every risk analysis includes, whether in the application of the technique or in the premises that its application is based on, elements of subjectivity to a greater or lesser extent depending on the type of technique employed. We remind the reader of the example mentioned of two passengers on a flight, one feeling "unsafe" and the other feeling "safe." This is obviously possible for both, regardless of being on the same aircraft, with the same crew, in the same weather conditions and

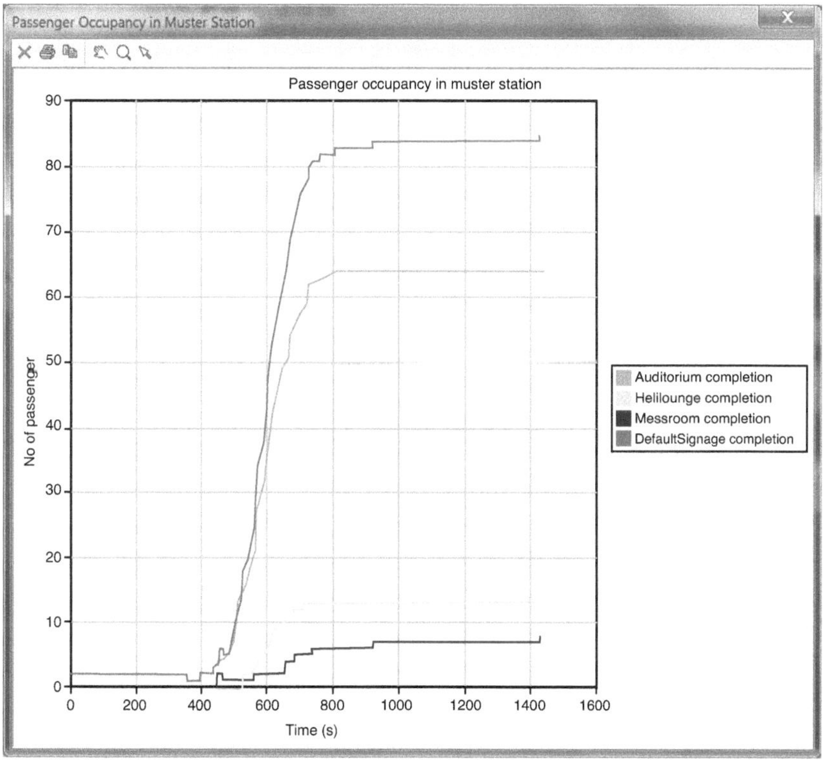

Figure 7.19 This graph shows the complete evolution of the occupation of the three FPSO meeting points under study. The abscissa axis indicates the evolution of time (in seconds) and the ordinate axis the number of agents. One of the curves shows the occupation of the meeting point in the restaurant (messroom); a second curve shows the occupation of the meeting point in the helideck waiting room (helilounge); and a third curve shows the occupation of the meeting point in the auditorium. A fourth curve shows the performance of the total occupation of all meeting points combined. The graph shows that the operation lasted between 1400 and 1500 s. It is also possible to note that the last agent to complete the movement occupied the meeting point of the restaurant (messroom). The graph also shows that the meeting point with the best performance, being occupied in the shortest period of time (between 600 and 800 s), was the one from the helideck waiting room (helilounge). *FPSO*, Floating production storage and offloading. *Courtesy of Safety-at-Sea, Glasgow UK, http://www.brookesbell.com/service/software.*

participating in the same flight experience. For one of the passengers, the knowledge and information available is sufficient to feel "safe" while the other is not. Keeping in perspective the comparison between the technical background of the passengers in the example and the experts who carry out risk analyses, organizations and professionals have different safety cultures and different positions regarding risk acceptance. In a qualitative analysis, a given scenario can be considered under acceptable risks for one expert and unacceptable for others. In quantitative analyses, an expert can accept the frequency values for a given scenario obtained based on international databases, while other experts may disagree and suggest other means to obtain frequency values of accidental scenarios. There is subjectivity in both cases. The subjectivity aspect can be more explicit and transparent in qualitative analyses, whereas in quantitative analyzes, it is hidden deep in the four-decimal-place accuracy. Certainly, even this paragraph can be considered controversial for those who admire the methods of quantitative analysis and for that reason "will more easily accept the risks" involved in the quantitative technique, which are not error-free. It is possible to imagine, then, how much discussion and divergences of opinions can happen during risk analysis studies, when diverse interests of the stakeholders exert great pressure on the results to be expected.

7.4.2 HAZOP chaos

The HAZOP technique is recognized as one of the best tools for organizing complex group work, especially when applied under the leadership of a competent professional. But when the safety culture is not going well and when risk management quality is not the highest, the application of HAZOP technique can turn into a real drama, with all the typical elements of a melodramatic story, including special appearance from actors in the roles of villain, "good guy," hero, guilty, "country's savior," "knows it all," innocent, in short, a stage of vanities, frustrations, and personal achievements—a little hell.

One of the mistakes that lead to this ultimate failure in the application of the technique is the execution of a project with insufficient communication among the disciplines, where each designer works independently behind closed doors trying to guess what the other designer is doing, but without a healthy open communication channel that encourages continuous information exchange. When it happens, the designers start to see in HAZOP a great opportunity to push their own strategies, solutions and visions and their way of executing the project, in direct confrontation with the other team members. But this is not the HAZOP purpose by any means. It is a tool designed to identify hazards and risks associated with operations that will be performed in the future, during the operational life of the technological enterprise.

HAZOP's objective is not to "improve the project" or "to extend operational comfort." These are constant objectives throughout the normal project work and not only on days planned for gathering of experts to reduce the unpredictability of accidents. But it is exactly this special circumstance—the meeting of several experts—that can create an expectation of a "golden opportunity" to expose their ideas and express their dissatisfactions, for those who are not familiar with the use of the HAZOP technique.

On one occasion, in a large semisubmersible (SS) offshore rig project, about 40 specialists (plus support staff) were gathered to carry out the final HAZOP of the basic design. The environment was not being controlled and the climate was quite unpleasant, with the expectation of several confrontations among the stake holders. And indeed it happened. A well-executed HAZOP can last from 1 day for very small projects, up to 2 or 3 weeks for complex projects or more than a month for large, highly complex projects.

HAZOP lasting longer than 2 weeks deserves special attention for the application of the tool to be productive. But in this particular case of this SS rig, after three days since the start of the event the first node had not yet to be analyzed. For the most the first two days were spent on discussions on the premises that should have already been agreed upon with a certain level of consensus. But on the third day, after more objective discussions about the process flowcharts started, the discussions got heated to the point where the event turned into a near real chaos with exchanges of accusations and fights related to the project that had been lurking for months. To make it worse, at each round of provocation exchange the experts called their managers as if they were de facto generals of a technological war, capable of supplying arguments to the theories defended by each expert. The situation deteriorated to the point that, although HAZOP was being held outside the company's office environment, at the company's headquarters, everyone kept talking about the discussions that were taking place at the HAZOP and even at a distance parallel fights started also in the headquarters where meetings were being called, becoming "the generals" parallel HAZOP." This complete mischaracterization of the event turned the scenario into a real "HAZOP CHAOS." During those days, professional alliances were formed and broken up, driven by an environment completely affected by the inability of application of the technique. Apparently for that group, genuine technical knowledge had lost the importance it deserves.

Finally, the confusion reached the top management and, lo and behold, the event that included more than 40 people, with rental expenses, travel, consultancy, leader, and coordinator professional fees had an unexpected outcome. The HAZOP leader received a phone call from top management with the order to adjourn the event immediately. Almost two days had passed without any activities, until new orders were received stating that out of the 40 or so participants, only 9 should remain at the event. The remaining participants would return to headquarters. New premises were also sent by the headquarters, considered based on the opinion of risk management experts at the headquarters, which simplified the whole process and subsequent replanning and drastic reduction of the event's schedule. From then on, the application of HAZOP flowed normally, based to good engineering practices, normal discussions aligned with the search for technical results, free from personal vanity, and disputes among groups.

As lessons learned, we can say that behind technological enterprises and all their complexity are people with their idiosyncrasies. And behind all the risk analyses that support the viability of technological ventures, there is always a major component of subjectivity, even when all the mathematical models and computer simulation tools employed are taken into consideration. Only operational experience, deep technical knowledge disseminated by all the specialists involved can lead to high-quality risk management engineering results. Experience and knowledge help professionals to deal with this delicate web of human and phenomenological relationships. This task needs to be executed in a productive and efficient for the fulfillment of the final goal: a safe technological enterprise. Therefore a great deal of study, technical knowledge, operational experience, in addition to an adequate level of team work skills are the requirements for success in the application of qualitative and quantitative risk analysis techniques.

7.5 Exercises

1. Identify the technique associated with each topic:
   A—Quantitative Risk Analysis
   B—Qualitative Risk Analysis
   C—PRA
   D—HAZID
   E—HAZOP

F—Brainstorming
G—Checklist
H—FMEA
I—Swift
J—LOPA
K—Fire Propagation Study
L—Study of Dispersion of Gases and Smoke
M—Explosion Study
N—EERA
A—Analysis of Loss of Liquid Containment and Environmental Control
P—Stability and Condition Studies of Naval Damage
Q—FSA

1. () Technique used to identify the effects of shock waves on the structural components of a facility, as well as the efficiency of existing safeguards.
2. () This technique is recommended for analysis and simple activities in terms of mental complexity.
3. () It is only possible to apply this technique when the project is at a maturity level that allows the analysis of pipeline and instrumentation flowcharts.
4. () Based on the deep knowledge on the phenomena related to the operation and the project of the technological enterprise.
5. () A more complete risk analysis technique, which allows to combine in a single analysis tool the results obtained from several previously applied risk analysis techniques, including grid (point clouds) from CFD study.
6. () It makes possible the use of a simpler methodology regarding HAZOP, albeit with limitations.
7. () This type of analysis provides information about the effects of temperature on the structure and on the equipment of the rig under study.
8. () Analysis that evaluates the efficiency of escape and abandonment systems in a simple but limited way.
9. () Simple technique recommended for situations where the information available for analysis is very scattered and poorly organized.
10. () The results of this type of study allow the verification the availability of safety systems such as FWP pumps in specific accidental scenarios of offshore rigs.
11. () Systematic and rigorous technique capable of producing detailed and audit-ready records regarding hazards identified in the processes.

12. () Technique that identifies scenarios, their causes and consequences, generating basic information for further studies.
13. () Used to assist in providing supplemental information for the location of gas detectors.
14. () Quantitative technique that allows the determination of theoretical risk reduction values required for a facility or system to be adjusted to an acceptable risk level.
15. () This type of technique makes it possible to investigate the effects of a dispersion spill in a body of water.
16. () Hazards of accidents of source external to technological enterprise are considered, such as natural disasters.
17. () Techniques that "attempt" to present risks through numbers based on mathematical modeling.
18. () Methodology based on the concept of protective layers.
19. () During the application of this technique the leader should try to avoid hindering the creativity of the experts involved.
20. () Study that uses spreadsheets and checklists to assess difficulties during escape and abandonment operations.
21. () Analysis that uses CFD to show the influence of temperature in oil spill scenarios with ignition.
22. () Also known as the "What if" technique.
23. () Uses checklist for easy and quick application.
24. () Mandatory risk analysis technique in most projects related to hydrocarbon process facilities.
25. () Tool capable of considering human factors aspects and safety culture to obtain improvement results in risk management.
26. () Analysis that uses computational tools and indicates values of overpressures on equipment and structural components of the facilities.

7.6 Answers

1M; 2G; 3E; 4B; 5Q; 6I; 7K; 8N; 9F; 10P; 11H; 12C; 13L; 14J; 15O; 16D; 17A; 18J; 19F; 20N; 21K; 22I; 23G; 24E; 25Q; 26M.

7.7 Review questions

- Explain the component of the risk management strategy line: "unpredictability reduction?"
- What are the differences between qualitative and quantitative risk analysis techniques?
- What are the main limitations of risk analysis techniques?
- At what stages of the technological enterprise are PRA, HAZOP, and Consequence Analysis techniques recommended respectively?
- Explain the methodology adopted in the PHA technique?
- What is a risk classification matrix and provide examples of techniques that use this resource?
- What does the term ALARP mean?
- What are the main differences between HAZID and HAZOP techniques?
- What documents are required to apply HAZOP?
- Is HAZOP an instrument for the improvement of operational comfort? Why?
- What are "guide words?" Provide examples.
- What are the main categories of failures identified in HAZOP? Provide examples.
- What precautions does a HAZOP leader need to take regarding control failures?
- What are the main control failures? Provide examples.
- Explain the process of the establishment of premises that precedes the application of the HAZOP technique?
- What is a "node" and what is its purpose in the HAZOP application strategy?
- In what cases can brainstorming be used as a risk analysis technique?
- What is the checklist validation process and when should it be carried out?
- What are the main advantages and disadvantages of the checklist technique?
- Explain the FMEA methodology and the situations where its application is recommended?
- What are the main advantages and disadvantages of the FMEA technique?

- What aspects need to be evaluated in each subsystem studied by the FMEA technique?
- What are the main advantages and disadvantages of the SWIFT technique?
- Provide example of a strategy for applying the SWIFT technique and describe its steps.
- Explain the concept of the protective layer?
- What are the advantages and disadvantages of the LOPA technique?
- Explain the term "consequences analysis?"
- Provide examples of scenarios analyzed in fire propagation studies.
- Provide examples of the application of computational fluid dynamics in consequences analysis.
- What are the limitations of studies that utilize CFDs and how should they be circumvented?
- Provide examples of scenarios analyzed in the studies of dispersion of gases.
- What are the expected results from the explosion studies?
- Explain the meaning of the term EERA.
- What premises need to be made before applying the EERA technique?
- What are the limitations of the EERA technique and which new technique allows more complete results for the same subject matter?
- List the items that should be evaluated in the checklist included in the application of the EERA technique.
- What are the most important results expected in the analysis of loss of liquid containment and environmental control?
- What type of mathematical model is used for spill analyses on bodies of water?
- What types of environmental impact are investigated in the environmental risk analysis for water bodies?
- What accidental scenarios can be described as examples for the investigation of spills in bodies of water?
- What are the possible results based on the stochastic model for analysis of environmental risks resulting from spills in bodies of water (with respect to the number of events and in the time domain)?
- What are the possible practical results from studies and analyses of environmental risks resulting from spills in bodies of water?
- What is the purpose of the studies of stability and condition of naval damage regarding the safety systems of an offshore rig?

- What are FSA studies?
- How is the approach to risk management problems based on FSA studies formulated?
- How are 3D models for FSA studies built?
- Why are FSA studies based on escape and abandonment systems?
- How are Human Factors and Safety Culture aspects considered in FSA studies?
- How can agents be positioned in a 3D model for an FSA study?
- Explain the incorporation of special tasks related to POB components in FSA studies.
- How are the physical effects on agents measured in an FSA study?
- Explain why agents can be considered a encapsulated set of mathematical functions in an FSA study?
- How are accidental scenarios defined in FSA studies?
- What is the margin of error of the results and what results can be obtained via simulations performed based on FSA studies?

CHAPTER 8

Human–system interaction

Human–system interaction is becoming increasingly important for risk and safety management. Accident investigations (Section 9.8) often point to deficiencies in the human–system interaction process as causes of accidents. The investigation of some accidents identifies human error as the root cause, whose consequences have been aggravated by a human error–inducing environment. As already discussed in Chapter 2, Fundamentals of Risk Management, the entire technological enterprise in some way creates an environment of human–system interaction, which can be susceptible to human error to a greater or lesser extent. Dealing with the aspects that influence this error-inducing environment is fundamental to the success of technological enterprises.

Projects based on the technological state-of-the-art require to take the subject matter into consideration from the conceptual phase to decommissioning (end of the technological enterprise's operational life). One of the difficulties encountered in the implementation of improvements related to human–system interaction is the subjectivity associated with human behavior, which needs to be considered within a scientific methodology compatible with engineering projects and activities. The book *Risk Management Based on Human Factors and Safety Culture* (2014, Elsevier, by the same author) addresses the topic in detail and includes a practical example of the application of the risk analysis technique full safety analysis in a floating production, storage, and offloading system. This type of analysis provides the framework for the systematic consideration of the influence of Human Factors and the safety culture in risk analyses.

In the following section, we will present some basic concepts related to human–system interaction, to offer some guidance to experts in risk management. They should look for practical ways to reduce the error-inducing environment in their technical activities, taking into consideration the influence of Human Factors and the safety culture as part of setting up the environment.

Risk Management in the Oil and Gas Industry.
DOI: https://doi.org/10.1016/B978-0-12-823533-1.00012-5

8.1 Human error

Regarding risk management we can say that human errors are the failures made by people regarding their responsibilities in the activities of interaction with systems, machines, and equipment. From the point of view of risk management engineering, human error is not associated with guilt or deceit, as it may seem to a layman or to specialists in other fields of knowledge, such as Criminal Law. Making mistakes is a natural phenomenon associated with human uncertainty related to the perfect execution of future tasks, and is no longer considered uncertainty after a task is not perfectly executed. Inaccuracy in the execution of tasks and functions is present in several other situations encountered in nature, even if not directly linked to human beings. In other words, human being makes mistakes and it is part of human nature. Even the best trained and skilled operator will eventually make a mistake. This is not just limited to professional activity, but is part of one's own life. However, exactly because making mistakes is something natural for human kind, engineering can contribute by minimizing the error-inducing environment, or at least reduce the consequences of these errors when they happen. Making mistakes is a natural trait of people. The evolution of the consequences of these errors leading to a catastrophe, however, is not a natural failure, but rather an escalation of events that constitutes an engineering failure.

As a didactic example, we can understand how human nature is closely associated with error through one of the world's most popular sports: soccer. Much of the interest in sports is motivated by the parallel between sports competitiveness and the real situations of people's daily lives. In a soccer match, unlike most other sports, three different results are possible: the victory of either team or the tie. But some matches cannot end in a draw, such as a championship's finals. When despite the established rules the game ends without a winner, then the final decision of the soccer match is obtained through a system known as penalty-kick shootout. These are the famous direct, close-range free kicks to the goal, with remote chances of defense by goalkeepers. Even with the great advantage for penalty shooters, there is no record to our knowledge of decisions of this type that have not ended up in a tie-breaker. Even at the top-level championships, such as at soccer World Cup finals, with the best players specialized in penalty kicks, a fact is uncontested: at some point, some of the penalty shooters will make mistakes, resulting in the defeat of their team. Note that similar situation is replicated in several other sports, and sports are interesting because they take place in an environment free of serious consequences, unlike the risk situations that we are all exposed to in our daily professional and personal lives. The occurrence of human error is expected even for the best experts in their fields.

There are several ways to classify human errors. With respect to the occurrence mechanism, the error can be classified as:

- Lapses: Failures based on *skill* requirements, for example, skipping a step.
- Lack of Training: Failures based on *noncompliance with technical rules*, for example, improperly opening an equipment cap.
- Lack of Knowledge: Failures based on *cognitive errors*, for example, misdiagnosing a problem.
- Violations: Failures by *deliberate action in violation of the prescribed procedure*, for example, disabling an alarm.
- Socio-technical: Failures caused by the *influence of culture and operational climate*, for example, declining confidence to make decisions in the midst of an emergency.
- Managerial and Organizational: Failures attributed to the *wrong managerial decisions and actions*, for example, failure to implement safety policies.

The study of human error, from the risk management perspective, is important because the high likelihood of occurrence of human errors in the processes, since people are the key components of the processes. By studying human errors, it is possible to develop methods to assess the risks of occurrences of these errors, as well as making it possible to develop systems aimed at reducing the vulnerability of technological enterprises regarding human error and its consequences.

The results of accident investigations point to human errors as one of the most frequent causes. Desiring safety is a matter of common sense, everyone wants safety. But to achieve safety, more than common sense is needed, technique is needed. Many human errors are made as a result of an error-inducing environment, whether due to physical issues or behavioral issues such as stress and managerial pressures. It is also necessary to consider the value of safety regarding to the organizational culture or the society in which the organization is inserted. Numerous factors can be misleading which makes the Human Factors theme multidisciplinary and complex. Even the lack of technological knowledge can be an error-inducing factor when we are inserted in a highly technological society, which requires the knowledge of technological tools. Those who are unable to keep up with the speed, with which technology evolves, end up standing apart, without the minimum knowledge about all necessary processes. This creates, in some circumstances, a group of people without the minimum training to deal with all the new technology that appears every day. Unfortunately, those who fall into this situation end up being considered as a group limited by the lack of technical knowledge to perform the tasks of their minimum routine. Some experts refer to this phenomenon as "technological illiteracy."

8.2 Human factors

Many experts consider that the majority of industrial accidents can be attributed to human error. Human failure analysis deals with the failures that people can make in their interactions with engineering processes. The earlier the human error analysis is conducted, the greater its efficiency in reducing the likelihood of human error, so it is important that an approach based on Human Factors (error-inducing factors) analysis be adopted since the design phase.

Human failures and their consequences are directly influenced by the Human Factors Project of the technological enterprise as a whole. We call Human Factors those that can increase or decrease the possibility of mistakes by humans, where such factors are established as a result of a technological project or enterprise. That is, human error may or may not happen depending on the Human Factors involved in the human–machine interaction created by the technological project or enterprise.

Human Factors are those that make up the human–system interaction environment that affect the likelihood of people making mistakes. As examples of Human Factors we can mention communication, indicators on computer screens, equipment identification tags, the use of color markings, alarm announcements, presentation of procedures, aspects related to the ergonomics, culture and personal life, among others.

Every machine or facility designed, be it is an industry, a car, a building, a cell phone, a notebook, an airplane, or a video game, in short, any equipment, facility, or technological enterprise, always interacts with human beings in some way. The anticipation of human–machine interaction can studied and designed for aiming at maximum efficiency and safety, or it can simply result from the natural project progress without receiving any specific attention. The factors involved in this human–machine interaction are called "Human Factors," and the study and design of their adequacy allows protection against the human error–inducing environment, which is the main cause of accidents identified by official investigations.

With respect to the term ergonomics, it has been most commonly used to identify aspects of Human Factors that may result in impacts on people's health as a result of the interaction of each agent with the designed environment.

A "Human Factors" predecessor, Ergonomics emerged to address the design and operational problems that arose as the result of the technological advancements of the 20th century. Its precursors were scientific management developed by Taylor and the study of the work developed by Gilbreth. It is a hybrid discipline, which had its origin when scientists started to work in tandem to solve complex and multidisciplinary problems. The main scientific fields that contributed to ergonomics are: engineering, psychology, anatomy, physiology and physics (mostly mechanics and environmental physics). Ergonomics was also especially influenced by the following emerging disciplines: industrial engineering, industrial design and systems theory. Several trends can be identified throughout the Ergonomics development process. First organizations tried to improve productivity by introducing new methods and machines. In the era of pure engineering, this worked because there were many opportunities for technological development, given that mechanization processes was in its infancy. Subsequently, attempts were made to increase productivity by optimizing task designs and reducing unproductive efforts. After the First World War, a new movement stimulated the development of psychological tests to measure various human traits such as intelligence and personality.

In bibliographic references, it is possible to identify books including texts on the history of Ergonomics and Human Factors which we recommend as a complementary reading for those interested in more information about the origins of this knowledge. We can summarize some essential points of this history, since 1857 when Jastrzebowski produced a philosophical treatise on ergonomics: The Science of Work. In Great Britain the field of ergonomics started after the Second World War. In the early days, there was some confusion regarding the similarity of the term "ergonomics" to "economics." The word "ergonomics" was coined independently by Murrell in 1949, despite fears that people would mistake the term for "economics." Even today, many people confuse the two terms and there are those who do not understand the meaning of ergonomics. The emphasis of ergonomics was on the design of equipment and workplace. Ergonomics is a multidisciplinary, complex area that causes even more confusion in understanding the meaning of ergonomics. The relevant fields were anatomy, physiology, industrial medicine, design, architecture, and engineering. In Europe, ergonomics was more closely associated with biological sciences. In addition to the physical issues associated with human work, there are cognitive issues that also influence the performance of each work activity. In the United States a similar discipline known as "Human Factors" emerged, but its scientific route was anchored in psychology (experimental and applied psychology, psychological engineering, and human engineering).

Undeniably Human Factors and Ergonomics have always had significant similarities, but their developments have followed different paths. Ergonomics has been more associated with the development obtained in Europe and Human Factors is more associated with the development obtained in the United States. Human Factors place much more emphasis on the integration of human aspects into the global systems design process. In the United States, this approach has gained recognition from reputable organizations. It achieved remarkable success in the design of large systems in the aerospace industry, especially through NASA and the US Space Program. The recognition of such scientifically respected institutions helped to validate the methodologies developed in the area of Human Factors. European ergonomics is more fragmented and has traditionally been more associated with basic sciences, biological and anthropometric themes, being limited to a specific subject matter or specific area of application and less multidisciplinary-oriented.

It is a big mistake to understand that there are conflicts between Ergonomics and Human Factors. Both approaches are perfectly aligned and in some cases complement each other. Despite some differences, there should be no concern regarding the use of both terms: "Ergonomics" and "Human Factors." In the United States, the HFS (Human Factors Society) changed its name to HFES (Human Factors and Ergonomics Society). It can be inferred that this change was made to signal the affinity between the areas, justifying a single society to represent the interests of those who identify themselves as advocates in both areas.

Each year, with the development of both the areas of Ergonomics and Human Factors, it promotes greater alignment, greater technical consistency between the two forms of approach. Currently, both Human Factors and ergonomics areas adopt the AWH (Adapt Work to Human) approach in replacement of the old and outdated AHW (Adapt Human to Work) approach, that is, both approaches seek to develop a work environment that is more receptive to the variability of human behavior as well as more receptive to different biotypes. AWH establishes that work must be suitable for people rather than other approaches that, although close to this concept, do not consider that the greatest efforts should be devoted to modifying the environment and the workplace and not people. This is an important foundation of his philosophy.

There is a relative correspondence between Human Factors analysis and human error analysis, although they are definitely different things. The first investigates the factors that influence the interaction between people and systems (including equipment and processes). It also considers people's interaction. In a different way, the latter investigates the possible failures that people can make in their interfaces in the engineering processes.

These failures and the frequencies of their occurrences are directly influenced by the Human Factors that make up the environment.

Some of the types of approaches and studies related to Human Factors are:

- *Human error analysis*: identification and systematic assessment of the possibilities of errors by engineers, technicians and other professionals familiar with the facility.
- *Human reliability analysis (HRA)*: involves the evaluation of the impact of human action on the processes of an industrial plant in terms of the likelihood of occurrence of errors.
- *Human Factors Engineering*: analyzes the interface between people and processes and the impact of this interaction on the systems operation.

The role of human factor engineering is to attempt to make changes to the "things" that people use and the environments where these "things" are used. It aims at a better way to combine these "things" and the "environments" where they find themselves with people's abilities, limitations, and needs.

Human Factors engineering increases the effectiveness and efficiency with which the jobs and tasks are performed to enable fatigue and stress reduction, Human factor engineering promotes the reduction of mistakes which improves safety, health, productivity, operability, besides the overall improvement of the technological enterprise quality.

Some principles can be established regarding the application of Human Factors in the field of engineering:

- Equipment, facilities, and technological developments are built to serve human kind and need to be designed with the users in mind.
- There are differences between individuals in terms of capabilities and limitations and this has important implications for the projects.
- Equipment design and crafting of procedures directly influence the error-inducing environment.
- Equipment, procedures, environments, and people do not exist separately and therefore they are required to be treated as a single system.

The scope of Human Factors studies covers all stages of a technological enterprise life cycle and its processes. Before any Human Factors study can be conducted, the boundaries and life cycle stages of the technological enterprise need to be specified. Likewise, the accidental scenarios that can be generated by human errors need to be defined.

There are two major trends that are present in human-based treatment approaches: human performance improvement and modification to facilities. Regarding people's performance, it is possible to achieve individual improvement, mainly through training, rewards and penalties associated with compliance to procedures, and through systematic audits and inspections. But the results obtained from techniques related to human performance improvements are very limited and reliant on subjective aspects when compared with the results of the techniques associated with facilities modification.

Facilities modification is the main strategy for the reduction of problems related to Human Factors and human errors and the most important aspects of this strategy are:

- Design for people by removing possibilities of human error by changing the task execution situation (in terms of equipment and procedures).
- Creation of recovery opportunities in cases of human error.
- Design of mechanisms that limit the consequences of human errors.

Regarding Human Factors engineering, the facility modification strategy is much more efficient in terms of results than the strategy of human performance improvement. But one of the most important aspects of the first strategy is the capture of modifications from the beginning of the design phase, when associated costs have considerably less impact.

A strategic failure related to Human Factors engineering is attempting to treat people's performance using methods similarly to those used for machine performance. This is very common especially in less skillful use of human reliability tools. The human performance highlights are creativity, adaptability, and flexibility. Machines are good at processes where human performance is poor, such as continuous surveillance capability, the level of precision in the execution of tasks and the memory capacity. Machines and people must be envisaged in technological enterprises as complementary elements and not as competing elements.

8.2.1 Main influences related to human factors

The Human Factors subject matter is multidisciplinary that transcends the boundaries of exact sciences. But this is not an impediment for the development of engineering-based means of significant technological enterprises improvement based on Human Factors. In this context it is important to understand that, due to its multidisciplinary and broad nature, there is not a single strategy or single model for approaching the subject matter, including engineering activities. As an example of a systematic approach to engineering, we reference one more time the content of the book *Risk Management Based on Human Factors and Safety Culture* (2014, Elsevier, by the same author). This book contains a methodology based on the dissemination of concepts aiming at the reduction of the error-inducing environment and presents a practical case study in the oil and gas industry, in which scientific tools based on mathematical modeling provide the framework for qualitative and quantitative risk analyses, taking into account Human Factors aspects. But despite the different approaches, the matters related to Human Factors involve the consideration of some essential elements to a near consensus as the correct to approach to be adopted.

8.2.1.1 People

All people who have direct or indirect relationship with the technological enterprise are considered as important influence to the error-inducing environment, and consequently in terms of Human Factors. They contribute to such influence, including designers, operators, managers, contractors, maintenance technicians and many other categories. The main aspects related to people to be considered for addressing Human Factors are: physical and intellectual capacities, personal skills, technical and general knowledge, individual behavior, level of training, education level, communication skills, among others. Regarding physical capacity the important aspects to be considered are strength, vision, hearing, and anthropometric parameters. With respect to the intellectual capacity, the ability of understanding the processes, diagnosing the problems and the reasoning speed stand out. Ability to use tools and to written language fluency demonstrated by clarity and accuracy should be considered among the required skills. Knowledge requires the understanding of the hazards and abnormal operations that may be related to processes. The most important behavioral aspects are responsibility and motivation. Communication should be characterized by contact availability, ability to adapt to the environment, receptivity, understanding ability, attitude, and adjustment to available time.

8.2.1.2 Procedures

The most important aspects regarding Human Factors related to procedures are: conciseness and writing accuracy, the level of credibility among users, accessibility, treatment given to revisions and updates, and ease of practical use. Procedures may exist in written or unwritten form, printed or in electronic format, conventionally archived or on computer networks. The credibility of the procedures is related to the continuous improvement through the contributions of users and also with the evidence of reliable means for incorporating users' feedback without loss of information. A procedure can be considered easy to use if users do not express any difficulties in adopting it. Basically this means that good procedures are those that allow the user to read, understand, and act. Reading is a complex activity and procedure technical writers should consider it as simple as possible to reduce the cognitive demand of their users. Most importantly, safety documents need to be extremely concise, without the use of "difficult" words and complex sentences. Safety-related texts should use the minimum number of words necessary to convey the ideas ("the minimum number of words is the maximum content"). Procedures are technical texts. Therefore the writing style and vocabulary may not appear very polished to those who are not professionals in the fields to which the procedures are applicable. Some less experienced procedure documentation writers, in an attempt to improve writing form and language style, may prepare some more esthetically pleasing texts, but with compromise to the procedures' technical objective, which require conciseness, objectivity, and content. A high-level operating procedure to be used by high-level technical operators do not require nontechnical texts. For example, an operation can be naively described as follows: "When the level of the pressurizer [pressure relief system (PZR)] reaches a value greater than or equal to 2.26 m (2 m and 26 cm), then the operator must open the relief valve of the third redundancy (VSA-3) of the set of safety valves (VSA-1/2/3) of the PZR." The same text, when written in a high-level technical and operational knowledge environment, can result in a simple language with a command such as: "If the PZR level ≥ 2.26 m opens VSA-3." When an operator has some difficulty in reading and understanding a procedure, several undesirable reactions can occur. The operator may decide that it is not worthwhile to follow the procedure and disregard it. Another reaction is seeking alternative, unofficial information sources, and in general not appropriate, to circumvent the procedure deficiencies. A very common and dangerous reaction is to continue the actions established in the procedure, even without understanding some of it (skipping the related part of the procedure), which can cause a catastrophe.

A technical writer responsible for a procedure has no way of controlling the cognitive demands that may arise from some extraordinary circumstance that is established during its application. Obviously, these circumstances can interfere with the execution of tasks. It is also not possible for those who write procedures to control the stress that may be generated by the procedure's own deficiencies or by an emergency associated with the fulfillment of tasks. But procedure writers can define under what circumstances each procedure can be applied. Certain procedures may require a high cognitive demand for its complete reading, understanding, and execution. This demand needs to be compatible with the scenario under which it is expected to be executed.

Motivated operators, even when having to deal with poorly written procedures, can obstinately try to comply with them without its full understanding, whereas less committed operators may decide to abandon the procedures when facing the first failure identified in them. Both cases pose threats to the facility safety, which increases the responsibility of those who in charge of the preparation of procedures. Well-prepared procedures will be used by both groups of operators, because they are in fact recognized as facilitators and therefore are effectively used to reduce the demand on operators from both behavioral profiles.

Poorly prepared procedures can pose problems and threats to safety for years to come in an organization, because local culture often makes it difficult to review procedures. Therefore, during the preparation of procedures, engineers need to pay attention to several factors that are decisive for their technical quality, such as: organization, structure, format, content, adequate level of detail, adequate level of numerical information and calculations, indexing clarity, clarity of references, consistency of figures and tables, attention to conditional statements, emphasis on complementary notes, attention to the records to be generated, clarity of announcements and warnings, cross-references consistency, graphs simplicity, and especially special emphasis on steps that are critical to safety. Another very important aspect is to differentiate the steps that require action by the operators from those that should only need monitoring (e.g., automation-related actions). It is essential to discard overly explanatory, literary or elaborate texts. The most positive review a procedure can receive is "thanks for the information you didn't provide because I didn't need it." Conversely, a good procedure should include enough technical content so that operators will never need to prepare parallel summaries, field manuals, etc. These artifacts are extremely detrimental to safety as they are not part of the evaluation and review process, becoming parallel procedures outside of the control of the facility's risk management. If we could summarize the characteristics of a well-designed and prepared procedure, it would be that a good procedure should be easy to understand and to read, as concise as possible, and detailed only to the exact extent needed by the operator.

Poorly designed procedures are one of the main sources of problems related to operational stress because they significantly increase the operators' cognitive load and, conversely, they can lead operators to a dangerous mechanized practice.

8.2.1.3 Stress

Excessive stress is just as bad as low stress. Operational activities require a normal stress level that must be maintained due to the criticality involved with the processes. Low stress level, as in a situation of mechanical fulfillment of procedures, can lead to lack of attention and carelessness by the operator. Too much stress can push the operators' cognitive ability to the limit. Stress has both positive and negative influence on operators and therefore it needs to be controlled. The effects of stress vary from person to person. The same type of stress source can trigger different reactions to different people. The main stress sources are work overload, physical environment (heat, noise, light, cold, physical discomfort, air quality, etc.), managerial demands, fear of poor performance, fear of punishment, simultaneous pressures of different sources, health problems, substance abuse, domestic, and personal problems. High stress level reduces the cognitive resources available to meet demands other than by the source of stress. An artificially low stress level can reduce attention up to the complete depletion of cognitive resources (falling asleep). This can lead to a situation of generalized lack of attention, both for the execution of operational tasks in the control room and for maneuvers and alignments in the field. An artificially maintained very low stress level can impair the understanding of procedures by reducing the concentration ability. Some other factors that also directly influence operational stress are night work, team or individual work, work while fatigued or dissatisfied about work.

8.2.1.4 Fear

Fear is a natural reaction associated with adversity and the knowledge of how to react to it.

Regarding Human Factors engineering, we can say that adversity can be represented as dangers, risks or crises, and always bring with it some component of ignorance, so every adversity generates, to a greater or lesser extent, a fear reaction. The mere expectation of adversity is sufficient to provoke the fear reaction. The best way to reduce fear is to reduce lack of knowledge. Therefore fear can be overcome by acquiring as much knowledge as possible about dangers, risks, and crises. When we manage to reach a high level of knowledge about an adversity, it may no longer be considered an adversity, and instead it can turn into a situation that no longer generates fear.

Fear is a natural biological reaction, which is not limited only to humans, but is part of nature. The effects of fear directly affect human × system interaction. Controlling fear in operational environments requires improving the overall level of technical and operational knowledge. The consolidation of knowledge also contributes to the reduction of fear through clear, organized, comprehensive strategies, divided into well-defined steps. When operators have a strategic plan for each crisis situation that they may face the technical and operational knowledge level will be sufficiently high and, consequently, the effects of fear will be reduced. As a biological reaction, depending on its intensity, fear can directly impact the error-inducing environment, making correct diagnoses difficult during crises and consequently impairing an effective response reaction.

8.2.1.5 Equipment

The considerations presented herein are related to the items that promote people's interactions with equipment and machines. They are the monitors of the process computers, control panels, the human–machine interface devices such as push buttons, valve handwheels, levers, visualization screens, tools etc. Computer screens or displays involve aspects such as operational parameter indicators, lights, alarms, video scenes, numerical values, control spreadsheets, graphs, and a wide range of means of presenting data and information.

The information presented on the display can be dynamic or static. Dynamic information is updated continuously or in time intervals, such as temperature and pressure indicators. The static information does not change during the operational routine, for example, the identification codes of each equipment and system. The information presented on displays can be quantitative or qualitative. They can be presented in the form of a status, or notices. A screen can provide representative information (flowcharts), symbolic, using colors, among other representations.

Regarding Human Factors engineering, the main aspects to be observed when using displays are location, identification of the display function, visibility, clarity, screen layout, level of complexity, and reliability.

8.2.1.6 Controls

Controlling equipment and processes can be carried out mechanically or via data input. The main aspects related to controls are location, identification, layout, ease of use, format, and type. Colors are also very important aspects in human–system interaction in control activities. Similarly, alarms also include color as an element of visual attention for reducing the error-inducing environment, in addition to sound. The alarms can have visual or audible announcement, they can be local (related to a part of the process) or general (related to the entire facility). Alarms can also be indicative of normal operations or can indicate emergency situations. They can indicate process parameters or simply the operational status.

Alarms can also signal the release of a system to start its operation or the detection of gas, flame, etc. To reduce the error-inducing environment, alarms need to be designed to be easily recognized during operation, visible and audible, easy to identify and understand. It is not recommended that a wide variety of alarm sounds or a wide range of colors be used. Operators should work with only two or at most three colors for alarm indicators, and at most two different sounds. The colors should indicate two statuses (e.g., on/off and perturbed/normal), and a third color should be used to signal the failure of the indicator's consistency, which means that the alarm indicator invalid. Alarm sounds should be split into two types: a more intense sound for alarms that require immediate and urgent action by operators, and a less intense sound for other alarms.

Complex facilities such as nuclear power plants, aircraft and offshore platforms should also not have elaborate and heavy alarm systems in terms of the operator's cognitive demand. On the contrary, the more complex the facility, the greater the need for simplification for the alarm architecture. Thus the entire complexity of the facility should be adjusted as part of an alarm announcement strategy compatible with the operators' human capacity. The so-called alarm storm is inevitable for severe transient situations such as accidents and emergencies in complex facilities. However, as it is a predictable phenomenon, both operators and the alarm architecture should be prepared to respond to an "alarm storm." The alarm architecture needs to include filters to associate existing alarms to each type of known accident. For example, if the set of alarms that characterizes a lift gas line breakdown accident can be represented by a filter, the operator, upon suspecting that this accidental scenario is in progress, may subject the "alarm storm" to that filter and thus confirm if all the typical alarms for that accident are present. From another viewpoint the operator should have a level of training and technical knowledge high enough to recognize the "alarm storm" situation and use the available filters as part of an attempt to make an accurate diagnosis of the developing situation.

Manual tools and manually operated equipment are also part of the error-inducing environment related to control activities. Quick and easy access to tools avoids improvisations and accidents, so their location needs to be defined to be user-friendly. The tools should be identified with labels and designed to be as user-friendly use as possible.

8.2.1.7 Computers

The main physical interfaces (hardware) of human–system interaction related to computers are screens or displays, the data input sources and other peripheral devices. Cognitive interfaces (software) related to computers are application programs, operating systems, and software systems. Computers can have complex interfaces through screens or displays, generating special reading requirements. Computers also demand cognitive activities related to the data organization and storage that need to become traceable and protected against losses. The interaction with computers also requires a high degree of adaptation in regard to the continuous software changes. Even when these changes are controlled and limited, the same version of software can be configured in different ways, which requires flexibility and technical skills for software migration or upgrade without information loss and time wasted.

8.2.1.8 Tasks

Are all the activities performed by people within the productive system of technological enterprises. Tasks are very diverse and with different physical and cognitive workloads. They can be part of normal activities such as, operational, maintenance, administrative, as well as critical components of emergency and crisis response systems. Tasks need to be assessed with respect to the balance of physical and mental workloads, the number of people required to perform them, the required communication facilities, the time constraints, the steps in which they are subdivided, the hazards identified in the activity, the protection safeguards against such hazards the degree of automation established in the execution of the work.

8.2.1.9 Workstation

The human–system interaction needs to take into account human comfort for work where the tasks are performed, the characteristics of the operational consoles, the general working conditions in the industrial areas and in the office areas, as well as the ergonomic conditions. It is important to perform anthropometric assessments of the task, considering aspects with high degree of influence such as lighting, stress sources present, physical and environmental conditions, opportunities for distractions, conditions of human comfort, movement of people related to local equipment, location, warning signals and visual communication.

8.2.1.10 Company or organization

Organizational aspects can exert influence in the error-inducing environment, such as managerial systems adopted, distribution of responsibilities, people's decision-making power, compatibility of the assignment of activities and positions, policies and organizational practices. The organizational culture has a great influence on the Human Factors (source of the error-inducing environment). Organizations generate expectations and adopt symbols, colors, practices, standards, and values, besides their own history that in itself influences the error-inducing environment. In order for the organization to contribute to the reduction of the error-inducing environment, it needs to provide the means to permanently deal with issues stemming from the assignment of responsibilities, assignment of authority, supervision activity, sources of resources and cost indexing. Other aspects to which the organization needs to maintain permanent attention are related to the level of commitment, the mechanisms of delegation of authority, the decision-making speed, the work shift scheme, the turnover of professionals and the imposition of policies and procedures.

8.2.1.11 Environment

The environment is established by the conditions in the areas surrounding the facilities, processes and physical workplaces. Important error-inducing environment characteristics are associated not only with physical matters such as noise and light, but also with psychological matters (work relationships, nervousness, tension, boredom, etc.), architectural matters (walls, lighting, space availability, noise, colors, decoration, equipment, design, etc.), and matters related to climate (rain, wind, fog, sun, temperature, air quality, cold, heat, etc.). An unfavorable environment can negatively impact working conditions, increasing the induction of human error and its consequences.

8.2.1.12 Safety culture

Within the organizational culture is the influence of its own pyramid of values. The position of safety-related matters in this pyramid determines the level of safety culture in the organization. Note that there is the official pyramid of values, which is expected to prevail throughout the organization. And there is also the actual value pyramid, which reflects the importance and value that is actually attributed to safety. There is always some disparity between what is intended and what is actually achieved regarding safety culture, but the organization needs to monitor and address such differences so that there is no loss of organizational identity and no degradation of the safety culture. Conflicts between safety and productivity are common, as well as conflicts between social expectations and corporate objectives.

The reinforcement of values makes it possible the development of a robust safety culture capable of maintaining a high level of risk management including in the midst of conflicts that are part of the life cycle of large organizations.

8.2.2 Human factors analysis

One means of analyzing Human Factors is the Human Factors Engineering Review (HFER) methodology. This methodology uses a checklist to evaluate a project under development or a facility already in operation. HFER is conducted by a small team of analysts or even by a single expert, depending on the complexity of the technological enterprise to be analyzed. A spreadsheet is also used for guidance and for recording results, such as observations and recommendations. Basically, the HFER consists of the following steps:

- Technology enterprise data acquisition.
- Subdivision of the technological enterprise into subsystems.
- Identification of the main categories of Human Factors to be investigated.
- Preparation of a questionnaire for each category.
- Application of the questionnaire and recording of responses.
- Refinement of information related to identified problems.
- Recommendations for improvement of aspects related to Human Factors.

The collection of information can be obtained through interviews, on-site observation and evaluation of procedures and records. Information and data of interest often are not formally recorded. In such cases, high level of experience is required by the experts to be able to carry out inspections on the premises and to evaluate the performance of operational tasks.

Facilities in the oil and gas industry are very large and need to be subdivided for an organized work to be carried out. Thus facilities need to be subdivided into systems within previously established criteria, such as units, sections of units, specific equipment, and physical areas, based on the type and complexity of each facility to be studied.

All aspects related to the error-inducing environment need to be considered in a HFER analysis. So, all aspects of Human Factors need to be investigated. The best organization format for this investigation is a checklist, with questions and points of interest to be assessed, covering all categories of Human Factors. The focus of the checklist is to investigate the quality of the human × system interaction, that is, how people relate to the facility. The categories of Human Factors often considered in this type of checklist are people, procedures, equipment, computers, tasks, workstation, organization, and environment.

The questions must be prepared to meet all the interests of each category of Human Factors. In general, experts already have lists of standardized questions and adjust them to the reality of each type of technological enterprise. We can mention some examples of questions that can be included in checklists of this type:

- Are all controls accessible?
- Is any special tool required to perform the task?
- Can operational set points be modified without official authorization?
- Are the required tools available and identified for immediate use?
- Can the isolation and safety shut-off valves have their position changed from the control room?
- Do safety-related controls need a key to be changed?
- Can the identification and grouping of controls cause any confusion?
- Do the panels and consoles have distributed push buttons and indicators aimed to avoid reading and operational errors?

Analysts use the information collected at the facility to answer each checklist question for each subsystem. For this purpose, the specialists conduct visits, interviews, field inspections and review of applicable procedures and standards. When specialists confirm the existence of Human Factors problems, they need to be investigated with a greater degree of refinement. This means that it is not sufficient answering the checklist questions with the acknowledgment of the problem, but it is also necessary to explain in detail the identified vulnerability.

When problems are identified, experts should assess which recommendations should be recorded for the purpose of correction. Thus the final deliverable of the HFER will be a report containing a summarized spreadsheet with the identified Human Factors problems, observations and detailed description of these problems and the recommendations for the necessary corrections.

An analysis of Human Factors will achieve effective results if it identifies, as much as possible, the factors that induce errors to those people who interact with the installation. This includes the factors that mislead designers, operators, managers, maintenance engineers and everyone else who participates in the entire life cycle of the installation, from the project to its operational routine.

An engineer specialized in Human Factors analysis can evaluate a technological enterprise through a check list where all the main items related to Human Factors can be observed. The adoption of a check list helps to ensure that no relevant factors are overlooked during the analysis, since the subject is multidisciplinary and complex, which can hinder the concentration and focus of specialists. For each Human Factor listed in the check list, the engineer specialized in Human Factors must evaluate the compliance verified in practice. In case of noncompliance, the engineer specialized in Human Factors must register recommendations to correct the problem, indicating responsible and deadline. The Human Factors to be considered by the Human Factors Analysis are subdivided into four major groups as shown below.

8.2.2.1 Safety culture and working environment

- Working Conditions: Areas are clean and organized; Instructions for housekeeping are posted; Noise and Vibration are maintained at tolerable levels including, if noise and vibration are not maintained at tolerable levels, Personal Protective Equipment (PPE); Temperature and humidity are normally within comfortable bounds; Ventilation is adequate; Provisions are in place to limit the time spent in extremely hot or cold areas; Lighting is adequate to perform operations safely; The safety climate and culture is perceived positively by de organization.
- Workload (overtime and fatigue) and Stress Factors: Company policy that addresses working hours and overtime is in place; Guards against fatigue are in place; Length of a normal shift is appropriate given the degree of alertness required by operator; Restrictions are in place to limit employee overtime; Staff levels size is sufficient to handle all shifts' routine and nonroutine duties; Staff experience level is also sufficient.
- Shift Work Issues: Control room is attended while operations are running; Provisions for scenarios when the relief (next shift employee) fails to show up are clear (these provisions have been considered in writing); Shift turnover periods are sufficient to adequately communicate plant operating conditions from off-shift to on-shift personnel.
- Labeling: Equipment (i.e., vessels, valves, pipes, instruments, controls, etc.)is legibly and unambiguously labeled; Labeling includes equipment identification, contents, capacity and, National Fire Protection Association (NFPA) coding as applicable; Labels are located close to the items that they identify; Labels are in the local language and others used by the operators; The responsibility for maintaining and updating labels has been assigned; Remote startup/shutdown switchers are clearly labeled; Equipment labels are consistent with nomenclature used in procedures; Components mentioned in procedures (e.g., valves) are labeled or otherwise identified; Switch labels identify discrete positions (i.e., ON or OFF, OPEN or CLOSE); Pipelines and electrical conduit are clearly labeled at points where they become

invisible (e.g., routed underground); Signs posted to warn workers of hazards; Signs are visible and easily understood.

8.2.2.2 Equipment and facilities

- Accessibility/Availability: Access to controls is adequate; Restrictions consider routine and emergency operations or personnel; Access is adequate at valve manifolds (including elevated valves) for routine, emergency and maintenance operations; Gauges and indicators are easily read and access; Process control system displays present the process information; For similar equipment this is presented in a unique manner to avoid confusion; Controls are easy to identify; PPE and tool requirements are established in operating procedures; PPE is available for routine and emergency use/situations; Workers are able to perform both routine and emergency tasks while wearing the PPE; Adequate tools are available for work; Tool exchange is provided as needed.
- Workstation Design: Operator Physical limitations considered (e.g., height, width, reach); Other characteristics (e.g., color blindness, left-handed operators) are also considered; Maintenance jobs are evaluated as part of the workstation design; Tasks requiring the operator to perform nearly simultaneous actions can be accomplished without traveling large distances; Manually operated valves are positioned to allow proper operation without muscle strain; Valve manifolds have been arranged to reduce the likelihood of misalignment.
- Control Room Design: Displays are visible from relevant working positions; Control system display targets (touch screens) are spaced adequately to prevent accidental operation; Board-mounted shutdown switches are sufficiently distinguishable from alarm acknowledgment buttons; Design of control room minimize the possibility of operator error or tripping during stressful situations; The control building air conditioning and pressurization is adequate to protect electronic instrumentation.
- Computers (Hardware and Software): Human/Computer interface; Operators can safely intervene in computer controlled processes; When values are entered manually by operators, the computer checks that values were provided within a valid range; If operator uses gloves to operate control systems, tactile feedback is not affected; Feedback/Displays adequate view of the entire process and details of individual systems is provided; Feedback/Displays labeling in control systems correspond to the P&ID's; Normal and upset conditions are clearly

displayed in the control room; Separate displays present information in a consistent manner; Glare on display monitors do not reduce feedback; The process control system displays provide feedback to operations personnel to confirm operator actions; Alarms are displayed by priority; An alarm summary is permanently on display or otherwise available; Alarms and signals are clear and distinguishable (Critical Safety from Control alarms); Alarm indicators (e.g., lights, horns, or whistles) adequately alert potentially affected areas and employees; The cause of "nuisance alarms" (repetitive alarms that operations personnel ignore or acknowledge without investigating) are determined and repaired in a timely manner.

- Emergency Equipment: An emergency shutdown panel exist and it is properly located; Panels are properly located, Emergency shutdown switches are guarded against inadvertent operation; Exit and response (e.g., evacuation route) signs are adequately visible; Emergency lighting is adequately installed; Emergency lighting is properly maintained and tested; Location of emergency equipment (e.g., first aid kits, fire gear, self-contained breathing apparatus) allow for rapid access and use; Equipment do not present additional hazards to personnel.

8.2.2.3 People

- Communications: Communication equipment is available for routine and emergency use; Communication equipment is intrinsically safe where required; There is an effective two-way communication between peer during an operation; Mechanisms are in place to encourage an effective communication between management and operators; operators and design engineers; Operating crews communicate unusual equipment or instrument status (bypassed or out of service) in writing.
- Individual/Personal Aspects: Calculations performed by personnel are documented in a consistent manner and checked periodically are independently for correctness (tables and formulas may be available); The facility enforce a drug and alcohol testing program; Program includes prevention, detection, and counseling.

8.2.2.4 Management systems

- Safe Work Practices: Permit to work systems including remote operations, lock out—tag out procedures, hot work (e.g., welding), confined space entry, opening process equipment or piping; Control over entrance into a facility by maintenance, contractor, laboratory or other causes.

- Incident Investigation and Analysis: Employees report all accidents (injuries or near misses); Investigations are carried out by multifunctional teams, including operators; Root causes are identified; Action items are determined to eliminate hazards as the priority; Results from investigation and analysis are documented; Records are available for future reference and follow-up.
- Management of Change: An effective tracking mechanism is in place to follow up open items from audits, preliminary hazard analyses, incidents investigations, Preventive/Predictive Maintenance (PM's), recommendations, procedure changes, new projects, etc.; Procedures are accurate and updated as needed, Procedures are audited effectively, Procedures are available to personnel even in case of an emergency, Procedures are written in the local language and others used by the operators, Procedures exists for all operating phases (e.g., star up and shutdown), Procedures prevent changing alarm set points, process control system or safely shutdown system control or logic (software) without proper review and authorization, Procedures require verification that instruments that are deliberately disabled during operation (e.g., shutdown interlocks bypassed to allow testing) are placed back in service, Procedures specify proper response to alarm indicators (e.g., lights, horns, or whistles) during emergency situations, Procedures steps are described in logical sequence, Procedures preventions are taken for taking steps out of sequence, Operators believe that procedures are easy to follow and understand warnings, cautions and notes are stand out from procedure steps.
- Competence Assurance and Management: Staff is recruited and selected against defined criteria for the job; Training Program (Employees and Contractors) includes an overview of the process and steps for each operating phase, operating limits, consequences of deviations and steps required to avoid deviations, safety, and health considerations; Accidents, near misses and hazards are considered to update training material and modify delivery frequency; Training is delivered after a change in the process or procedure and it includes an explanation of why and how safety can be affected; Instructors are trained in how to deliver courses; Courses include evaluations of performance for attendants and instructors; New employees, visitors and contractors are trained, at a level that will be required by their task, in the processes

hazards before exposure; Employees receive adequate training in safely performing their assigned tasks (even those infrequently performed); Employees receive refreshment trainings as often as necessary; Training material consider different learning styles (visual, aural, read, write, and kinesthetic); Training materials, course evaluation and attendance is documented and filed for future reference.

- Organization: Involvement of all organization levels is evident; Safety critical roles and responsibilities are clearly defined; Operators participate in the writing procedures process.
- Maintenance Program: Instruments, displays and controls are promptly repaired after a malfunction; Plant instrument indicators are routinely checked for accuracy; Program recognizes that there are maintenance of higher priority; Valve misalignment has been evaluated.
- Emergency Preparedness and Response: A written on-site emergency plan exists and includes a designated and recognized chain of command; Critical valves or equipment close to the process can be shut off from a safe location and this can be accomplished in a timely manner; Emergency procedures (including shutdown) have been developed and employees are trained accordingly; Emergency procedures are presented in a clear, step-by-step format to reduce the "panic" factor during upset conditions; On-site plan designate a specific entity or individual responsible for contacting outside bodies (i.e., local authorities, fire department, and press); Additional operators (e.g., from other areas or from off-site) can be called in quickly to help during an emergency; Emergency drills and evacuation maps are periodically conducted with participation of all areas and departments, and also, drills performance is evaluated and improved; External impacts is considered by an emergency response plan and consider steps to respond to external events such as power outages, winds flooding and, releases from neighboring facilities or processes; Backup power is readably available in areas where a power outage could cause a higher risk; Guard are in place to prevent internal or off-side vehicles to hit process equipment; Operators practice emergency response under extreme environmental conditions (e.g., at night, cold, rain).

8.2.3 Programs for consideration of human factors

Simple process safety management systems do not always address matters related to Human Factors. This happens even though it is known that most of the accident investigations results point to causes related to human errors. Higher level process safety management systems do address Human Factors matters and establish a program for monitoring and implementing measures to improve and control the error-inducing environment. Human Factors treatment programs include:

- Human Factors Design.
- Review of Human Factors Engineering (RHFE).

Each opportunity for human–system interaction is influenced by Human Factors. The actions and operational practices need to be studied and dealt with, considering each type of facility, thus defining the Human Factors design and the revisions of engineering of Human Factors (RHFE).

8.2.3.1 Human factors design

Designing for Human Factors requires the assessment of tasks, people, procedures, workplace, equipment, computers, organization, and environment. It is necessary to identify the tasks to be performed and the requirements for the people's adequate performance (considering the physical and cognitive aspects). Determining these requirements requires a significant studies and analyses of tasks and can be time consuming. The main objective is to verify the compatibility between people and the tasks they perform. People can act as filters related to the requirements needed to perform tasks. Only requirements compatible with people shall be met. Periodically, tasks need to be analyzed and compared with the characteristics of the people assigned to them. Feedback from the operators is very important for the success of this type of evaluation. The company or organization should provide the means to ensure that channels are opened so that the feedback can be used productively.

Procedures are essential for all operational activities, in addition to ensuring compliance with international requirements. Procedures should be used in training. They need to be elaborated with focus on the operator's daily routine and not as a record of didactic knowledge. The workplace should have all hazards addressed and assessed. Human Factors design guidelines, when defined, should always be applied and not just acknowledge verbally. Equipment should meet reliability and ergonomics requirements. Computers need to use reliable computing and control systems. The company or organization needs to remove elements that are harmful to a good safety culture, such as feeling of guilt, adoption of punishment and reward instead of promoting project improvements, excessive tolerance, alienation from people regarding risks, creating the appearance of safety practices, productivity over safety even when safety slogans are used, etc. The environment needs to be adjusted to provide the most favorable conditions for people to maximize their performance in the execution of tasks.

8.2.3.2 Human Factors Engineering Review

Each facility needs to undergo a HFER (Section 8.2.2), and all issues related to Human Factors should be identified. HFER recommendations must be implemented, and HFER should be revalidated periodically. In addition, Human Factors treatment programs interfere in the sense of adding value to Human Factors aspects in traditional elements of process safety management, such as

- operational procedures.
- training,
- safety review in prestart up conditions of systems,
- physical integrity of mechanical equipment,
- change management,
- accident investigation,
- emergency response plan,
- compliance with audit requirements.

8.2.4 Human factors in the life cycle of technological enterprises

The Human Factors engineering approach is important throughout the life cycle of technological enterprises. Human Factors influence the error-inducing environment from the design stage through construction, commissioning, operation, maintenance, testing, inspections, emergency response, critical startup and shut down operations, decommissioning, and including demolition and end of the technological enterprise operational life. But the best time for human factors knowledge to be applied is in the design phase of technological enterprises. The application of such techniques after the design phase is used as a "remedy" to reduce the problems of a patient with a chronic illness: lack of treatment from the inception (design) of issues related to Human Factors.

The design phase provides the best possibilities for the elimination or at least reduction of the likelihood of occurrence of most human errors. In the design phase it is possible to provide the enterprise with the resources required for an adequate response to the consequences of human errors throughout the life cycle of the facility.

Considering that designers also make mistakes, a good strategy is to have independent experts conduct a project review regarding the adequacy of Human Factors. The most critical design errors involving Human Factors are lapses and failures in training and knowledge. Typically, these errors occur due to technical education deficiencies, training and operational experience of the designers. The design phase is the best time for the principles of intrinsically safe installations to be applied.

8.2.4.1 Design errors

An accident in a steam generator (pressure vessel) has two fatal victims. The vessel exploded due to the water level being decreased below the minimum level specified. The equipment had been designed with two level monitoring displays, two low level alarms, a low-level feed water pump startup interlock and a low-fuel level interlock to shut down the heaters. The equipment could to be isolated by two valves (for nonexclusive use) of the steam generator. For some reason, these valves were closed. The causes of this closure are unknown, but that was the reason for the low water level and the fatal accident itself. There might have been many possible causes for this undesirable closure, but that is not relevant for this example. The design of the facility should not allow the valves to isolate the steam generator by being easily shut off, including when the equipment is in operation. Some of the Human Factors aspects that are important in this accident are each safety-related equipment needs to have its own exclusive isolation valves.

Isolation maneuvers should generate clear and audible alarms, so that operators can immediately be aware of such status change.

8.2.4.2 Construction and assembly errors

Construction and assembly errors are quite common. The most common mistakes are training failures, knowledge failures, and lapses. The most frequent causes for this type of error are lack of motivation and training. Through checks and inspections, during and after construction and assembly activities, it is possible to reduce errors of this kind and verify that the design specifications are being followed, as well as best engineering practices. A good strategy is to include designers and members of the commissioning team among the experts who conduct the tests and inspections. These tests and inspections should verify that the contractors are interpreting the design documents correctly and whether modifications should be included based on problems identified in the field during construction and assembly.

Examples of construction and assembly errors are debris and foreign materials that are often found inside pipelines. Other frequent problems are related to failures in structural components designed for supporting pipelines and equipment, resulting in vibration and fatigue. Another failure is caused by poorly finished sections of a pipeline that impair circulation and can result in stale corrosion products and dirt. Incorrect use of materials can also cause serious consequences for facilities.

8.2.4.3 Commissioning and startup errors

Commissioning and startup are very risky phases in the life cycle of a technological enterprise and deserve special care by engineers. During this phase, new actions are being carried out, people are under heavy workload and demand, operators still lack experience with the facility, more people are involved in the activities, increasing human errors risks and the exposure of people to risks, besides being a period where many changes occur all the time. The most common mistakes are lapses and failures in training and knowledge. Examples of errors during commissioning and startup of the installations are provisional pipeline supports authorized without proper technical validation with risks of associated accidents, incorrect loading of process substances resulting in unexpected reactions, valves left in out of alignment positions can cause loss of containment (leaks) with fires and explosions, etc.

8.2.4.4 Maintenance errors

Maintenance errors also occur frequently. The most common mistakes are lapses and lack of skills in performing tasks. A constant source of concern is the balance involving cost reduction, maintenance level, equipment reliability, and expected consequences in the event of failures. As an example of maintenance failures we can mention:

- Engineers at a facility are attempting to remove an incorrectly installed actuator from a motorized valve and during that operation a fire broke out causing with six fatalities. The engineers inadvertently skipped some isolation tasks and this caused leakage of flammables and fire.
- A reactor temperature controller fails and a batch reaches a very high temperature. The cause of the identified fault is a terminal disconnected from the temperature sensor. An analysis finds that during the

last maintenance operation, the screw had been lost during assembly and an inadequate screw was provisionally used to secure the terminal, causing the accident.

- A shaft was incorrectly installed in the bearing. The difficulty in disassembling the unit leads the engineers on duty to cool down the shaft and heat up the bearing to proceed with disassembly. During the operation, engineers used gas (hydrocarbon) at low temperature as the coolant and at the same time used an open flame (oxyacetylene torch) to heat up the bearing. A big explosion occurs with four fatalities.

8.2.4.5 Emergency response error

Human behavior is very unreliable under the stress of an emergency situation. Operators may be afraid of making hasty decisions, there is overload of tasks, excessive complexity in tasks, elements of unpredictability and surprise in the ongoing scenario, alarms, noise and discussions can disturb the environment, and emergencies can occur in the end long shifts, besides they can last long and tiring work hours. There are several attempts to classify human behavior when exposed to a stressful situation and these efforts are based on aspects related to the design of the facility, formal procedures, training received, and organizational structure.

Human behavior under stress is prone to several tendencies of undesirable reactions, among which we highlight are as follows:

- Defensive Escape: this reaction can assume various forms of behavior, but basically the person can adopt a selective inattentive behavior in an attempt to avoid thoughts related to the seriousness of the situation, hazards, and risks. This type of reaction can lead the person to try to engage in activities that distract them as a way to escape from the stress that the actual risk scenario imposes. The person can try to "pass the baton" to another team worker or call other people to avoid responsibility for the decision.
- Forged Consensus: it is the tendency of a group of people under high stress to protect a consensual decision during an emergency. The group itself acts by putting pressure on other people who disagree with their consensual decision. The group also reacts by censoring external information that may override consensus on the decision.
- Acceptance of Risk Beyond Reasonable: people tend to accept more risks when making group decisions. The arguments employed to achieve the objective of accepting a high risk are: diffusion of responsibility among several people, use of persuasive people to convince

others regardless of the facts, increased familiarity of people with the risk situation by simply insisting on discussing the problem (similar to "winning by persuasion").

- Fixation in Situations of the Past: this reaction under stress is characterized by the focus of arguments during an emergency based excessively on events that have occurred in the past but that do not have the perfect correspondence with real developing facts during the present emergency.
- Tunnel Vision: using an analogy, the human beings are able to analyze the scenarios in which they are immersed, observing them through a 360°field of vision. To achieve this we can move our head to the right, to the left and back. However, our comfort region is only 20° wide, because we only need to look forward toward the immediate objective. When this objective receives an exaggerated value, the phenomenon of "tunnel vision" occurs, which is a much more radical reaction, where we become so concentrated and focused on a certain objective that everything else around us can disappear like the walls of a tunnel. In a crisis when we are focused on a single objective in an exaggerated manner, we can lose sensitivity to everything else that happens around us. It is as though we are crossing a tunnel where the exit on the other side is the sole focus of our attention. Moreover, when we make the decision to focus our decisions to achieve a single objective, it can generate decisions based on only one side of the developing crisis scenario. Every time we make decisions, other hidden decisions are also made as a result. For example, if our analysis focus is set exclusively on the comfort and flexibility of air travel, we can opt for a state-of-the-art business jet.

However, when we make this decision, other unconscious decisions are being made in parallel as well. Depending on the situation and the flight plan, the decision may not be only for a business jet, but also for an aircraft with looser safety requirements, for landing at an airport with less infrastructure, for a crew more influenced by our flight schedule requirements. All of these factors will then become part of the flight scenario, and it is not always done consciously by those whose decision is based on a "tunnel view."

8.2.4.6 Shutdown, decommissioning, and demolition errors

The shutdown, decommissioning, and demolition phases are extremely vulnerable and neglected from the point of view of risk and safety management. Although the technological enterprise life cycle is coming to an end, the hazards remain present and need to be taken into consideration. In these phases there is more people's engagement in the processes and, consequently, more chances for human errors and evidently greater exposure of people to risks. The main types of mistakes made in these phases are the lapses and the lack of knowledge and training. Several examples of serious commissioning errors can be provided. For example, when certain installations are decommissioned and demolished, there can be cases of hazards being kept in place due to serious failures. Unfortunately, this has happened with radioactive sources left in the rubble of disabled facilities. These sources got into the hands of people ignorant about the risks associated with radioactivity, resulting in a nuclear accident of grave consequences.

8.2.5 Intelligent identification of systems and equipment

The need for objectivity and conciseness is so pressing in communications during a crisis that some protocols based on acronym codes can be adopted to simplify the exchange of information between operators. These codes are not created in the same way as labels, tags, or equipment tags are created although they replace them in some ways. An example is the KKS ("Kraftwerk-Kennzeichen-System") code which is used as a standard for identifying systems and equipment in power generation installations. The engineering used in the most modern power generation plants in the world, established a common language for all technical disciplines associated with this type of installation, recognizing the current needs for human and machine interaction. Therefore this code can be used by different engineering specialties, different companies and countries, regardless of the local language as it does not have a direct link between the letters of the codes and the names of the equipment. In reality, KKS codes replace traditional equipment names in conversations, facilitating communication between operators regardless of nationality, language, and company.

This language allows a communication protocol involving civil engineering, mechanics, electrical applications and instrumentation and control systems. Through KKS coding, operational reliability and efficiency are considerably increased due to the high level of communication conditions during planning, maintenance, parameter setting, and operation activities. The continuous increase in the complexity and automation of state-of-the-art facilities presupposes greater efficiency in the technical communication language. KKS provides the standardization of all types of identification and all types of operational processes for power generation plants. The use of protocols such as KKS increases the resources and possibilities for detailing the identification of equipment, components, systems and structures during conversations and alarm announcements. Although KKS originated in the power generation industry, it can also be applied in other industries that also adopt the high operational level as a value.

8.3 Limitations of quantification techniques related to human reliability

There are still no standards related to Human Factors for the oil and gas industry. Some oil companies have developed their internal guidelines for addressing the matter. But the American Petroleum Institute (API) recognizes the importance of the reduction of human errors aiming to increase the safety, productivity, and quality of the processes in the oil industry. However, to develop human performance, managers need specific technical guidance on the prevention of human errors and to reduce the likelihood that such errors can lead to problems or accidents in processes of the oil and gas industry. The API has prepared a guide for an audience of managers with different levels of knowledge on Human Factors engineering. This guide has become the API 770 standard and was created to provide managers with a basic understanding of human errors causes and to recommend ways to reduce such human errors from a person's individual viewpoint. API 770 also describes means to incorporate HRA into risk management and process safety activities.

However, API 770 provides much more basic and generic guidance on Human Factors subject matter than a methodology for the effective reduction of human errors. Maybe this is so because the most effective way to work in the Human Factors field is through the development of concepts that make up a solid safety culture. The techniques of human reliability have major limitations regarding the results of the reduction of human errors. In fact, human reliability techniques are very efficient in identifying the most critical points related to human errors in processes. But human reliability techniques lack a direct approach for specific treatment to Human Factors that make up the error-inducing environment in processes. Human reliability techniques contribute to identify and quantify human errors, but do not provide direct solutions for reducing them.

API 770 states its objective as: "helping to understand the principles of Human Factors engineering for practical use." It is our understanding that such objective is not fully met with respect to the meaning of the term "for practical use." Despite the numerous practical examples of operational cases presented in the standard, a "for practical use" methodology it cannot be established only through these examples. API 770 is an "Executive Summary" of important concepts on the subject matter (human performance, human error). The text is more academic than operational, that is, it delves into theory and what is presented as practical is actually still theoretical in nature. Examples of more mature regulations on the subject matter are from the nuclear and aeronautical industries, for which Human Factors have been integrated as part of their routines longer.

Section 3 of API 770 presents "strategies for improving human performance."

The most important part of this text is related to the "examples of situations of probable error," summarized as follows:

- Deficient procedures
- Inappropriate, Inoperative, or Misleading Instrumentation
- Insufficient Knowledge
- Conflicting priorities
- Inadequate signaling
- Inadequate feedback
- Discrepancies between Policy and Practice
- Disabled Equipment
- Poor Communication
- Poor layout
- Violations of Population Stereotypes
- Excessive Sensitive Controls
- Excessive Mental Tasks
- Opportunities for Error (excessive repetition of an event)
- Inadequate Tools
- Poor maintenance
- Computer Control Failure
- Inadequate Physical Restrictions (inadequate door handles, etc.)
- Appearance at the expense of functionality (two "mirrored" control panels, beautiful but with error risk due to duplicity).

Despite being titled "strategies," the referenced section presents only three generic lines as an improvement approach:

- Hardware change
- Procedures changes
- Policy changes

Finally, Section 4 of API 770 standard presents the Managerial Use of HRA as an EVALUATION instrument, not directly IMPROVING human performance. But the standard itself outlines the main limitations of HRA:

- Perfection: it is not possible to predict all influences on human behavior.
- Validity/specificity: Probabilistic models cannot be completely verified, they do not represent specific tasks of a process plant adequately.
- Precision/uncertainty: lack of data on the probabilities related to human behavior.
- Reproducibility/propension: numerical precision is elusive. In reality, the most important input data for the application of the HRA technique are subjective, resulting from the analysts' premises and assumptions.
- Traceability/scrutability: HRA analyses produce lengthy documents and tedious and immense tasks, making the understanding of the overall results difficult.

8.4 Rapid Entire Body Assessment

Regarding ergonomic aspects, several types of analysis contribute to significant reduction of the error-inducing environment. The most accurate analyses with the most effective results are those performed during the design phase. However, the day-to-day operations eliminate and create new tasks as the processes get improved. In addition, wear and tear and replacement of parts and equipment do naturally generate new tasks, new operating practices, new physical configurations with associated ergonomic problems to be solved. How to make quick and practical assessment, if the performance of new tasks can cause some damage to those who working on them? The Department of Industrial Engineering at the California State University recommends the tool Rapid Entire Body

Assessment (REBA) for basic tasks. This simple and easy-to-use tool allows a quick assessment of the safety of people in the execution of tasks, without excessive associated costs and without the need for sophisticated computational tools. There are variations in the application of the REBA technique, but we will present below an application example of the technique for a very common situation in our times: the use of a laptop computer in a workplace.

8.4.1 Example of the application of the REBA technique

The use of laptops is increasingly widespread in all activities. In some companies, desktops have been completely replaced with laptops. An employee can work from several physically separate units, establishing a workstation at each of these locations. The main equipment is the laptop, which connects the employee with the entire corporate environment and with the rest of the world. This example will consider the use of laptop by a worker in a control room on an offshore oil exploration and production rig, during its construction phase.

The main control rooms of offshore platforms follow ergonomic standards aiming to provide the best possible working conditions for operators. Consulting companies are often hired for the specific design of control rooms. However, although being very close to an "ideal workstation," sometimes an operator or another employee uses a laptop in the area. The control room environment does not always have space for the laptop without jeopardizing the tasks in progress. Adequacy difficulties are even greater during the construction and assembly phases, where working environments are improvised, which can mean ergonomic problems with consequences for workers' health. The following example will assess the conditions associates with a laptop on the consoles in the main control room during the construction phase.

8.4.2 Human body mechanics during execution of tasks

Let's consider a 47-year-old man, weighing 70 kg, working with a laptop on the console of a Main Control Room under construction. If the man is photographed performing the task in his most frequent posture, the REBA tool supports a general analysis of the task to be performed based on the photograph (see Fig. 8.1).

From the photograph (see Fig. 8.1), experts can observe:

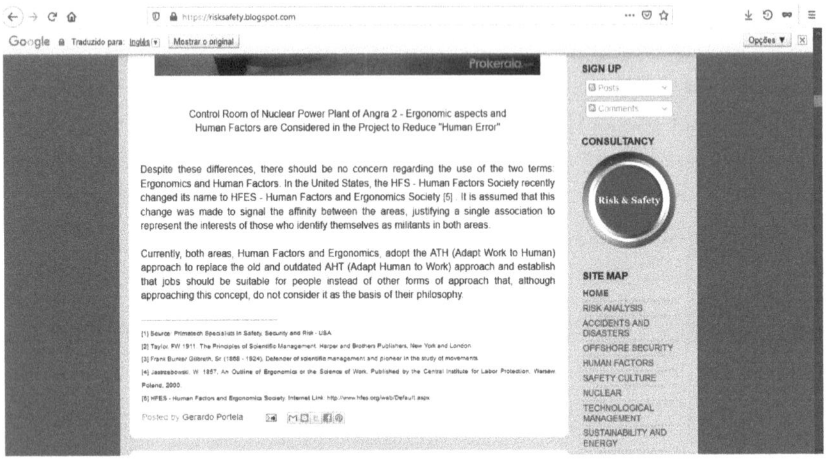

Figure 8.1 The website https://risksafety.blogspot.com/ offers a step-by-step execution of a REBA analysis. After accessing the site click the "Human Factors" link on the site map. All the references needed for application of the tool are described in the content, including pictures, tables and calculation examples of calculation. *REBA*, Rapid Entire Body Assessment.

- *Workstation adjustment*: adjustment of posture, freedom of movement of the arms in their normal range.
- *Seat*: comfort and relaxation when sitting, freedom of rotational movements and the spine angle, for example, flexed about 30 degrees.
- *Posture*: poor posture. Loads considering that the sitting position can represent approximately 180% more compression in the elements of the spinal column in comparison with the upright position.
- *Hands*: typing posture at the time the photograph (see Fig. 8.1) was taken. The hands posture related to other tasks. Number of equipment within the reach zone.
- *Loads*: based on the information in the photograph (see Fig. 8.1), specific tables can be consulted to obtain data that allow the calculation of loads on bone and muscle structure.

8.4.3 Anthropometry

The sample photograph (see Fig. 8.1) was taken during the construction and commissioning of a new oil rig. The man in the photograph (see Fig. 8.1) is Brazilian who weighs 70 kg, and his anthropometric data was obtained through a questionnaire from an anthropometric data collection sheet. Through statistical tools, the data collected about the employee are compared with the data from anthropometric information databases. This comparison ensures the best use of the tables for the assessment of the risks in the performance of the task.

8.4.4 Static work

The employee posture was analyzed accurately as depicted in the photograph (see Fig. 8.1). Although the employee is not typing at the time of the snapshot, he is interacting with the other colleagues who are doing the assembly work. We can consider the employee spends most of the time exchanging information with colleagues and checks additional information on the laptop to guide them. Therefore we can consider that the head position is facing the colleagues, as shown in the photograph (see Fig. 8.1), as this is probably his most frequent posture. In addition to using the laptop, the employee being analyzed also uses a printout that is placed in the space between the laptop and the end of the console that serves as a table. As the space is insufficient, part of the printout is outside the edges of the table. A power cable is also incorrectly positioned between the legs and the console/table. The information obtained thus far is sufficient for the REBA analysis to assign a score. In this example, the score result was 9, which means that the task is considered "high risk" and requires investigation and implementation of changes.

8.4.5 Repetitive work, cumulative trauma, and use of hand tools

The employee being analyzed works in a control room under construction. However, it does not mean that the employee's work situation is temporary. When the construction of the control room is complete, the employee will be assigned to another construction and assembly work where he will perform his activities under very similar conditions. Thus we can consider that for the employee being analyzed this is their routine work environment. Therefore working in the conditions shown in the photograph (see Fig. 8.1) can, over time, cause illnesses due to cumulative trauma, recurring body strain, and muscle/bone diseases. The typing hand position does not favor "the natural line." It also does not favor the "wrist oscillation" position, which is more correct. The employee posture is also unfavorable regarding the use of other manual devices and accessories such as mouse, mouse pad. This is partly due to the disorganized workspace.

8.4.6 Rapid Entire Body Assessment evaluation

The situation depicted in the photograph (see Fig. 8.1) establishes a paradox. The main control room of an oil production and exploration rig is one of the most carefully studied locations regarding ergonomics. All details are considered by the designer so that operators have the best working conditions possible. However, despite all the level of quality of the design of control rooms of this type, some groups of employees working in the construction and assembly of these rooms work under completely unfavorable conditions. These employees build a perfect room for their fellow operators and yet over the years they work in different control rooms still under construction under conditions that can cause cumulative trauma, repetitive strain illnesses and muscle/bone diseases.

8.4.7 Rapid Entire Body Assessment recommendations

Although the task is being performed in a unit still under construction, it needs to be redesigned based on the results of the REBA analysis. Although the workstation is temporary, it needs to be considered as a "Visual Display Terminal" and get as close as possible to the associated standard of comfort and safety.

This example is calculated in its entirety on the website: https://risksafety.blogspot.com. On this website (Fig. 8.1) the Human Factors link presents the step-by-step execution of the REBA analysis described above, as well as the photograph and all the tables and references required for the application of the REBA technique. For quick access use the link https://risksafety.blogspot.com/p/fatores-humanos_2.html.

8.5 Lessons learned

Human Factors are essential for the improvement of risk management and operational performance of all activities that involve a human—system interface. From the operation of nuclear power plant control rooms, aircraft control rooms and offshore rig control rooms, human—system interaction, when studied, evaluated and properly designed, can significantly reduce the error-inducing environment and its consequences. But it is not limited to environments that involve complex operations, but also in daily tasks, human—system interaction remains essential for improving risk management and performance in the execution of tasks. We have chosen the automotive GPS (Global Positioning System), a very common equipment that is part of people's daily lives, to exemplify the importance of Human Factors in risk management.

GPS is an electronic device that has been increasingly used to assist drivers in location tasks. Its operation is based on a network of 26 satellites positioned in orbits in such a way that at least the signals from 5 satellites are always within reach of a ground station. Only 4 satellites are sufficient to determine the altitude, longitude, and latitude of the receiver's antenna. With this system, a GPS device user has a safe reference to their position and, through maps and navigation software, they can drive to any address without requiring prior knowledge of the area. It is a complex system that requires reliable software to interact with the driver quickly, clearly and accurately, without causing distraction from the safety functions while driving a vehicle.

There are several types of GPS devices that can be used in a vehicle. Some drivers also use cell phones with GPS or other nondedicated devices. Only devices sold for specific use in vehicles shall be considered herein. We will discard the others for their total inadequacy, for lacking proper mounts and their use can result in an unsafe condition and, consequently, cause accidents.

8.5.1 Influence of Global Positioning System on the driver

The dashboard of an automobile has been the subject of studies of Human Factors and ergonomics for decades. A driver needs to split his attention between traffic and the instrument indicators on the vehicle's dashboard. Failure to pay attention to traffic can lead to an accident. But a failure to monitor instrument readings can also cause many problems, including fatal accidents. If we add a GPS navigator to the dashboard of an automobile, such problems will be maximized, and a failure in the understanding of a GPS instruction can also result in accidents.

8.5.2 Global Positioning System position in the dashboard

The overwhelming majority of new vehicles sold do not yet have a GPS navigator built into their dashboard. Only some models include it as standard equipment. Even for these vehicles, the presence of a GPS built into the dashboard significantly increases the driver's demand for attention to the controls. The driver using a GPS needs to pay more attention to the dashboard than a driver in a car without it. This is even more critical when the GPS is sold as aftermarket accessories.

When the GPS is integrated into the original vehicle dashboard, the car manufacturer has previously studied the best position and the most adequate interaction of the driver with the GPS. However, in most cases, this does not happen, because drivers purchase a GPS unit and without lacking proper knowledge, attempt to install it in an perceived ideal location in the vehicle. Regarding the original vehicle dashboard design, the GPS device can be considered a "foreign body" for which it was not accounted for. GPS device is often poorly positioned and hinders the driver's view. At other times the power supply cable interferes with the driver's movements. In other cases GPS mounts cannot withstand heat and get loose while the vehicle is in motion. Attention is demanded to the driver needs to demand attention while reattaching the GPS unit.

8.5.3 Audio information

During GPS navigation, the device provides audio information related to the actions to be executed by the driver.

Often times the driver, in addition to using GPS, keeps the radio on, talks to a vehicle passenger, and may even answer a cell phone call. Some GPS units are capable of centralizing various audio functions and send a Bluetooth signal to the vehicle's radio device. This causes other audio functions to be interrupted when the GPS sends an instruction. However, configuring the system correctly is time-consuming and it takes willingness before starting each day's drive.

8.5.4 Visual information

The audio information is not sufficient to guide the driver, so the driver's visual attention needs to be split between the road and the GPS screen. In most cases the GPS screen is too small, making it difficult to read, and at other times, the screen is so large that it partially obstructs the view of the driver. There may also be a lag between the audio information and the visual information. This implies more visual attention demand from the driver.

8.5.5 Software and configurations

Despite all designers' efforts, not all software for managing maps used in GPS is user-friendly. Some are slow and get locked-up while performing its functions, requiring a restart that is often performed while the car is in motion. There is also some level of complexity in the software configuration. There are different options for route calculations and recalculations when a driver error is detected. The use of GPS requires reasoning activities in parallel to the driving action, requiring mental effort and, in some cases, added stress.

8.5.6 Driver's GPS knowledge

To get the best performance from a GPS navigator, it is necessary to read the instructions manual and knowledge of navigation software and configuration requirements. As not all drivers perform these tasks, many GPS navigators are underutilized or do not work properly, and may even cause accidents.

8.5.7 Definition of the GPS position in the dashboard

Lack of knowledge about mounting the GPS unit on the dashboard can cause the device to fall during travel. It is necessary to mount the device in a position capable of proper signal reception from the satellites. If the device is positioned incorrectly, it may stop functioning intermittently due to poor signal.

8.5.8 Limiting the level of audio information

It is necessary to configure the volume, the language and most importantly the audio and communication signals that will be generated. Some poorly configured GPS devices emit so many sounds that specific guidance related to the direction to be followed can hardly be understood. Other GPS devices, just as poorly configured, remain mute and do not provide the minimum audio information required by their functions.

8.5.9 Visual information overload

Some GPS screens seem extremely polluted with excessive information as the result of improper configuration by the user. Others lack the bare minimum information, also due to configuration failure by the user. Users who require reading prescription glasses experience a conflict because they cannot wear the glasses for driving, but cannot read the information on the GPS screen without them. For these users GPS is bad, but in reality is the user's visual impairment that makes the system bad in the human–system relationship.

8.5.10 Human–system interface

Even though the driver is an expert in the GPS users manual and its navigation software, other anthropometric driver characteristics can make a big difference in term of system performance. Some GPS navigators have a touchscreen. To interact on such a small screen and drive at the same time, it is necessary to have specific physical skills and characteristics. A person with very large fingers can completely unconfigure the device in the midst of a 100 km/h trip, adding stress and increasing the risk of an accident. Usually the driver does not have the ability to make quick decisions during the GPS setup, while simultaneously maintaining the coordination of the movements required for driving. Such an ability has a great influence on the proper functioning of the GPS–driver system.

8.5.11 Conclusion

GPS is a very useful device for driving directions. However, the human–machine interfaces need to be improved and developed to reduce the error-inducing environment. The use of GPS affects the driver's attention in traffic and poorly installed and configured devices can cause serious accidents. The GPS user need to have good knowledge of the use of the equipment for the system to work properly. The choice of the GPS device should be appropriate for the cognitive and anthropometric characteristics of the driver, taking into consideration the vision and hearing abilities, besides other physical characteristics.

Transit authorities should assess whether it is time to develop a standard for audio and visual information on GPS. Each software and configuration generates its specific alarms and commands. A minimal standard could facilitate the drivers' understanding of the commands of GPS devices. The driving task requires great attention by the vehicle driver, and its execution is tied to the safety of human lives. Most GPS devices today are distracting to drivers. Authorities in collaboration with specialized technicians should seek appropriate standard solutions and the use of technologies such as the Heads Up Display (HUD), so that GPS can be used with maximum safety. HUD is a type of technology that enables the projection of GPS information directly on the vehicle's windshield, thus maintaining the position and orientation of the driver's field of vision compatible with the driving task.

As illustrated by this example, the simple addition of a small piece of equipment to a system familiar to the user can significantly overload the cognitive and physical demands. This increase in demand to perform the task contributes significantly to accentuate the environment for inducing human error, with increased risk of accidents and losses. This finding is not applicable to the GPS example only, but it also serves for any human–system interaction system, especially those involving high risks such as those present in the tasks performed in the facilities of the oil and gas industry.

8.7 Exercises

1. Evaluate the following statements related to Human Factors and fill in the blanks with "T" for true and "F" for false:
   ( )1. There is no difference between the meanings of the terms "Human Factors" and "ergonomics."
   ( )2. The HFER (Human Factors Engineering Review) tool includes the application of questionnaires in the form of checklists.
   ( )3. Human Factors issues applicable to computers are all hardware-related.
   ( )4. The best way to deal with problems related to human error is to improve the performance of people in executing tasks.
   ( )5. Tasks should be studied regardless of the workstation.
   ( )6. The most important factor for people not to make mistakes is training and education.
   ( )7. Most aspects related to the task execution environment cannot be modified.
   ( )8. In general, little can be done to change a culture.
   ( )9. Human Factors analyses can be contemplated in management routines by adding typical elements related to Human Factors, such as items to be evaluated in existing process safety programs.
   ( )10. Problems related to Human Factors can be identified and most appropriately treated in the design phase.
   ( )11. Before implementing a Human Factors program, it is necessary to develop standards and procedures for addressing the topic.
   ( )12. Most accidents are caused by human errors.
   ( )13. Human Factors analyzes consider people's characteristics and also the characteristics of the facility to be analyzed.
   ( )14. Recurring human errors can be avoided through a system of punishment for operators who systematically make mistakes.

( )**15.** Organizations have less impact on problems related to Human Factors at the facilities.

( )**16.** Displays and control screens have less influence in the study of Human Factors.

( )**17.** Human Factors analyses should be focused on the processes operations.

( )**18.** Human Factors is a technical term for human errors made in industrial activities.

( )**19.** The human–system interaction is treated only partially by traditional design safety management systems.

( )**20.** Human Factors are those that exert influence on the error-inducing environment related to the human–system interaction.

2. The following list contains some problems that can occur in facilities and process plants. Fill in the blanks with "E" for human errors, "F" for Human Factors, and "EF" for problems applicable to both concepts. Justify each answer.

( )**1.** A valve was not opened by an operator.

( )**2.** A nut was not fastened.

( )**3.** Project deficiencies.

( )**4.** Attention failures.

( )**5.** The pump identification label is incorrect.

( )**6.** An isolation valve was closed by an operator during normal operation.

( )**7.** The procedure text is confusing.

( )**8.** Operator's distractions in the control room.

( )**9.** Alarms intentionally disabled.

( )**10.** A storage tank overflows.

8.8 Answers

1. 1—F, 2—T, 3—F, 4—F, 5—F, 6—F, 7—F, 8—F, 9—T, 10—T, 11—F, 12—T, 13—T, 14—F, 15—F, 16—F, 17—F, 18—F, 19—T, 20—T.

2. 1-E: The operator was unable to fulfill his responsibility regarding the task.

3. E: The responsibility for executing a task has not been fulfilled.
4. F: Design deficiencies contribute to increased error induction.
5. F: Something in the environment has distracted the operator.
6. F: An incorrect indicator on the equipment label increases error induction.
7. F: The project allows a dangerous operation to be carried out, which increases the error-inducing environment.
8. EF: Human error when writing the procedure and the procedure contributes to expand the operator's error–inducing environment.
9. EF: Human error because a control room operator is trained not to be distracted, and there is something in the control room environment that can distract him, which is a contributing factor for increasing the error-inducing environment.
10. EF: Human error made by improperly disabling alarms and error-inducing environment due to the disabling of alarms being possible under conditions established by the project.
11. EF: Human error due to lack of monitoring by the operator and increase of the error-inducing environment due to the lack of overflow protection interlock.

8.9 Review questions

- Define "human error" from the viewpoint of risk management engineering.
- How are human errors classified with respect to the mechanism of occurrence?
- What is the meaning of the term "technological illiteracy" and how does this problem affect security?
- Define "Human Factors" from the viewpoint of risk management engineering.
- What is "design for Human Factors"?
- Explain the differences between the terms "ergonomics" and "Human Factors."
- What is HFES (Human Factors and Ergonomics Society)?
- Explain the difference between an AWH design and an AHW design.
- What are the types of approaches and studies related to Human Factors?

- Name the main influences related to Human Factors.
- Explain the REFH methodology and its steps.
- How are programs for treatment of Human Factors developed?
- At what phases of technological enterprises do the tools for improving Human Factors apply? Provide examples.
- What are the most likely reactions for operators under the influence of stress during emergency response?
- Explain why quantitative human reliability analysis techniques are considered limited to reduce human errors.
- What is API 770?
- What are the limitations of applying the ACH technique according to API 770?
- What is REBA and in what situations does this technique apply?
- Explain the meaning of the term HUD, "Heads Up Display."

CHAPTER 9

Risk management systems

Companies and organizations establish the distribution of responsibilities for conducting programs and risk management methodologies in their organizational structure. Positions for directors, managers and supervisors are created for work on risk management and safety-related matters. In this process, other closely related fields such as health and the environment can be added to safety, which has its advantages and disadvantages. As explained in previous chapters, the risk management strategic line (Figure 2.5) singles out operational technical knowledge as the most important item.

Therefore the purpose of this chapter is not to describe methodologies or the organizational structure for safety management in organizations. There are several models of organizational structure, but the models with directors, managers and supervisors per se do not ensure the technical and operational knowledge essential for effective risk management. Discussions on risk management model, organizational charts, job descriptions and organizational structures are very common. But efficiency requires administrative simplicity so that it does not cease to be the main focus, hidden within bureaucratic organization charts. Risk management is a technical engineering activity, not a generic management activity. When the word management is used in this field, it necessarily needs to be accompanied by another word forming the most appropriate term: Technological management. One of the main problems with organizational structures is the lack of technical and operational knowledge by occupants of positions established in the organizational charts. Unfortunately, organizational models allow this to happen, to a great extent because of over reliance on the organizational structure. It is a very frequent mistake to confuse a dedicated organizational structure with a solid technical operational knowledge base team. In this case, a good organizational structure is not sufficient to provide good risk management practices, on the contrary, it becomes a dangerous way to hide the lack of knowledge. The purpose of this chapter is to help experts identify the relevant technical and operational aspects, within the formal risk and safety management structures created by organizations, which are often bureaucratic and tedious.

Risk Management in the Oil and Gas Industry.
DOI: https://doi.org/10.1016/B978-0-12-823533-1.00009-5

If the most important item related to accident prevention is the knowledge associated with the phenomena that are part of the processes, then formal systems need to highlight and value such knowledge. The corporate routine generates meetings, minutes, procedures, documents, and historical data that need to be recorded but that have no intrinsic value if they are not supported by sound technical knowledge of operational activities. Nonspecialized managers can be familiar with accident occurrence charts the same way that they are familiar with cost charts. Technological management training is essential, that is, the manager needs technical knowledge related to the operational activities and to have technical authority to carry out the risk management activity effectively. It is not enough to know risk management tools such as Hazard and Operability Study (Hazop), Hazid, Fluid Dynamic Computing, etc. It is essential to know the operational practice of the activities to be protected. Leadership skills or to be well advised by specialists are not sufficient for working in this capacity. It is essential to have technical authority to make independent, critical safety decisions.

In this chapter, we will highlight some important characteristics of risk management systems formalized within organizations and companies. But, regardless of a good risk management system in place, or lack thereof, at an organization, experts in the field need to know how to act in any case, both exploring the strengths of established management, and avoiding existing traps in weak and bureaucratic risk management systems.

9.1 Risk management in the corporate environment

Every human activity generates risks related to these aspects and for that reason organizations need to develop a structured method for dealing with such risks, keeping them at acceptable levels. Risk management provides decision makers with a systematic approach to cope with uncertainty. This systematic approach needs to be established in compatibility with the risk management strategic line presented in Figure 2.5, in which the element of highest priority is operational technical knowledge. Risk management makes use of the organization's existing processes to identify, assess, and monitor risks. Total risk elimination is impossible. Risk management provides a means to reduce threats and, consequently, risks by keeping them at acceptable levels. This allows the organization to consciously move forward in its technological endeavors with respect to the associated risks, and not naively in a trial-and-error approach, without a technical foundation to avoid most disasters.

The corporate world creates a well-established environment of rules and practices, which gets consolidated as an organizational culture. Risk management also finds space within this culture. The experts need to pay attention to elements genuinely capable of offering an effective response to threats and risks (Figure 2.5). The corporate world often includes indicators, audit routines, inspection routines, but do not preserve technical knowledge as the risk management's highest value. Risk management systems in organizations are not built overnight, or rebuilt just as quickly after catastrophic accidents. Experts cannot use deficiencies of risk management systems as a reason to stop doing preventive work to avoid an accident.

On the contrary, risk management and safety experts need to understand the level the corporation's management system and to act to prevent accidents, regardless of the level in place. Obviously, for each level of corporate risk management the most appropriate methods, techniques, and tools should be employed.

Improving the level and quality of an organization's risk management can occur slowly, but nevertheless, objective actions can be taken at any time by managers aiming to avoid catastrophic accidents.

9.2 Centralization and decentralization of risk management

Risk management practice can be more centralized or more decentralized. Many organizations prefer it centralized and this is partly due to an attempt to mimic the organizational structure. However, risk management is an activity with multidisciplinary characteristics, which makes centralization limited type of management. No one has a better perspective to manage risks than those who operate the systems on a daily basis that are meant to be protected. External managers can offer tools and provide general guidance, but the knowledge required for understanding and making the best decisions belong to those who operate the systems routinely. Too much centralization of activities can create an abyss between the operation reality and its perception by the managers.

In reality, every engineer and every professional in the countless technical and scientific fields need to manage risks as part of their routine. Everyone needs basic training in this regard. The corporate system can act in a centralized fashion to monitor the safety of each field, but safety itself is totally dependent on the operators and specialists responsible for each field. No professional is more capable to identify threats in an electrical system than the electrician. Likewise for chemists, physicists, biologists and all other specializations. Excessive centralization of risk management systems can de facto reduce the liability of specialists and operators each field. This in turn ends up requiring consolidated and easily manipulated data without the ability of exposing more serious problems on the operational fronts. In his type of corporate scenario, activities may overestimate universal items such as the use of PPE (Personal Protection Equipment), occupational health problems, absence frequency due to accidents, among others. It does not mean that these aspects are not as important as the others, but due to the generalist view imposed by centralized risk management models, severe thermodynamic transients, dangerous operational maneuvers and inappropriate behaviors in control rooms, among others, become less likely to be detected so as to receive the right attention at the right time.

The ideal system is fully decentralized or decentralized enough so that the responsibilities of the operator and experts of each area are valued.

9.3 Association of different technical fields

Risk management is a multidisciplinary field, which is one of its main characteristics. Organizations often make mistakes when trying to meet administrative objectives by associating multiple disciplines with a single system. If this logic were correct, then the entire corporation should be subordinate to the risk management manager, because all corporate activities involve risks to a greater or lesser extent. It is clear that the objective of this strategy is to streamline corporate life, simplifying the organization's structure, reducing management, eliminating peers. The problem is that the specialist is required to have a multidisciplinary vision and training, but the management structure cannot include such different fields without conflicts. Doctors can take action to prevent an epidemic disease from spreading on an offshore rig, and the tools and means for that can involve risk analysis tools used in other areas.

However, engineers working on the design of fire fighting systems, despite using (in some cases) the same risk analysis tool as doctors, they are engaged in a completely distinct work with a different scientific bias. This can be extended to issues related to the environment, social impact, quality, property safety, population health, etc. It is our understanding that if a single manager accumulates responsibility for decisions in all these fields, in reality he will not make any decisions, but will simply confirm the decisions of each expert, which makes his activity totally meaningless and without credibility.

We recommend that experts in each technical field also become formally responsible for the risk management decisions associated with each field in their domain. Interdisciplinary interfaces can be dealt with in another way, according to tools that we will present later in this chapter. The risk management experts in each field require a multidisciplinary profile and training, that enables good dialog and a openness regarding the interface areas themes. However, this should not be confused with a risk management model made up of "generic managers," where decisions are made by someone regarding topics in which they lack the proper technical authority.

9.4 Historical data records and management by indicators

Some models are anchored in indicators. Indicators are extremely important resources to help identify improvement opportunities and help monitoring efficiency. But risk management cannot be limited to indicators. Although some indicators are mandatory even as required by the labor authorities, anchoring risk management in these indicators generates a poor form of management, vulnerable to accidents. The main components that influence the occurrence of accidents are the culture of safety and human factors. Indicators can hardly represent the real organization conditions regarding safety culture and human factors.

Data acquisition is essential to record, albeit in an incomplete fashion, the experience accumulated after each day's work. Based on data acquisition and formal records, several important and valuable indicators can be generated and several problems can be investigated. Lack of knowledge can lead to misleading data interpretation and the indicators results can be obtained a completely incorrect way. When there is deficiency in technical and operational knowledge, professionals tend to overestimate the value of statistical tools and mathematical models for data processing. This is due to the fact that statistics and mathematical data modeling are used in several fields of knowledge such as financial, accounting, administrative, etc., and for this reason they are easily accessible tools, which require no operational experience to be used. Risk management centered on technical and operational knowledge results in completely readings of the data and indicators, much richer and linked to the reality of the phenomenological activities and processes that make up the activity that is to be protected. Anchoring only in statistical data and indicators can be considered a naive approach to management, subjected to extremely poor interpretations, giving room for conclusions incompatible with the operational reality. Data, statistical results and indicators should be used, but always subjected to the eyes of experts with solid technical and operational experience to avoid typical distortions and naive interpretations that lead to incorrect conclusions.

9.5 Risk management, occupational safety and safety engineering

There is a recurring confusion related to the terms risk management, occupational safety, and safety engineering. Some other terms also add to the confusion: property safety, Health, Safety and Environment (HSE), safety, etc. The meaning of risk management, the central theme of this book, is presented in detail in Chapter 2, Fundamentals of Risk Management. In Brazil the terms operational safety and safety engineering are tightly associated with the regulatory requirements imposed by the Ministry of Labor. Regulatory standards by the Ministry of Labor is enforced by law and their noncompliance can generate fines and penalties for companies and organizations. Thus in Brazil, Safety Engineering training was developed as a graduate program, with the objective of training professionals to advise organizations and companies for compliance with the regulatory standards of the Ministry of Labor. Some of the best safety engineering graduate courses include a good technical foundation on general safety-related matters such as combustion, fire, ergonomics, fire extinguishing methods, first aid, etc. But what drives the institutions and professionals who take safety engineering courses are the legal requirements involving the need for the official "safety engineer" job position and compliance with the regulatory standards of the Ministry of Labor.

Therefore we need to warn that the "safety" training adopted in Brazil does not necessarily enable professionals with this title to make decisions about the risks of any technological enterprise. "Safety engineering" training in the end is very generic and, therefore, is focused on more general aspects such as occupational safety and compliance with the Ministry of Labor regulations. An engineer with only a graduate degree in engineering safety is not qualified to assess the risks of a transfer of Liquefied Natural Gas between ships and terminals, not qualified to assess the risks of fire and explosion in offshore platforms, or not qualified to assess the risks of accidents with loss of radioactive coolant in nuclear power plants. This does not devalue the training in safety engineering, but only clarifies that this training is specifically designed to prepare professionals to help organizations and companies to comply with the regulatory standards of the Ministry of Labor at an elementary technical level. In other words, the safety engineer trained in the Brazilian model is not a specialist in risk management, but a facilitator in the application of the Ministry of Labor, standards that are generic, of an elementary technical level and which are applicable to all professional activities, including those unrelated to engineering. Risk management depends on specific knowledge ofthe field that is to be protected and requires professionals with solid practical experience, making the so called "safety engineering" training in Brazil being completely unnecessary. In most other countries, there is no equivalent training as a "safety engineer" as in Brazil, aimed specifically at facilitating the application of the Brazilian Ministry of Labor regulatory standards. In the United States, for example, at the California State University, the safety course for engineers has a much broader scope and is focused on the physical and chemical phenomena that are part of industrial processes. This type of training comes closer to what may be basic initial training for the professional who wants to become a specialist. In reality, the risk management specialist needs to be trained from as a wide breadth of experience as possible in the operational area where one intends to work. An engineer of any specialization who accumulates significant operational experience in a given field can start an academic training for specialists in risk management. This can be achieved by complementing the training in their original fields with graduate courses available at the top universities.

These trainings are focused on risk and safety management for specific activities in the oil and gas industry, such as offshore, onshore, upstream, downstream, etc.

9.6 Risk-based design

Risk-Based Design (RBD) is a high-level risk management strategy that allows greater freedom for technological innovation regardless of the established international safety rules and standards that may pose limitation to new technologies. The strategy is to make massive use of in-depth technical and operational knowledge to prove that the risks of a given project are acceptable, even if the technology adopted in the project is so new that it has not yet been supported by international rules and standards. It is necessary that the standards applicable to the project recognize the RBD as a methodology. An example of such recognition is the openness that International Maritime Organization (IMO) offers to the subject matter. Many investments in new maritime safety technologies have been avoided over the years due to the certainty that the newly designed equipment would not find support in the official standards for its approval. Although the latest equipment and methods are known to be safer, at the time of the writing of the standards, there was no knowledge about these new technologies, and therefore these standards offer no means of supporting the application of these innovative technologies, moreover, waiting for updates/revision of the rules is almost always unfeasible within a project timeframe. As this is a well-known problem, designers and operators prefer to rely on well-established technologies, albeit at a lower safety level, since following this approach ensures the immediate certainty of the project's adherence to IMO standards. As a result, technological development in the maritime area was impacted by the lack of technological innovation for many years. But with the growing challenges related to building large passenger ships and other demands of maritime transport, IMO standards currently open up opportunities for innovative projects that are conducted within the RBD.

The technological advances attributed to the RBD strategy are currently made explicit by the rational methodology for dealing with safety matters, within an effective cost—benefit ratio. For best results the RBD strategy requires full adherence to the following principles:

- A consistent safety measurement means needs to be employed. In addition, a formal quantification procedure based on risk analysis needs to be adopted. The feasibility of this type of work taking into account the complexity of what is considered "safety" requires the establishment of a precise focus on the key points related to the specific safety of each technological enterprise (selection of the most important accidental scenarios to be studied). There are, in different contexts, numerous procedures and guidelines for conducting quantitative and qualitative risk analyses, and for risk management activities. Experts cannot get lost in the available procedures, but they need to focus all their attention on the following aspects: project safety objectives; identification of project hazards; analysis of risks associated with these hazards; risk management that is intended to be adopted throughout the operational life of the technological enterprise.

- The risk analysis procedures and methodologies to be adopted have necessarily to be integrated with the project's execution process in order to allow the interaction between safety and other important factors related to the project, such as performance, costs throughout the lifecycle of the enterprise and functionality. Consequently, advances in safety system design decisions need to be made available to all other designers, allowing the optimization of the decision-making process that each project discipline needs to practice.
- Taking into consideration the level of computational tools that may become necessary to manage all information relevant to safety, as well as the effect of changes related to safety decisions throughout the project that have implications for functionality and cost, it becomes necessary the use of a differentiated approach to handle the volume of information required for all project disciplines. Parametric models of shared understanding can allow a good interaction among all the parties involved in the project, through an easy, fast, and accurate access to essential basic data for an initial approach to the problems of each discipline.

9.7 Safety peer review

- Safety Peer Review (SPR) is a technique developed by us recently, already tested and effectively applied with success in facilities in the oil and gas industry. This technique had its origin in the nuclear industry, but its methodology has been adapted to the oil and gas industry reality with proven results for improving risk management and safety. The technique combines elements of risk analysis with objective actions. The major advantage of this technique is that it allows the organization of all the previous safety documentation, identifying recommendations of studies that have not been met. At the same time, the technique makes it possible to identify new safety recommendations in a practical way, based on the information exchange between operators who perform similar functions in equivalent, but distinct installations. This means that at least two facilities can be studied concurrently, as one of the principles of the SPR is the comparison between similar operating units. The technique consists of the following steps:

- Identification of safety studies conducted in the facilities' project and postproject phases, such as Preliminary Risk Analysis (PRA), Hazard Identification Analysis (HAZID), HAZOP, Quantitative Risk Analysis (QRA), Consequence Analysis, Layers of Protection Analysis (LOPA) and others;
- Technical evaluation of the recommendations arising from the analyses and studies conducted for the facilities under study regarding their effectiveness, feasibility, and regulatory compliance;
- Confirmation of the current compliance status with the recommendations of the analyses and studies, based on interviews with the technical and operational teams, field trips and inspection of the project documentation and reports issued;
- Issuance of a final report with updated status for each recommendation, based on objective documentary and field evidence and a "corrective action plan" to eliminate any pendencies that might have been identified.
- Holding of a SPR event for each facility under study, with the participation of the peers from both facilities in each capacity: operational, technicians and managers, based on the following routine:
  - Establish, jointly with the responsible managers, a calendar contemplating two events (one for each facility) gathering operators, technicians and managers with the objective of equalizing and updating information, comparing operational practices, identifying opportunities for improvements and sharing of operational experience, all these activities related to the safety area;
  - Identify topics relevant to safety for discussions with the operators and the technical support staff of the facilities under study;
  - Prepare a Safety Check List for application in visits to each facility, to be used as a field tool during the event, considering the best

operational practices recognized by the standards and international operating companies;

- Prepare a schedule contemplating for each facility: a meeting for the description and preparation for application of the "Safety Checklist"; field trip and "on-site" application of the Checklist; and meeting to prepare safety recommendations as the final result of the event;
- If necessary, assess the participation of an independent consultancy specialized in risk management to coordinate the event, which shall be considered coresponsible for the technical quality of the recommendations, adherence to the schedule and the final recommendations report;
- Suggest a list of similar facilities, as a reference for operational safety ("benchmark") for possible opportunities for future experience exchanges.

The success of the application of the technique relies on the proven track record in projects and operation of the type of facility under study. Another prerequisite is the participation of experts with proven experience in risk and safety analysis in facilities similar to those under study. Experts need to be familiar with the use of risk analysis tools such as HAZOP, HAZID, PRA, fire propagation studies, explosion and gas dispersion, with a history of services provided related to these topics.

The professionals involved in the coordination of a SPR should have at least 10 years' experience in risk and safety analysis, area classification, fire detection and fire fighting systems, escape and abandonment, HAZOP leadership and/or HAZID and other risk management tools.

SPR is one of the most complete and efficient tools to combine risk analysis techniques, operational experience, and practical actions. Due to the high importance of the requirement related to the technical profile of the coordinators of the use of the tool, it is possible for expressive and reliable results to be achieved. The action plans resulting from the application of the SPR technique are very realistic, since they are submitted upfront to operators who work daily at the facilities involved in the review.

9.8 Accident investigations

Accident investigations make it possible to understand the causes of accidental events, the extent of the losses and opportunities for improvement so that accidental events do not recur. The most important aspect for a successful accident investigation process is not the investigation methodologies or techniques, but rather the domain of the researchers' knowledge about the activity being investigated. But it is evident that good practices and investigative tools also contribute to the organization of data, reports and objective evidence, which makes the investigation more productive and accurate. As the domain of knowledge about the activity is the most important element of accident investigation, there is a wide variety of investigation routines, according to the industry type. Investigators are specifically trained to conduct investigations in their fields of technical expertise. Each field has its typical routine for conducting the investigation, taking into consideration the most relevant aspects, and the specific characteristics of the activity being investigated. But some common basic concepts applicable to various types and methodologies of research, regardless of the type of industry or activity. Next we will present some tools and concepts with such characteristics.

9.8.1 Systematic cause analysis technique

This technique is also known by the acronym SCAT (Systematic Cause Analysis Technique) and can be described by a typical sequence of steps as follows:

1. The first step of the technique is to describe the accident event, identifying the losses that occurred and evaluating the potential loss involved.
2. The second step is to frame the event described within a specific type of accident, for example, contact with some energy source, contact with some substance, collisions, falls, fires, explosions, leaks, radiation, system overloads, equipment failures, environmental damage, etc.
3. The third step is the identification of the immediate causes, more easily and directly associated with the description of the event. Immediate causes can be nonstandard practices such as unauthorized equipment operations, protection failures, resource allocation failures, operations at inappropriate speeds, deactivation of safety devices, use of defective equipment, lack of PPE, inadequate load movements, ergonomic failures, commissioning and operating failures, inappropriate human behavior, drug influence, equipment misuse, failures in the use of procedures, failures in the assessment of hazards and risks, failures in verification, reaction failures, communication failures, detection

failures, failures to correct nonconformities, etc. But other failures stem from conditions noncompliant with the expected standard, such as inadequate protections, inadequate safety equipment, defective tools, congestion, disorganized workplace, presence of hazardous materials without proper care, inadequate indicators and operational data, planning failures, climatic conditions, lack of knowledge of rules, design failures, lack of definition of technical requirements, failures in technological management systems.

4. The fourth step seeks the identification of the (basic) root causes, which are those that contribute with greater weight to the accident occurrence. Root causes can be associated with personal factors such as physical limitations, physical training for the task, mental and psychological training, physical and mental stress, lack of knowledge, lack of qualification, motivational problems, substance abuse, etc. Root causes can also be associated with typical work factors, such as problems of inadequate leadership or supervision, engineering failures, purchase failures, maintenance failures, equipment failures, inadequate norms and standards, communication problems, excessive amount of equipment wear, inefficient planning, inadequate emergency systems.
5. The fifth step seeks to identify specific control failures, including inadequate control systems, texts of inadequate control standards, inadequate compliance with control standards and also problems related to the overall management system. Relevant items such as leadership, planning, risk assessment, human resources, compliance assurance, project management, training and competence, communication and promotion, risk control, asset management, contract management, purchasing, emergency preparation and control, risk monitoring, etc.

9.8.2 5 Whys technique

It is a technique that uses the brainstorming tool (Section 7.1.5), but centered on the repetition of the question "why did the accident occur?" in order to identify root causes considered "below the surface" that are less visible in an initial approach. It is also known as the 5 W (5 What) or 5W1H technique (when the question "How?" is included at the end). The main characteristics of this technique are as follows:

- Easy to learn and apply.
- It can be a quick application method depending on the team and the development of the application of the technique.

- Dependent on the subjective judgment of the participants.
- In the sequence of questions, the trend is that the final answers are the most likely root causes, although this is not always the case.
- If the application of the technique is repeated by another team, the sequence of questions may arrive to different conclusions.

9.8.3 SHELL method (Software, Hardware, Environment, Liveware 1, Liveware 2)

This method consists of evaluating the interfaces between software, hardware, environment and liveware, paying attention to the points of elevated risk, until the relationship considered the most dangerous and responsible for the accident can be identified. It is a method widely used in the investigation of air accidents. Storms and winds are natural phenomena that contribute to accidents occurring mainly when the human element fails. Procedures, equipment, other agents and the environment as a whole can also add to a human error and trigger the process of accidental scenario development. The method studies the relationships among these elements seeking to identify the most dangerous relationship. The basis of the method is the five-interaction elements described below:

- Software (S): operating manuals, instructions, rules, regulations, drawings, flowcharts, maps, computer programs, etc.
- Hardware (H): equipment, physical installations, tools, instruments, etc.
- Environment (E): operating environment, natural influences, climatic conditions, visibility, sound, noise, vibrations, physical and mental stress, local safety culture, general culture, external pressures, etc.
- Liveware 1 (L): refers to the human element and relates to all other elements of the SHELL method. It includes aspects about the individual's physical and mental health, education, skills, decision-making ability, etc.
- Liveware 2 (L): refers to other people who interact with Liveware 1. If, for example, Liveware 1 is a control room operator, Liveware 2 can be a field operator, members of the maintenance team, managers, etc.

9.8.4 Causal tree or fault tree technique

It is a technique based on Boolean Algebra and the creation of a logical diagram, built using a deductive process of ordering causes and effects. A part of an accidental scenario is postulated (top event) based, for example, on a previous accident or a threat to future operations. In a deductive exercise, the possible causes of such an event are sought, as well as the possible combinations between these causes. The logical diagram is developed from logical symbols to represent the connections and interconnections between the causes that led to the top event. In analogy to a tree, the top event can be represented by the top of the tree, and each basic cause can be represented by a branch. The branches are interconnected through the trunk and it is fed through the tree roots, which in this analogy represents the most important root and basic causes, and that had the most influence on the occurrence of the accident. *The technique also performs the so-called "cut set" and "minimum cutting set."* A cut is a combination of causes that leads to the top event. A minimum cut is the minimum combinations that lead to the top event. The steps of an accident investigation based on the fault tree method are as follows:

1. Survey of event information: its objective is to obtain records of all available information including times of events, reports and observations collected by researchers at the event site. Hypotheses considered and all obscure facts that deserve further investigation are also surveyed.
2. Definition of the top event of the fault tree: accurate description of the accidental event that occurred so as to define it as the top event.
3. Assembly of the fault tree: based on the observation of the top event experts ask: what was necessary for the event to occur? Was the identified cause really necessary for the event to occur? Besides being necessary, was it sufficient?
4. Evaluation of the set of minimum cuts and their importance: the experts seek to identify the minimum combination of events that can explain the occurrence of the top event.
5. Analysis of the results obtained: the experts compare and evaluate the most relevant minimum cuts.
6. Recommendations and conclusions: the experts gather objective evidence in a conclusive discussion based on the results of the fault tree technique and present recommendations and conclusions to prevent the accident recurrence.

For the sake of the technique's greatest efficiency, it is important to sort the facts chronologically and clearly identify facts and hypotheses. Also essential is the accurate description of the top event. The recommendations to be recorded should be selected based on criteria that suppress recommendations with little added value.

9.9 Surveillance system

Surveillance System (SS) is a technique for the systematic execution of safety tests adopted in high-level risk management systems, characterized by rigorous safety tests execution frequency and treatment of the results obtained in each test. This tool had its origin in the nuclear industry, which has a system in place with the same principles for performing out the safety tests of nuclear plants. The technique can be used in any other industry (including oil and gas) with a high-level risk management system. The first step toward the application of the SS technique is the definition by risk management specialists of the list with all the important safety-related tests of the facility, the establishment of the test frequencies, those responsible for executions of the tests and those responsible for the availability of equipment/operating systems associated with the tests. All these data need to be input into a surveillance and control software which, from the first execution of each test, begins sending messages and information to the professionals involved, ensuring that no safety test is missed or deferred without an approved justification on record.

Surveillance software is a simple program that works as a database, and can be purchased off-the-shelf, or developed internally by organizations using simple and traditional computational tools in their engineering offices. The implementation of the SS requires specific technical preparation for each test that will be performed for the first time in the system. Before the official execution of the tool, experts need to ensure that the professionals involved in each test meet the technical requirement to enforce the strict compliance with the execution frequency to be established. If all the conditions for the execution of a weekly test are met to make it feasible, this might not mean that the organizational structure will still be prepared to repeat the same test the following week. Therefore it is essential to understand that the tests performed using a SS required planning with a high degree of technical maturity, thus allowing the fulfillment of the continuous execution within the established frequency.

As the number of safety tests included in the system grows, the complexity of the SS activities increases accordingly, requiring the continued attention by those involved with the execution of the tests and with the control of the software. Messages are issued reminding each tester of the upcoming scheduled safety tests that are under their responsibility. For each test, information about the testers, the operational conditions and, most importantly, the results obtained are recorded. For each nonconformity identified in the results, in addition to warning messages for those interested, the SS penalizes the facility and imposes deadlines for correcting the nonconformity. Penalties and deadlines vary according to the severity of the identified nonconformities are and can vary from a simple recommendation for the next preventive maintenance operation, to requirement of production reduction as a preventive and compulsory measure to find the solution to the problems. All penalties are associated with deadlines for the elimination of nonconformities.

In a SS, for example, testing a set of three Fire Water Pumps pumps at 50% can be included. Suppose, as an example, that after the test, one of the pumps fails. Depending on the characteristics of the facility, the SS may penalize it by reducing production to 75% of maximum capacity, for the first 24 hours after the test failure. After 24 hours, the SS penalizes the facility again, this time with a 50% reduction in production capacity. Finally, if the equipment has not been repaired within 48 hours, the SS penalizes the facility with its complete shutdown until the repair is completed. Evidently, this is only a didactic example and each equipment has to be studied and evaluated for the risks that they may represent should they fail. Through this analysis, penalties are established and, consequently, deadlines are set for solving the problems.

The SS is an extremely effective tool for keeping operational structures in continuous state of readiness for critical maintenance. It also contributes to the maintenance of stock levels ready for the replacement of key parts for the safety of the facility. People involved with the tests then can count on the support of the tool keeping the operational mobilization for the fulfillment of the execution of important scheduled safety-related tests. Some SSs have the limited capability of working with no more than ten tests, while others support tens and hundreds of tests. Several industrial facilities, with a high level of risk management, have used this system for decades with proven successful results.

9.10 Capillarity of concepts and principles of risk management

In this section, we want to note that there is a wide range of application of the concepts of risk management and safety. One of the difficulties in this area is its multidisciplinary nature and the extent of the level of details required to solve safety problems of each discipline. As we have shown in previous chapters, risks are part of the reality of life, and human existence is impossible without living with risks. In other words, life is a succession of acceptance and rejection of risks and therefore the concepts of risk management are applicable to almost anything. Many tools and concepts presented in this book, with a focus on the oil and gas industry, can also be applied in other industries and even in other human knowledge activities. More than that, the risk management concepts, especially those related to human factors and safety culture, are useful for our personal lives. It is possible to apply such concepts and tools for risk assessment in the economic, medical, and commercial fields. More than techniques, concepts and tools, the main element for achieving success is to master the knowledge about the activity to be protected.

9.11 Risk and safety management in the energy industry postpandemic COVID-19

As presented in the previous chapters, the area of risk and safety management is multidisciplinary and complex. Risk and safety management experts find application of their expertise in virtually all human activities. Part of the themes and problems addressed in the area of risk and safety management is mathematically modelable. Another part is subjective, influenced mainly by the interference of human behavior. At the end of 2019 China began to spread a new disease and this caught the attention of the whole world. But in the beginning of 2020, what would be another viral outbreak became a global pandemic with economic, environmental consequences, losses to the image of countries and organizations, in addition to many human losses. The occurrence of a global pandemic has always been a scenario feared by risk and safety analysts and managers. However, until then this was a scenario considered to be of low probability, especially when we raised the hypothesis of a virus becoming lethal and contagious to the point of reaching the entire planet.

But this scenario identified as "low probability" became real, confirming that, despite all the safeguards recommended by risk analysis techniques, there was no way to guarantee absolutely that an unlikely scenario could not happen. In reality, we cannot include unlimitedly all possible scenarios in professional risk analyzes. We need to establish qualitative and quantitative criteria to select the most likely and relevant scenarios in order to create technical safeguards that reduce human, environmental, property, and image losses. However, those with very low probability of occurrence scenarios should not be completely despised by those working in the area of risk management and safety. As far as possible, these remote scenarios should be at least the target of consequences analysis, because in the case of the unfortunate occurrence of these apparently unlikely scenarios, at least we will have a mental exercise previously performed, increasing the chances of dealing better with the adversity of catastrophic scenario such as the one established by the pandemic COVID-19.

The fact is that, since March 2020, the world has changed its form of social coexistence, work, transportation, communication, production, which has affected all human activities, planning, and future perspectives. The impact of COVID-19 pandemic exceeded all expectations and will bring consequences and definitive changes for humanity. Experts worldwide try to assess possible consequences of the scenario of COVID-19 pandemic, but what really prevails so far is the uncertainty about the consequences of this pandemic and the uncertainty about the possibility of other pandemics arise in a global world in constant interaction is this face-to-face or virtual interaction.

All areas of human activities were directly or indirectly impacted by the pandemic. But what would be the main points to be noted by risk and safety management experts in relation to the energy industry in the world, especially the oil and gas industry?

The changes in the energy matrix that were already underway before the pandemic were directed to reducing carbon emissions. There will be new trends of change after the pandemic? What about changes in the behavior of postpandemic? That is transitory? What change is definitive? What will change in human activities? Can we go back in the search for cleaner energy sources? The world will proceed fastest in sustainable energy matrices after COVID-19 pandemic? How should these risks be considered in risk and safety management work in the oil and gas industry? We will seek answers to these questions at a time when it is not possible to confirm definitive and absolute answers. But the oil and gas industry continue to produce. Despite the dramatic changes triggered by the COVID-19 pandemic, how should we deal with the need for continuity in risk management and safety activities going forward?

The COVID-19 pandemic can trigger an acceleration of process changes in the global energy industry, including disruptive changes in the energy matrix, generating a direct threat to the oil gas industry that we know today. However, precisely because of the multidisciplinarity that characterizes the area of risks and safety, specialists in this area may consider easier to face the transition between the energy matrix that prevails today and the new energy industries that are emerging as promising trends for the coming years and decades. We are in a transition to a new world that needs to change quickly, after being surprised by an unimaginable pandemic. There are changes that we can already consider as definitive and others that can promote an acceleration of disruptive changes in the area of power generation. It is the arrival of the "new" without the right to "rehearsal," without the right to "preparation." It is the urgent arrival of the "new" that is necessary, in some cases, necessary for survival.

There is no doubt that the current trend is that the leading role in the area of energy generation will change over the course of this decade, until 2030. Our proposal is to be prepared to use what we have learned so far to protect life, property, the environment and the image of organizations and countries in a world with a new energy matrix. We are in a process of change that may become greatly accelerated by the surprising effects of the pandemic COVID-19.

9.11.1 "World energy outlook": International Energy Agency

Based on the "World Energy Outlook 2020" Report issued in October 2020 by the International Energy Agency (IEA), we will exercise in this item our ability to identify "new paths" for risk and safety management. In this exercise, we will build on the postpandemic scenarios COVID-19 postulated by experts from the IEA. They indicate the probable new protagonists for the generation of energy in the world, replacing the forms of energy with high CO_2 emission and low efficiency, as is the case of the oil and gas industry. Those who dedicate themselves to study and knowledge in the area of risk and safety management should not fear the future, but rather anticipate it by further improving their current concepts and knowledge, contextualizing them to the new forms of energy generation that will probably take center stage by 2030.

The IEA's "World Energy Outlook 2020" Report analyzed in detail the possible impacts of the COVID-19 pandemic on the energy sector over the next 10 years. One of the points of attention is the verification of whether the transition to cleaner forms of energy generation will be stagnant or accelerated. In the year 2020, due to the reduction of productive activities and the need for social distance, there was a record of global drop in carbon emissions. But despite this, the world is far from ensuring a steady period of declining emissions. This is because this record fall in emissions in 2020 was not the result of a change in culture or concept, but it was a change forced by the circumstances created by the pandemic. The corona virus has caused more disruption to the energy sector than any other event in recent history, and the consequences of this will be felt for years, perhaps decades.

The IEA's expert assessment is that global energy demand fell by 5% in 2020, and this drop caused a 7% drop in CO_2 emissions and an 18% reduction in investments in the energy sector. Analyzing the issue from the point of view of the type of fuel, it is estimated that the drop in demand for oil was 8% and for coal 7%. This contrasts with the estimate of a slight increase in the contribution of renewable energy sources to the global energy matrix. With regard to natural gas, experts estimate that there was a 3% drop in demand in 2020, while the drop in demand for electricity appears to have been less, falling by only 2%.

The analyzes point to a decline of 2.4 gigatonnes (Gt) in annual CO_2 emissions in 2020. This means that the world is back to where it was a decade ago in terms of CO_2 emissions. Despite the reduction in oil and gas production, experts concluded that there was no drop in the same proportion in 2020 in methane emissions, a powerful greenhouse gas. Despite the experts' efforts, uncertainties prevail amid so many changes and possibilities for developments. At the beginning of 2021, uncertainty remains about the duration of the pandemic, its economic and social impacts. And the responses that society and industry are producing in relation to the pandemic open up a wide range of possibilities for the direction of the power generation industry. In the midst of so many uncertainties, the "World Energy Outlook 2020" Report issued by experts from the IEA, evaluates the four main postpandemic scenarios for the power generation industry until 2030:

- The Stated Policies Scenario (STEPS), in which Covid-19 is gradually brought under control in 2021 and the global economy returns to pre-crisis levels the same year. This scenario reflects all of today's announced policy intentions and targets, insofar as they are backed up by detailed measures for their realization.

- The Delayed Recovery Scenario (DRS) is designed with the same policy assumptions as in the STEPS, but a prolonged pandemic causes lasting damage to economic prospects. The global economy returns to its precrisis size only in 2023, and the pandemic ushers in a decade with the lowest rate of energy demand growth since the 1930s.
- In the Sustainable Development Scenario (SDS), a surge in clean energy policies and investment puts the energy system on track to achieve sustainable energy objectives in full, including the Paris Agreement, energy access and air quality goals. The assumptions on public health and the economy are the same as in the STEPS.
- The new Net-Zero Emissions by 2050 case (NZE2050) extends the SDS analysis. A rising number of countries and companies are targeting net-zero emissions, typically by midcentury. All of these are achieved in the SDS, putting global emissions on track for net zero by 2070. The NZE2050 includes the first detailed IEA modeling of what would be needed in the next 10 years to put global CO_2 emissions on track for net zero by 2050.

A dark period will be inevitable following the pandemic. Processes and behaviors will be questioned, some reduced and others definitively replaced. Of course, there will be losses for some and gains for others. An avalanche of changes initiated in an unplanned way, at an unimaginable speed, generating reformulation of concepts, ways of working, ways of producing and living.

Considering the STEPS scenario, the demand for energy would recover to the prepandemic level in early 2023 but that date would be delayed until 2025 if the pandemic continues beyond 2021 as established by the DRS scenario. Before the pandemic, the estimated growth in demand for energy was 12% between 2019 and 2030. The most optimistic scenario, STEPS, this percentage drops to 9% while in the most realistic scenario, DRS, the growth in energy demand in the same period it would not exceed 4%.

The DRS scenario would generate cascade effects, some very surprising. If the pandemic continues, experts foresee an inevitable aggravation of the global economic crisis, with income reduction mainly in developing countries. This would cause a reduction in the size of houses, houses, equipment, vehicles, all motivated by the drop in income caused by the economic crisis. IEA experts estimate a reduction in the average household size of 5% by 2040 associated with a reduction of 150 million refrigerators in use, which would further reduce energy demand. An estimated reduction of 50 million fewer cars is also expected in the DRS scenario when compared to the STEPS scenario.

Results obtained from efforts to bring access to electricity to poor countries would also end up lost, especially in Africa where in 2019 approximately 580 million people still did not have access to electricity. Several projects were being developed with the aim of reversing this situation, however, they should hardly be put into practice if the scenario that prevails is the DRS.

In the midst of conditions contrary to the growth in world energy demand, solar energy seems to have the tendency to become the new "King" of electricity. In all four postpandemic scenarios defined by the IEA, renewable energies appear with rapid growth, but at the center of the constellation of new possibilities for technologies for electric generation is solar energy. Support policies and technological maturity are making it easier and easier to access solar energy. There has been a big reduction in the costs of producing electricity by solar energy in the last decade. This has made solar photovoltaic energy consistently cheaper than new coal or natural gas plants in many markets. Currently, solar projects are able to offer the lowest electricity costs ever seen.

In the STEPS scenario, it is estimated that renewable energy is expected to account for 80% of the growth in energy demand by 2030. Hydroelectric plants remain the largest renewable source of electricity generation, but solar energy is expected to become the main driver the growth of renewable energies in the world, as it is expected to set new implantation records every year after 2022. Wind, onshore and offshore energy will also play an important role. The advancement of renewable sources of electrical generation, especially solar generation, is much stronger in the SDS and NZE2050 scenarios. The accelerated pace of change foreseen for the electric sector places an additional advantage for robust energy matrices that have diversified sources and that promote the flexibility of the system through the use of renewable generation means. Despite the growth of renewable sources, the crisis caused by the pandemic COVID-19 created risks for the global power generation structure.

According to the IEA Report, electric power generation networks could become the transformation link in the electric sector in the next postpandemic years, with implications for the safety and reliability of electricity distribution. The need for expansion projects for new transmission and distribution lines worldwide in the STEPS scenario is 80% greater in the next decade than the expansion seen in the past 10 years. When we go through a process of rapid transition from sources of energy generation (such as the one caused by the pandemic COVID-19), the importance of electricity distribution networks increases significantly as it is necessary to connect new and diversified sources of energy to the system. But the economic crisis has also affected the finances of many companies and countries, especially in developing economies. There is no balance between the resources needed for the design, construction and assembly of intelligent, digital and flexible electricity networks in relation to the expected revenues for the transmission network operating companies. This creates a very high risk for investors in the postpandemic scenario.

Traditional energy sources such as coal had a significant drop in world demand as a result of COVID-19. Experts believe that in the STEPS scenario, demand for coal will not return to precrisis levels and that the share of coal in the world energy matrix may reach, for the first time since the Industrial Revolution, values below 20% by 2040. Utilization of coal for power generation is expected to be strongly affected with the sequence of reductions in global demand for electricity. Policies for phasing out coal, increasing renewable energies and competition from natural gas are expected to lead to the elimination of 275 gigawatts (GW) of coal-fired electricity generation capacity by 2025 (13% of 2019 total), including 100 GW in the United States and 75 GW in the European Union. Increases in coal demand in developing economies in Asia are expected to be significantly smaller than expected and insufficient to offset declines in other markets. Coal's share in the global power generation mix is expected to decline from 37% in 2019 to 28% in 2030 in STEPS and to 15% in the SDS scenario.

Now let's analyze the point that most interests us: the possible decline of the oil and gas industry. According to the IEA Report, unless there are any additional policies aimed at rapidly changing the global energy matrix, current policies are insufficient to promote a rapid decline in the oil and gas industry. In other words, this means that the era of growth in global oil demand is expected to end at the end of the next 10 years, but the way the economy is expected to recover during this period creates fundamental uncertainty. In both STEPS and DRS, demand for oil is expected to stabilize in the 2030s. However, a prolonged economic slowdown would take more than 4 million barrels per day (mb/d) of oil demand in DRS, compared to STEPS, keeping it below 100 mb/d. The behavioral changes resulting from the pandemic have a dual effect. The longer the routine interruption, the more ingrained changes that affect oil and gas consumption become, such as working from home or avoiding air travel. However, not all changes in consumer behavior would hurt oil. The oil and gas industry would benefit from a short-term aversion to public transport, the continued popularity of SUVs and the late replacement of older, inefficient vehicles.

In the absence of a more radical advance in current policies for changes to a renewable energy matrix, according to the IEA, it is still too early to predict a rapid decline in oil demand. The natural increase in income over time in emerging markets and developing economies creates a strong underlying demand for mobility, offsetting reductions in oil use in other more traditional markets. But transportation fuels are no longer a reliable factor in ensuring the growth of the oil and gas industry. The use of petroleum products in passenger cars is expected to peak in both STEPS and DRS, reduced by continuous improvements in vehicle efficiency and robust growth in sales of electric cars. The use of oil products for chartered vehicles and long-distance transportation, according to the IEA Report, is expected to vary according to the prospects of the global economy and international trade because if the economy is bad, long-distance and cargo transportation in general it is also negatively impacted. The upward pressure on oil demand will increasingly depend on its increasing use as a raw material in the petrochemical sector. Although there is a forecast for an increase in recycling rates, there is still much room for increased demand for plastics, especially in developing economies. However, as the oil used to make plastics is not burned, the scenarios studied by the IEA show a peak until 2030 (and the subsequent decline) in total oil-related CO_2 emissions.

With regard to natural gas, until the pandemic COVID-19 the main highlight was the abundance and availability of this resource. In recent years, natural gas has performed better than other fossil fuels, but different political contexts can produce strong variations. Based on the scenarios foreseen in the IEA Report, in the STEPS scenario there should be a 30% increase in the global demand for natural gas by 2040, an increase that is concentrated in South and East Asia. Political priorities in these regions—notably an impulse to improve air quality and support industry growth—combine with lower prices to support the expansion of gas infrastructure. In contrast, this is the first IEA Report in which STEPS projections show that gas demand in advanced economies will decline slightly in 2040. An uncertain economic recovery also raises questions about future prospects regarding the record number of new facilities exports of liquefied natural gas approved in 2019.

The IEA Report also points to a trend toward greater transparency in the calculation of methane emissions with implications for the environment, by tracking different sources of emissions of this gas. In economies based on high carbon energy sources, natural gas continues to benefit from producing lower emissions when compared to coal. However, this is less important in countries planning a path to zero net carbon emissions, where coal is generally already on the decline. Methane emissions along gas supply chains—as highlighted in the IEA's Methane Tracker—remain a crucial uncertainty, although better data from companies and from aerial measurements, including from satellites, should soon improve the understanding of emission sources methane across the energy sector. In Europe, considering the STEPS scenario and, in all parts of the world in the SDS scenario, the challenge for the gas industry is to retrofit for a different energy future. This can come through demonstrable progress even with reduced demand for methane, through alternative gases such as biomethane and hydrogen with low carbon emissions, and technologies such as carbon capture, utilization and storage (CCUS).

What are the biggest dilemmas for the oil and gas industry and what are the risks associated with these dilemmas? According to the IEA report, lower prices and falling demand, resulting from the pandemic, reduced the value of future oil and gas production by about a quarter. Many oil and gas producers, especially those in the Middle East and Africa, such as Iraq and Nigeria, are facing acute fiscal pressures as a result of the high dependence on oil revenues. Now, more than ever, fundamental efforts to diversify and reform the economies of some of the major oil and gas exporters seem inevitable. The United States shale industry has met almost 60% of the increase in global demand for oil and gas in the past 10 years, but that increase has been fueled by easy credit that has now run out. By 2020, major oil and gas companies had reduced the reported value of their assets by more than $50 billion, a palpable expression of a shift in perceptions about the future. Investment in oil and gas supplies fell by a third compared to 2019, and the extent and timing of any resumption of spending is unclear. The same is true of the industry's ability to respond to a possible resumption in a timely manner: this may portend new price cycles and risks to energy security.

Low-cost energy sources that generate low carbon emissions and the diversification of energy sources are becoming the keywords for many economies and for oil and gas companies. Falls in the production of existing fields create the need for new upstream projects, even in scenarios of rapid transitions from energy sources. However, investors are increasingly skeptical about oil and gas projects due to concerns about financial performance and the compatibility of companies' strategies with environmental objectives. Some of the financial concerns may ease if prices rise and projects start to offer better returns, but doubts about the industry's contribution to reducing emissions will not go away.

Some experts from the IEA consider that, given the current situation, the world has become unprepared for a definitive decision to reduce carbon emissions. According to the IEA report, global emissions are expected to recover more slowly than after the 2008–2009 financial crisis, but the world is still a long way from a sustainable recovery. CO_2 emissions at STEPS would exceed 2019 levels on the way to 36 Gt in 2030. Emissions are lower in the event of a delayed recovery, but a weaker economy also drains the momentum from the change process in the energy sector. Lower fuel prices, compared to precrisis trajectories, mean that the payback periods for efficiency investments are extended, slowing the rate of improvement in overall efficiency. The pandemic and its consequences can suppress emissions, but low economic growth is not a low-emissions strategy. Only an acceleration in structural changes in the way the world produces and consumes energy can break the emission trend forever, says the "World Energy Outlook" Report of the IEA.

But experts point to sustainable path options for overcoming the crisis caused by the pandemic COVID-19, in the hope that the air will become cleaner than in the period of lockdowns that occurred in 2020. A radical change in investment in clean energy, in line with the IEA's Sustainable Recovery Plan, it offers a way to drive economic recovery, create jobs, and reduce emissions. According to the IEA Report, this approach has not been highlighted in the plans proposed so far, except in the European Union, United Kingdom, Canada, Korea, New Zealand and some other countries. In the SDS scenario, full implementation of the IEA's Sustainable Recovery Plan, published in June 2020 in cooperation with the International Monetary Fund, would put the global energy economy on a different postcrisis path. The additional investment of US $1 trillion per year between 2021 and 2023 in the SDS scenario is directed toward improvements in efficiency, power generation networks and electricity distribution with low emissions and more sustainable fuels. That would make 2019 the definitive peak for global CO_2 emissions. In 2030, emissions in the SDS scenario would be almost 10 Gt less than in the STEPS scenario.

According to IEA experts, cities see major improvements in air quality by 2030 if the SDS scenario prevails, without interruptions in economic activity or in the lives of people who changed their routines and cleared the air in 2020. The IEA states that in the next 10 years, lower emissions from power plants in urban areas, lower emissions from residential heating units and industrial facilities should be recorded in the SDS scenario. According to IEA experts, the drops would be 45%–65% in the concentrations of fine particles in cities, and cleaner transport would also reduce other pollutants on the streets. The main pollution reductions in developing economies also come from improving access to an emission-free kitchen. The SDS scenario, as defined in the IEA Report, would not completely eliminate all sources of air pollution, but while the number of premature deaths from poor air quality would continue to increase in the STEPS scenario, in the case of the SDS scenario, more than 12 million premature deaths in the next decade.

Avoiding new emissions is not enough: if nothing is done about emissions from existing infrastructure, climate targets will certainly be out of reach, say IEA experts. According to the IEA Report, a new detailed analysis shows that if today's energy infrastructure continued to operate from now on as it did before the pandemic, it would lock itself in by causing a temperature rise of 1.65°C All of today's power plants, industrial facilities, buildings and vehicles would generate a considerable level of future emissions if they continued to rely on massive combustion of fossil fuels. If all of these assets, as well as the plants currently under construction, were operated for similar lifetimes and in ways similar to those of the past, they would still be emitting around 10 Gt CO_2 in 2050. That is why the SDS scenario does not include only the much faster deployment of clean energy technologies, but it also foresees the operation of existing assets, which were previously large emitters of carbon, in a technologically very different way than it would be in the STEPS scenario. The existing coal-fired power plants, for example, would need to be adapted, reused or taken out of operation in case the SDS scenario prevails. This, according to IEA experts, could cut coal emissions in half by 2030.

In reality, according to IEA experts, to transform the energy matrix, it would be necessary to extend the changes to sectors other than the electricity generation system. The energy sector would take the lead, but a wide range of strategies and technologies would be needed to deal with emissions in all parts of production activities. The IEA Report indicates that emissions from the energy sector have fallen by more than 40% by 2030 in the SDS scenario as a result of the annual growth in photovoltaic solar energy production, which would triple compared to today's levels. Electricity would play an increasing role in overall energy consumption, as increased production of renewable energy would help to reduce emissions from sectors—such as passenger transport—whose electrification has become economically viable. The most difficult tasks for the transformation of the energy sector lie elsewhere, particularly in industrial sectors such as steel and cement, in long-distance transport, in balancing various changes that occur in parallel considering a complex energy system where we would still have to consider factors such as population acceptance and public interest. Maintaining a strong pace of post2030 emissions reduction requires a focus on energy efficiency, and increased electrification. Important space opens up for the use of liquid and gaseous fuels with low carbon content. Low-carbon hydrogen and technologies such as CCUS would increase significantly, based on a decade of rapid innovation that would be implemented in the 2020 s.

IEA experts believe that the vision of a world with zero net emissions is coming into focus. Ambitious actions would be needed in the coming decades. The ambitious path set by the SDS scenario depends on countries and companies reaching their announced zero net emissions targets, within the expected deadline and fully fulfilled. These are primarily targets for 2050, although there are countries that have set earlier targets and, more recently, China announced a date for carbon neutrality in 2060. Achieving these targets is important not only for the countries and companies involved, but also for accelerate progress elsewhere, reducing technological costs and developing regulations and markets for low emission products and services.

The IEA Report states that reaching net-zero emissions globally in 2050, as predicted in the NZE2050 scenario, would require a set of additional dramatic actions over the next 10 years. Achieving a 40% reduction in emissions by 2030 would require, for example, that low-emission sources provide almost 75% of global electricity generation in 2030 (compared to less than 40% in 2019), and that more than 50% of passenger cars sold worldwide in 2030 were electric (2.5% in 2019). Efficient electrification, significant energy efficiency gains and behavioral changes would play important roles, as would accelerated innovation in a wide range of technologies associated with the production of hydrogen by electrolysis. No part of the world economy could be left behind to achieve a net zero of global emissions.

Experts understand that to achieve zero-net emissions, governments, energy companies, investors and citizens would need to be involved and everyone would have unprecedented contributions to make. The changes that would provide the emission reduction envisaged in the SDS scenario are far greater than many imagine and would need to happen at a time when the world is trying to recover from the COVID-19 pandemic. These changes would need to rely on the continued support of politicians around the world, while they would also need to meet the development aspirations of a growing global population. Achieving zero-net emissions globally by 2050 goes far beyond that, both in terms of actions within the energy sector and those that would be needed in all other sectors of the economy. For any path toward net-zero emissions, companies would need clear long-term strategies backed by investment commitments and considerable impact. The financial sector would need to facilitate a significant development of clean technologies, help transition fossil fuel companies, grow important energy businesses, and bring low-cost capital to the countries and communities that need it most. Involvement and choices made by citizens would also be crucial, for example, in how they heat or cool their homes or how they travel.

According to the "World Energy Outlook 2020" Report issued in October 2020 by the IEA, governments have a decisive role in this whole process of change. At a time when the COVID-19 pandemic creates extraordinary uncertainty, governments have unique capabilities to act and guide the actions of other sectors of society. They can lead the way, providing strategic vision, stimulating innovation, incentives for consumers, policy signals and public finance decisions that catalyze the action of private actors as well as support for communities where livelihoods are affected by rapid changes. They have a responsibility to avoid unintended consequences for the reliability or accessibility of the power supply. Our sustainable and secure energy future is a choice for consumers, investors and industries, but above all for governments, concludes the Report issued by IEA experts.

Reference link for direct access to the "World Energy Outlook 2020" Report issued in October 2020 by the IEA: https://www.iea.org/reports/world-energy-outlook-2020.

9.11.2 Risk management in the postpandemic world energy future

Considering the world energy panorama presented in the previous item, in the next 10 years the activities of exploration and production of oil and gas should maintain their continuity, however at a less accelerated pace. From 2030, although some scenario options allow an expectation of a rapid decline in oil and gas exploration and production activities, this is realistically unlikely as a radical and global change in government policies favoring energy generation means would be necessary. renewable sources, accompanied by a perfect alignment of public opinion, companies and organizations worldwide, in addition to, of course, the full support of final energy consumers across the world economy.

In order for a rapid decline to occur in oil and gas exploration and production activities, it would be necessary that in all countries of the world, energy generation projects starting in 2020 adopt the zero emissions target. At the same time, the entire existing power generation industry today would need to be reformed and modernized to align with the goal of zero emissions. A perfect global alignment would be necessary, involving energy generation and distribution activities, transport modes, the behavior of consumers, investors and government officials. Virtually the entire planet would need to come together and prioritize the rapid development of technologies for generating zero emissions energy.

Such a scenario is, in theory, possible, but a rapid and definitive decline in the oil and gas industry is very unlikely, even with the effects of the COVID-19 pandemic favorable to the decline of the oil and gas industry. Such postpandemic effects are expected to cause a great acceleration in all changes that are directed toward a cleaner global energy matrix and less and less dependent on the oil and gas industry. In practical terms, the most likely is a slow, not a rapid, decline in the oil and gas industry, mainly from 2030 on. Until then, oil and gas exploration and production activities will continue to grow, and a slow decline in oil and gas activities will continue. exploration and production is expected to begin in 2030. This means that for several decades to come, it will most likely still be necessary to continue to apply and improve technology and knowledge for risk management in the oil and gas industry.

With the COVID-19 pandemic, its effects and the resulting drop in demand for oil and gas in 2020, many things will need to change faster than we could have imagined before the pandemic. The significant reduction in demand for oil and gas is expected to extend until at least 2025, with effects until the end of the decade. And this drop will promote a global environment more favorable to the evolution of new energy generation techniques. In a world that needs less oil and gas, other forms of energy will gain a greater share in the global energy matrix. The need for oil and gas is associated with an economic growth in which companies, countries and final consumers become "hydrocarbon eaters." This rampant "hunger" for oil and gas does not seem to make any more sense in the possible postpandemic scenarios, which are expected to start a downturn in global economic growth with radical changes in behavior in human activities.

Engineers and experts used to working on project risk management and operating oil and gas exploration and production units will have enough time to make the slow transition from oil and gas industry risk management to risk management of the new power generation industries that are expected to take center stage from 2030. If the oil and gas industry has its main danger liquid and gaseous hydrocarbons, the new energy generation technologies that appear on the horizon include their own risks.

The new energy generation technologies get stuck and often even stop evolving due to difficulties related to the new risks that these technologies generate. A classic example is nuclear energy, considered a form of energy generation with virtually zero carbon emissions. Nuclear energy has been present in the energy matrix for decades, but the risk of radioactive contamination by liquid and gaseous effluents is a challenge for specialists. Not to mention nuclear waste, an unsolved problem with consequences that can reach generations and even last for centuries, with a much higher and irreversible environmental impact when compared to the effects of carbon emissions. It is also necessary to consider the environmental impact of waste from batteries that are necessary for electric cars, power plants, and consumers of new sources of energy generation. The use of this equipment on a large scale, for example, in means of transport, creates risks very different from the current risks. They are equipment that concentrate a lot of energy, with chemical risks involving contamination, difficulties of disposal and recycling, maximization of fires and explosions. It is important to note that new sources of energy generation cause interference in the environment and the risks associated with this need to be managed, as much as was already necessary in the use of old sources of energy generation. New wind and solar plants, whether offshore or even onshore, will need to have their risks analyzed and mitigated by risk management experts. Certainly, the lessons learned from the long cycle of the oil and gas industry should not be wasted, but redirected toward managing the risks associated with these new technologies.

In practice, it is recommended that risk and safety experts working in the oil and gas industry follow the same strategy as companies and operators that lead, with good ethics, the gas—oil industry. Virtually all oil companies have a discourse of interest in new sources of energy generation, in a clear demonstration of alignment with new world trends in search of a more renewable, less polluting energy matrix. However, not all oil companies actually make consistent moves toward developing alternative energy sources.

Most companies need to show some interest in alternative energy sources in order to survive in an increasingly demanding market in relation to this issue. However, they lack technical knowledge and do not carry out significant research in the area of new energy sources. But among the true technological leaders in the oil and gas industry, there are companies that are genuinely transforming themselves in the direction of developing their own technology in the area of clean and sustainable energy production. They are companies that think and look to the future, without giving up the long transition period in which it will still be possible to operate productively in conventional oil and gas exploration and production activities. This is the only way that seems to be smart and acceptable, both for large oil companies and for specialists working in these companies, especially specialists in risk and safety management, because in times of paradigm changes, risk management will always be a theme highlighted.

Finally, for those who are today dedicated to the study and knowledge of risk and safety management techniques, there will certainly be many years ahead to apply this knowledge to the activities of the oil and gas industry. This will probably be the case for at least another three decades. However, there is no longer any way to dedicate to any area of technical activity in the oil and gas industry without keeping a close eye on new sources of energy generation, their new associated technologies and the progressive technical maturation. Most likely, the oil and gas industry will begin to lose its role in power generation from 2030, accelerated in part by the consequences of the COVID-19 pandemic. It is also more likely that the oil and gas industry will continue to play a significant role in the energy matrix until 2050. But in at least two other (less likely) scenarios identified in the "World Energy Outlook 2020" report issued in October 2020 by the IEA, a definite disruption and the rapid decline of the oil and gas industry could happen by 2050. In all these more or less optimistic scenarios, oil and gas industry risk and safety experts will find numerous opportunities to continue applying all the knowledge and technologies developed over the decades when the oil and gas industry was the protagonist of the power generation industry. The future awaits us and it will always be good for those who study and truly hold knowledge.

9.12 Risk and safety management and the potential of the new digital tools

Much is said about "digital revolution." It is undeniable that a huge change has already occurred and perhaps even greater changes are underway, based on the use of digital tools and associated with the power of communication and data transmission on the internet. But we also have to recognize that many tools, digital resources and applications, appear as "technological phenomena" or "marketing phenomena" but soon after they end up being overcome by more changes, more applications, more news always associated with the increasing digitization of processes and methodologies that relate to productive activities. In the midst of this effervescence of constant news that excites people, companies and organizations, what in fact could cause a major change in the conceptual way of managing risks, taking the oil and gas industry as the main reference? Will automation be the highlight of digitization processes aimed at improving safety? Or will the possibility of recording, tracking and disseminating data overcome even automatisms capable of predicting and preventing accidents? Would not it be more likely an evolution of computer simulation of accidental scenarios, including in the simulations the influence of the active presence of people in these scenarios? Or will the increasing digitization lead to a mix of these and other elements, raising the quality of risk and safety management? We have no ambition here to try to map all the conceptual and methodological changes that may occur from the evolution of digitalization associated with risk and safety management. We will try to highlight some promising ideas for risk and safety management, some more developed and others still to be developed. In this short text, we will seek to point out some possibilities of relevant changes that can become a reality with the help of the growing evolution of digitalization.

9.12.1 New risk analysis methods

The collection of digitized data is nothing new. But as digitalization progresses, it is possible to collect data in real time in larger quantities and transmit it more and more quickly. With the "5G" internet, the data transmission capacity will grow in a general way. The "internet of things" allows the equipment of a production facility to store and transmit a lot of information from each equipment about its normal operation, in emergencies, occurrences of failures and malfunctions. Everything that happens in an industrial unit can now be quickly collected, stored, and transmitted with an unprecedented degree of detail. These data can supply internal "operational databases" of each company and later can also be included in "operational databases" for comparisons and studies at national and international level. Even questions of confidentiality about information related to a company's internal failures could be circumvented since the data could be transmitted without identifying the source, just to compose the "operational databases" used for the purposes of calculating reliability and risk analysis.

In this way, we could have an enormous amount of data, in real time, on countless operational parameters and countless operating parameters of the equipment. Let us imagine that every industrial installation would collect data on the functioning of all its equipment and send it to a real-time analysis to be carried out on computers with high processing capacity. This would greatly facilitate the work of QRA, based on historical data and even real-time data. We could imagine risk analyzes going on every minute based on the data collected in the previous minute. More than an important logical interlock for safety, we could have "Artificial Intelligence" performing risk analysis every minute.

Such a large set of data collected, recorded and transmitted in real time would facilitate the evolution of risk management by exploring the concept of "Twin Design." This concept is based on the creation of a digital copy of a physical product, service or process. This "digital twin" would function as a detailed simulation of objects or models of operation, replacing the creation of real prototypes. This feature would be useful both in the project and in the day-to-day life of the unit after it starts operating.

In the design phase, we used Piping and Instrumentation Diagram/ Drawing (P&ID) drawings to perform a qualitative risk analysis, such as the HAZOP. In this type of risk analysis we seek to foresee possibilities for dangerous alignments and possible operational failures so that we can create safeguards to protect against accidents. This type of analysis is often carried out in the design phase as presented in Chapter 7, Reducing Unpredictability. In order to carry out a HAZOP, it is necessary to bring together experts who will spend days and even weeks analyzing P&ID and other documents in order to improve the safety of the project. As much as there is an effort to quantify the risk analysis process, there is always a large margin of subjectivity involved in establishing the study premises and in defining the scenarios to be analyzed. With digitization it is possible to store and consult a huge amount of data about previous projects. But with the recent progress of digitization, we will be able to go much further in the evolution of risk analysis tools. We are using only HAZOP as an example, but the massive use of data can be applied in several risk analysis tools. In the case of HAZOP, as we know it today, we can imagine a revolution if we consider digitization resources like "Deep Search." The concept "Deep Search" refers to a way of performing data search that is much more advanced than the searches that we normally do on the internet. With "Deep Search" it is possible to search for information within the files, without the need to open them one by one, whether text files, images, audios, etc. Researchers are already developing tools that allow, while the designer draws a P&ID, suggestions based on previous projects and HAZOP to be offered to the designer from data searched through "Deep Search."

The research could be carried out using the designer's own database, with all of his previous projects or even on the internet, searching for similar projects, suggestions for alternatives, previous accidents, etc. In reality, the search would be carried out not in a conventional database but in a "Data Lake." More than a simple database, the architecture of the "Data Lake" is simple because the data can be structured, semistructured, or unstructured. In addition, they are collected from various sources within the organization, while the traditional "Data Warehouse" stores them in files or folders. A "Data Lake" could be fed with project documents, accident investigations, films, audios, technical specifications, risk analyzes, 3D models and a multitude of media and digital documents. Through a tool that uses "Deep Search," a designer drawing a P&ID could submit his work to a survey to identify all the information that may show opportunities for improvement in the project, even before the drawing is issued.

In practice, with the evolution of "Deep Search" it would be possible that, while the designer designs (a pipe with valves, tanks and equipment, etc.) a digital tool would make a "Deep Search" search in an evolved database in real time., the "Data Lake," able to find in audios, videos, documents and drawings everything that relates to the project and can improve it.

That is, at the exact moment when the designer is designing the P&ID, a kind of risk analysis (similar to a "Digital HAZOP") would be performed through "Deep Search," preventing the designer from issuing the document with something wrong, that has caused accidents before. It would be possible for "Digital HAZOP" to suggest improvements in real time to improve what is being projected based on learning from previous projects on the same theme. With each project, this ability to correct and standardize would be improved through the concept of "Machine Learning." This is a concept associated with "Artificial Intelligence" which is another broader concept. In the concept of "Machine Learning" a system can modify its behavior autonomously based on its own experience. The objective is that the system can become increasingly efficient. As thousands of designers worldwide submit their designs to systems that use "Deep Search" and "Machine Learning," with each correction, the system can improve and suggest the safest, most successful design configurations, that have generated fewer failures and accidents. So there would be much less to analyze through the old form of risk analysis that we practice today. As the designer works, the system already analyzes the risk of what is being designed and suggests the necessary changes. Just an overview and no longer those long weeks of risk analysis, repetitive and not always productive.

9.12.2 Risk analysis including the human behavior

Have you noticed that engineering documents do not normally show people? Yes, that's right, we engineers in general design systems, oil platforms, refineries, marine transfer terminals, control rooms, processing plants, ships, airplanes, buildings, all as if there were no people in these places. Except when we present artwork or architecture, human beings are often omitted from design documents as if what we designed would work on its own, without constant interaction with people. This also applies to risk analyzes. We often carry out fire propagation analysis, explosion analysis, gas dispersion analysis, among others, with a lot of detail. Many quantitative risk analyzes use fluid dynamic computation grids with impressive realism, however, amid the representations of shock waves, temperature ranges and gas concentrations, human beings are not explicitly present in the quantity foreseen in the project. It is common to analyze the effects and damage to supports, equipment, pipes, floors, walls, but we omit the representation of human presence in these scenarios. This is a mistake that needs to be repaired.

I think that we should include human representation in drawings and design documents in the same proportion that people need to be present in the workplace. I believe that the representation should even be considering gender, age, height, weight according to anthropometric data of the operators and workers who will be interacting with the real project to be built. In some cases we have included decorative details in the design representation of a control room, but we have failed to represent the people who will be working on it. An interesting experience is to request that people be included in a typical engineering design document. When we evaluate a process plan without the representation of people, the process plan appears very organized and safe. When we include in the drawings all the people who will normally need to be in the projected environment, we can often be surprised by the agglomeration and lack of space for circulation. In addition, with the inclusion of people it is possible to notice how it is more frequent than many people think the proximity of people to areas at risk.

The difficulty in representing humans in design documents also has to do with the need to represent people's behavior, as individuals move around, react differently, walk at different speeds, some walk in groups, others alone, in addition to being much more fragile than the steel that makes up most oil and gas installations. With digitalization, new 3D models are being developed with the ability to include in the workplace not only "mechanical dolls" to represent people. Some computer simulations include what is being called "encapsulation of mathematical functions" capable of simulating some characteristics of people's behavior. In Chapter 7, item 7.3 we show a digital tool capable of simulating, among other things, the behavior of people in an escape and abandonment operation in an Floating, Production, Storage and Offloading (FPSO).

Until some time ago, the processing capacity of our work computers was limited and we could not simulate hundreds and even thousands of "encapsulations of mathematical functions" interacting in a 3D model of an FPSO. But with the evolution of digitization this is now quite accessible, even with a notebook. We then have a new door of possibilities described in Chapter 7 item 7.3 of this book: FSA—"Full Safety Analysis" capable of simulating an accidental event and including human behavior in this simulation. We can perform risk analysis with 3D mesoscopic models, including clouds, grids representative of fire propagation, explosion and dispersion of gases generated by fluid dynamic computation, while at the same time people on the site are represented with their reactions to the evolution of the accidental scenario. Thus we can identify improvements in the escape routes that do not work and have a more precise idea of the possible damage to people and not only to the structure and equipment. We could also test procedures, efficiency of meeting points, duration of abandonment operations, sizes of doors and stairs, different strategies for displacing individuals and groups. More recently, with the pandemic COVID19, this type of tool is being used to study social isolation on board an FPSO. There are endless possibilities. The limit, as always, is research, development, the evolution of digital tools and the creativity of engineers.

9.13 Applicable technical standards

Throughout this book, we have relied on numerous national and international standards for references. However, we have opted to centralize the references to these standards in the present section, to avoid repeated use of the acronyms of standards in the text, which could hinder the flow of the text regarding its greatest objective, which is to present of technical and operational knowledge. This centralization is also the result of a concept repeated throughout the book, which identifies technical and operational knowledge as the most important element. We didactically seek to give as much emphasis to technical knowledge as possible, so as not to allow knowledge associated with the handling of standards to be perceived as being superior to the knowledge that is recorded in the standards.

Moreover, good standards are quite dynamic, undergoing continuous reviews, updates, and amendments. It is this dynamics that ensures the technical value of the standards. The value of a standard depends on its ability to follow the technological evolution that happens on the field, be it design, construction, assembly, or operation. After all, as we have already said:

> *Standards and procedures are the documents with the best technical references, recorded by those who know how to do it, for those who still don't know how to do it don't repeat the same failures made in the past.*

Every risk management expert, who knows the complexity of the subject matter to which they are dedicated, will always feel when tackling each new task as a professional who still has some learning to do. As long as there is a new task, there is a lesson to be learned. Within this concept, technical knowledge is always superior to standards, because in addition to preceding standards, it is also the essence of standards. Technical knowledge is born before standards, this cannot be forgotten by those who intend to write standards. And that's exactly the way it is, following this logical sequence, that we have chosen to gather in this section a set of references to standards to facilitate experts in their search for the best records of technical knowledge. The objective is to facilitate the search for operational experience aiming to identify practical solutions to risk and safety management problems. Technical standards are continuously evolving living documents. It is part of the risk management expert's job to always check the validity and applicability of the standards, both from the viewpoint of technological evolution and from the chronological validity.

9.13.1 Reference standards[1]

- ABNT – Associação Brasileira de Normas Técnicas
- ANSI – American National Standards Institute
- API – American Petroleum Institute
- ASTM – American Society for Testing and Materials
- COLREG – International Conference on Revision of the Internation- al Regulation for Preventing Collisions at Sea
- CONAMA – Resoluções do Ministério do Meio Ambiente do Brasil
- FSS Code – International Code for Fire Safety Systems
- IEC – International Electrical Commission

[1] **Translator's Note:** the Brazilian Standards and Regulations were left in Portuguese for easier reference search.

- IMO – International Maritime Organization (Organização Marítima Internacional)
- ISGOTT – International Safety Guide for Oil Tankers and Terminals
- ISO – International Organization for Standardization
- LSA Code – Life-Saving Appliances
- MARPOL – International Convention for the Prevention of Maritime Oil Pollution from Ships
- MODU CODE – Code for the Construction and Equipment of Mobile Offshore Drilling Units
- MTE – Regulamentações do Ministério do Trabalho e Emprego – Normas Regulamentadoras –NRs
- NFPA – National Fire Protection Association
- NORMAN – Regulamentos Aplicáveis da Autoridade Marítima Brasileira (DPC)
- SOLAS – Convention for the Safety of Life at Sea

9.14 Lessons learned

9.14.1 False safety improvement plans

In the aftermath of a series of serious accidents in several of its facilities, a large oil company decides for the complete change of its management in the field of safety. Seeking to reassure the technical community that the procedures and practices were being radically revised, the company decided to create an impactful plan aimed at safety improvement. But productivity had the highest value in the organizational culture, which made all the difference in the results. Despite the company's trauma caused by the accidents and victims, due to its productivity culture, the plan for safety improvement was designed under the massive influence of their productivity culture. The title chosen for the plan itself made it clear that safety was not at the top of the pyramid of values. The plan designed title adopted was "Production Excellence Plan."

Deeply influenced by a culture of productivity for decades, engineers and risk management experts, even in a situation of evident risk management degradation, they were unable to free themselves from their productivity culture and were trying to resolve the safety problems the only way that their culture vision allowed: improving production conditions.

Engineers and experts were unable to see that the Production Excellence Plan was nothing more than the same problem-solving structure in place for decades in the company. This new structure, although it could contain variations, it was based on providing operational comfort, to increase production with fewer interruptions. In the managers' mindset, fewer interruptions can equate fewer accidents, but this was a critical mistake that will end up generating millions of dollars in material losses, in addition to human losses, in yet another serious accident.

Determined to take the company out of the oil and gas industry operating companies' black list, the managers went to great lengths and managed to get a budget approved so that everyone who was aligned with the plan could make it a reality. This contributed to increased expectations and added external interest in the activities that were about to unfold. There were large sums of money allocated for improvement projects, renovations, purchases of new equipment, consultancies, intensification of routines, etc.

But the old productivity culture was stronger than ever and with the availability of the new funds all the company's managers, almost unanimously, were waiting for them to put into practice dozens of changes and new projects that had been in the wish list for years. In fact, the Production Excellence Plan was understood by the entire organization as a plan for production improvement, creating more operational comfort for operators, more automation, more redundancies, and more equipment replacements. The company's managers were completely obscured by the bigger problem that was threatening the company. After a period of low morale, resulting from recent accidents, managers understood that safety problems would be resolved quickly with all these actions, but they were unable to see that in reality all actions are about improvements in productivity results and not exactly aimed at safety improvement.

The Production Excellence Plan is at its peak, when the scale of the failure of the management in progress begins to emerge. Many projects start lagging and all suffer an astronomical cost increase, which draws the international market's attention regarding could be causing the cost of projects in the area of oil and gas to be so overestimated. Managers' opportunistic actions to include automation systems and special equipment, projects quickly began to lose competitiveness. That was when the first large projects ended up becoming unfeasible, causing serious impact on the company's schedules and results, without adding any effective improvement to safety.

Even more serious is that due to the sophistication of the new automation systems, new equipment and activities, production has also become more prone to failure and more complex for builders and operators. That was when it hit rock bottom. The first completed project based on the new concept to begin operations suffers a major accident while still in the commissioning phase, resulting in property and human losses. It was only then in the wake of this catastrophic moment that the entire organization was able, tragically, to realize the error in its value pyramid.

Despite the genuine objective of improving safety being present since the beginning of the Production Excellence Plan, it was found that there was a cultural problem where safety was not at the top of the value pyramid as everyone in the organization had assumed. After that, the company and its managers did manage to move safety to the top of the value pyramid and understood that there was a lack of technical and operational knowledge in risk and safety management required for developing an improvement plan, legitimately based on risk management principles. The lesson learned was that management models, unrealistic plans, motivational actions, large budget, constructions, projects, will, determination, etc., none of it replaces the genuine knowledge of risk management engineering. Only such experience and knowledge can steer engineers, experts, companies and organizations toward effective improvements in risk management activities. Accidents always find the easiest ways to happen. Safety must always overcome the most difficult routes to get established.

9.15 Exercises

1. Can the risk management expert claim the deficiencies in the risk management systems to justify accidents?
2. What is the main problem caused by excessive centralization of risk management within the organization's structure?
3. What are the consequences of centralizing the decisions of the many disciplines related to risk management in a single manager?
4. Risk management systems anchored on indicators are vulnerable to what types of problems?
5. Is the safety engineering course a prerequisite for training the risk management specialist?

6. An oil and gas processing offshore rig designer has developed an escape and abandonment system with equipment that is an alternative option to lifeboats. What kind of approach should the designer take to add the equipment to the project, even if this equipment is not yet approved and recognized by current technical standards?
7. Two marine terminals have shown problems related to safety systems. The system is operated differently by each operator, although the equipment is the same for both terminals. What type of risk management approach should be applied to the situation to improve the risk management associated with these industrial units? Justify.
8. The community surrounding a refinery has complained about safety and has claimed that old technical studies have not been properly followed, with safety recommendations not being implemented. What kind of risk management approach should be adopted by experts, considering that the company has four other facilities with the same characteristics, and one of them has also undergone similar questioning 5 years ago. Justify.
9. A recurring accident has been happening in an oil and gas processing facility, but apparently the event has not causes losses to people or to the facility, except for brief few-minute long operation interruption for correction following each accidental event. For this reason, the operators involved are not too concerned about the problem, although they have sufficient technical knowledge to avoid it. Due to the increased event frequency and the possibility of more serious accidents, the company's risk management expert suggests conducting an investigation to identify the problem's root cause and prevent the accident recurrence. What type of technique is recommended for this scenario? Justify.
10. What could happen if a surveillance system were deployed without the company being organized to support the frequency of testing the security systems required?

9.16 Answers

1. The risk management expert needs to assess the state of development of the organization's management system where it operates and adopt

the appropriate measures for the situation in order to avoid accidents. Organizations are continuously evolving toward more rigorous risk management and safety. The risk management expert needs to understand the current level of safety of the organization and determine corrective actions compatible with the particular situation.

2. Excessive centralization in the organizational risk management structure can release local managers from their responsibility for technical actions aiming at preventing accidents. Excessively centralized systems tend to cause over reliance on indicators as central managers are unable to have direct access to risk locations and scenarios, becoming dependent on indicators to make decisions.
3. The multidisciplinary nature of the risk management field, where each discipline requires highly specialized knowledge to provide its risk management solutions. Therefore the centralization of all decisions in a single manager creates dependency on experts in each area.
4. The main disadvantage in anchoring risk management totally in indicators is the distancing from the management of the phenomenological activities and processes that make up the activity to be protected. This can lead to a naive risk management system, based on interpretation without the support that only be provided by operational experience.
5. No. Although the safety engineering course can add to the training of the risk management expert, this course is not a must have requirement. The prerequisite for becoming a risk management expert is specific knowledge about the activity to be protected associated with an academic background with emphasis on the depth of knowledge such as a specialization or graduate degree in a related field.
6. The integration of the new equipment could be justified by a RBD approach, where the designer conducts the applicable risk analysis and presents the technical arguments to demonstrate that the project is based on calculated risks to be maintained within acceptable levels.
7. The most recommended approach is Safety Peer Review (SPR). This technique allows operators to be gathered in a single event to discuss divergent technical opinions under the supervision of experienced experts, producing recommendations for the units' risk management improvement.

8. The most recommended approach is SPR. The SPR technique has the ability of including an initial phase for survey and verification of previous recommendations based on studies and safety analyses. In addition, the SPR technique can be used to organize a Peer Review event with the participation of the current unit and the unit that was questioned 5 years ago. Thus operators of both units will be able to exchange information, providing solutions, and improving the risk management of the facilities.
9. The 5 Whys technique is recommended. This technique is fast, simple to apply and recommended for accidents with minor complexity. The experts of the activity to be protected are directly involved in it and can be easily take part in the application of the 5 Whys technique.
10. Due to the strictness of the surveillance system, when tests cannot be executed, operation and production will be severely affected by penalties that can include shutdown of the production.

9.17 Review questions

9.16.1 Explain the difference between generic and technological management.

9.16.2 What are the consequences of over-centralized risk management systems?

9.16.3 Who should be formally responsible for risk management decisions in each field?

9.16.4 What are the consequences of risk management systems totally based on indicators?

9.16.5 Why is the safety engineering course offered in Brazil not a prerequisite for training specialists in risk management?

9.16.6 Explain the Risk-Based Design technique.

9.16.7 What are the principles that yield the best results regarding the application of the RBD technique?

9.16.8 Explain the Safety Peer Review technique and its advantages.

9.16.9 What are the steps for applying the SPR technique?

9.16.10 What are the prerequisites for applying the SPR technique regarding the experts involved?

9.16.11 Define accident investigation.

9.16.12 Provide examples of the main accident investigation tools and techniques.

9.16.13 Describe the steps that make up the SCAT accident investigation technique.

9.16.14 Explain the meaning of the acronym SHELL regarding accident investigation.

9.16.15 Provide definition for fault tree technique.

9.16.16 Describe all steps of an accident investigation based on the causal tree method.

- Explain the meaning and origin of the term surveillance system.
- List the main references for general standards applicable to risk management in oil and gas industry.

CHAPTER 10

Synthesis

Risk management is a multidisciplinary subject matter that requires experts to be trained to deal with various branches of knowledge. Safety culture and human factors in risk management are becoming increasingly important. Safety culture and human factors are areas related to human behavior, thus engineers who keep up-to-date with the latest technological trends need to be prepared to deal with subjective matters in parallel with the use of traditional objective engineering methods. The difficulty in organizing a multidisciplinary theme that associated mathematical precision with subjectivity requires a strategic line model for risk management practice. The model presented in this book consists of five elements, in the following order of importance: technical and operational knowledge, hazard reduction, removal of agents (people), emergency control, and unpredictability reduction.

Technical and operational knowledge is considered the most important element for the success of risk management strategies. Nothing protects a facility more than professionals who have deep knowledge of its processes, systems, and phenomenology. The basic hazard of the oil and gas industry is hydrocarbon, without which the industry would not exist. But at the time of an emergency, the quantity of hydrocarbons should be reduced as much as possible as part of the protection strategy. Hazard reduction in the oil and gas industry focuses on hydrocarbon inventories and is conducted through emergency shutdown systems, whose response occurs in a short period of time by automation.

To reduce the risk of human losses, agents need to be removed from accidental scenarios as soon as possible. Minimizing the number of people in the accidental scenario, the potential risk of victims is reduced. Thus once people are protected, the emergency can be diagnosed, assessed and, in some cases, controlled through systems and actions related to power generation, ventilation and air conditioning, inerting, gas detection, fire detection, fire fighting, passive protection, among others. Accidents always include some elements of unpredictability. Such unpredictability needs to be reduced as much as possible starting in the design phase through to the operation of the facilities of the oil and gas industry. For this purpose,

Risk Management in the Oil and Gas Industry.
DOI: https://doi.org/10.1016/B978-0-12-823533-1.00007-1

several risk analysis tools and techniques are used to assess possible emergency scenarios, assisting designers and operators in providing protection safeguards for these scenarios.

The evolution of risk management engineering has attested the importance of the safety culture and human factors (factors that make up the error-inducing environment) as frequent causes of accidents. Technology managers along with risk management experts should be mindful of opportunities for improving the safety culture of organizations and for reducing the error-inducing environment in oil and gas facilities. New tools, such as full safety analysis, rapid entire body assessment, safety peer review, surveillance system, among other techniques, have offered a new universe of opportunities with enormous potential for improving management risks and safety of oil and gas industry.

Bibliography

API RP 14J, 2007. Recommended Practice for Design and Hazards Analysis for Offshore Production Facilities. API RP, Washington, DC, USA.

Ávila Filho, S.P., 2011. Apostila de Confiabilidade Humana. Universidade Corporativa Petrobras, Rio de Janeiro.

Bega, E.A., Delmée, G.J., Cohn, P.E., Bulgarelli, R., Koch, R., Finkel, V.S., 2006. Instrumentação Industrial. Interciência IBP, Rio de Janeiro.

Brauer, R.L., 2006. Safety and Health for Engineers, second[a] ed. John Wiley & Sons, Inc, Hoboken, NJ.

Bridger, R.S., 2009. Introduction to Ergonomics, third[a] ed. CRC Press Taylor & Francis Group, Boca Raton, FL.

Capra, F., 1982. O ponto de mutação: a ciência, a sociedade e a cultura emergente, twenty fifth[a] ed. Cultrix, São Paulo.

Cardoso, L.C., 2005. Petróleo do Poço ao Posto. Editora Qualitymark, Rio de Janeiro.

Chadwell, G.B., Leverenz, F.L., Rose, S.E., 1999. Contribution of human factors to incidents in the petroleum refining industry. Process. Saf. Prog. 18 (4).

Chaffin, D., 2005. Improving Digital Human Modeling for Proactive Ergonomics in Design. Ergonomics. SAE International, Brooklyn, MI.

Cohn, P.E., 2006. Analisadores Industriais: No Processo, na Área de Utilidades, na Supervisão da Emissão de Poluentes e na Segurança. Interciência IBP, Rio de Janeiro.

Deming, W.E. 1982. Quality Productivity, and Competitive Position, first[a] ed. Massachusetts, USA.

Department of Energy, 1988. Comparative Safety Evaluation of Arrangements for Accommodating Personnel Offshore. Londres, UK, Department of Energy Section 9 + Appendix 7.

Descartes, R., 1637. O Discurso do Método, first[a] ed. Leiden França.

DET Norske VERITAS, OREDA-92, 1993. Offshore Reliability Data Handbook, (second[a] ed.) DNV Industries Norway ISBN 82 515 0188 1.

EVE/EVI, 2008. Evacuation Simulation Software. University of Strathclyde and Safety at Sea Ltd, Glasgow.

Frutuoso, P.F.M., 2003. Análise da Confiabilidade Humana (ACH) na AQR. In: 7° Congresso de Atuação Responsável ABIQUIM, São Paulo, 6. 8. Palestra disponível em: <http://www.abiquim.org.br/atuacaorespon-savel/7cong/segundodia/Paulo_Frutuoso.pdf> (acesso em 9 ago. 810.).

Gilbreth, F.B. Jr, Carey, E.G., 1963. Cheaper by the Dozen, Crowell, 1948, Expanded Edition.

Guarin, L., Majumder, J., Shigunov, V., Vassalos, D., 2004. Fire and flooding risk assessment in ship design for ease of evacuation. In: Proceedings of the Second International Conference on Design for Safety, Osaka, Japão out.

Guarin, L., et al., 2007a. Design for fire Safety. In: Proceedings of the Second International Conference on Design for Safety, São Francisco, CA, USA set.

Guarin, L., Majumder, J., Puisa, R., 2007b. Human Life Safety – Risk Analysis. Safedor Report.

Health and Safety Executive. 1995a. A Methodology for Hazard Identification on EER Assessments, HSE Consultants Ltd, OTH 95 466. <http://www.hse.gov.uk/research/othhtm/400–499/oth466.htm>.

Health and Safety Executive, 1995b. Prevention of Fire and Explosion, and Emergency Response on Offshore Installations, first[a] ed. Suffolk, HSE Books.

Health and Safety Executive, 1995c. Review of Probable Survival Times for Immersion in The North Sea. HSE OTO 95 038. Disponível em: <http://www.hse.gov.uk/research/otopdf/1995/oto95038.pdf>.

Helin, K., Viitaniemi, J., Aroma, S., Montonen, J., Evila, T., Leino, S., Maatta, T.O., 2007. Digital Human Model in the Participatory Design Approach A New Tool to Improve Work Tasks and Workplaces. VTT Working Papers, v. 83, Finland.

HFES – Human Factors and Ergonomics Society, 2012. Disponível em: <http://www.hfes.org/web/Default.aspx >, USA.

Hollnagel, E., 1993. Human Reliability Analysis Context and Control Computers and People Series, first ed. Academic Press Inc, San Diego, CA.

Hollnagel, E., 1998. January Cognitive Reliability and Error Analysis Method, first[a] ed. Elsevier Science, London,.

Hollnagel, E., 2004. Barriers and Accident Prevention, first[a] ed. Ashgate Publishing Company, Aldershot.

IAEA International Atomic Energy Agency, 1991. Safety Culture, Safety Series, n.75, INSAG-4, Vienna.

IMO (SOLAS), 1974. International Convention for the Safety of Life at Sea. London, UK.

IMO MODU Code. 2009. Code for Construction and Equipment of Mobile Offshore Drilling Units. London, UK.

International Association of Oil & Gas Producers, 2009. Vulnerability of Humans, DNV Report no. 32335833/14, rev 2.

International Association of Oil & Gas Producers, OGP, 2010. Evacuation, Escape & Rescue. London, UK. Risk Assessment Data Directory Report, n. 434, 19 March.

ISO 13702, 1999. Petroleum and Natural Gas Industries – Control and Mitigation of Fires and Explosions on Offshore Production Installations – Requirements and Guidelines. Geneva, Switzerland.

Jastrzebowski, W. An Outline of Ergonomics or the Science of Work. Warsaw, Central Institute for Labor Protection, 2000.

Jones, J.C., 2003. Hydrocarbon Process Safety. Whittles Publishing, Tulsa, OK.

Jordão, D.M., 2002. Manual de Instalações Elétricas em Indústrias Químicas, Petroquímicas e de Petróleo, third[a] ed. Qualitymark, Rio de Janeiro.

Lafraia, J.R.B., 2001. Manual of Reliability, Maintainability and Availability. Qualitymark Petrobras, Rio de Janeiro.

MSC.1/Circ.1238, 2007. Maritime Safety Committee – Guidelines for Evacuation Analysis for New and Existing Passenger Ships, 30 out.

MURRELL, 1949. Applied Experimental Psychology: Human Factors in Engineering Design. London.

Newton, I., 1687. Philosophiae Naturalis Principia Mathematica. London.

NORMAN 01, 2005. Normas da Autoridade Marítima para Embarcações Empregadas na Navegação em Mar Aberto. Rio de Janeiro, Brazil.

Pastura, F., 2000. Avaliação da Criação e da Difusão do Banco de Dados Antropométricos e Biomecânicos Ergokit, M.Sc., Rio de Janeiro, Tese – Universidade Federal do Rio de Janeiro, Coppe/UFRJ, Brazil.

Pavard, B., Mouton, C., Gourbault, F., 2008. Context dependant móbile interfaces for collaboration in extreme environment. In: M. Klann (Ed.). Mobile Response Interfaces, firsta ed. New York, Springer Verlag,.

Peters, G.A., Peters, B.J., 2006. Human Error Causes and Control, first[a] ed. Taylor & Francis Group, Boca Raton, FL.

Portela, D.A., Ponte, JR, G., 1998. Qualidade na Gestão da Tecnologia sob uma Visão Crítica da Ciência Racionalista, M.Sc., Rio de Janeiro, Dissertação – Centro Federal de Educação Tecnológica do Rio de Janeiro/CEFET RJ, Brazil.
Portela, D.A., Ponte, JR, G., 2014. Gerenciamento de Riscos Baseado em Fatores Humanos e Cultura de Segurança. Elsevier, Rio de Janeiro.
Porter, J.M., Freer, M., Case, K., 1995. Computer aided ergonomics and workplace design. In: Wilson, J.R., Corlett, E.N. (Eds.), Evaluation of Human Work: A Practical Ergonomics Methodology, second[a] ed. Taylor & Francis Group, London.
PRIMATECH, 2008. Understanding and Applying Human Factors for Process Safety. Primatech Training Institute, Las Vegas, NV.
Rasmussen, J., 1997. v. 27, n.2/3. Inglaterra Risk Management in a Dynamic Society: A Modeling Problem. Elsevier Safety Science, London, pp. 183–213.
Rasmussen, J., Rouse, W.B., 1981. Human Detection and Diagnosis of System Failures, first[a] ed. Plenum Press, Nova York.
Reason, J.T., 2003. Human Error, first[a] ed. Cambridge University Press, Cambridge.
Robertson, D., 1987. Escape III – The Evaluation of Survival Craft Availability in Platform Evacuation, first[a] ed. London, International Offshore Safety Conference.
Roger, L.B., 2006. Safety and Health for Engineers, second[a] ed. John Wiley & Sons, Inc.
Santos, V., Zamberlan, M.C., Pavard, B., 2009. Confiabilidade Humana e Projeto Ergonômico de Centros de Controle de Processos de Alto Risco, first[a] ed. IBP, Rio de Janeiro.
Sebzali, Y.M., Wang, X.Z., 2002. Joint analysis of process and operator performance in chemical process operational safety. J. Loss Prevent. Process Ind. 15, 555–564. Leeds, UK.
Swain, A.D.;, Gutmann, H.E., 1983. Handbook of Human Reliability Analysis with Emphasis on Nuclear Power Plant Applications. Sandia National Laboratories, Albuquerque, NM.
Sykes, K., 1986. Summary of Conclusions Drawn From Reports Produced by, or made available to, the Emergency Evacuation of Offshore Installations Steering Group, MaTSU.
Tanner, D.E., 2002. Ten Years Incident Reports Underscore Human Error as Primary Cause of Accidents, Bulletin: Summer, v. 57, n. 2. Disponível em: <http://www.Nationalboard.org/SiteDocuments/Bulletins/SU02.pdf> (accessed 15.11.09.).
Taylor, F.W., 1911. The Principles of Scientific Management. Harper and Brothers Publishers, New York and London.
TECHNICA, 1983. Risk Assessment of Emergency Evacuation From Offshore Installation, London UK, Technica Report F 158, prepared for DoE.
TECHNICA, 1988. Escape II – Risk Assessment of Emergency Evacuation from Offshore Installations, London, UK, OTH 88 8285, ISBN 0 11 412920 7.
UK Legislation, 2011. ISBN 0 11 073610 9Available from: http://www.opsi.gov.uk/si/si2005/20053117.htmThe *Offshore* Installations (Safety Case) Regulations SI2005/3117. The Stationery Office, Norwich.
UK Step Change in Safety. Loading of Lifeboats during Drills – Guidance. <http://step-changeinsafety.net/ResourceFiles/Lifeboat%20Lo-ading%20Guidance%20Final%20Copy.pdf>, 2003.
Vassalos, D., 1988. OTH 558, ISBN 0–7176–1595-2, 294pp, coauthor A. Kourouklis An Experimental, Theoretical and Full-scale Investigation on the Snap Loading of Marine Cables, vol. I and II. HSE Books.
Vassalos, D., 1999a. Shaping ship safety: the face of the future. Mar. Technol. 36 (2), 61–74. abr.

Vassalos, D., 1999b. Time-Based Survival Criteria for Ro-Ro Vessels, UK, Transactions RINA, Bronze Medal Prize, coauthors Jasionowski, A, Dodworth, K, Allan, T, Matthewson, B and Paloyannidis, P.

Vassalos, D., 2004. A risk-base approach to probabilistic damage stability. In: Proceedings of the Seventh International Ship Stability Workshop, Shanghai, China.

Vassalos, D., Hamamoto, M., Papanikolaou, A., Molyneux, D., 2000. In: Vas- salos, D., Hamamoto, M., Papanikolaou, A., Molyneux, D. (Eds.), Contemporary Ideas on Ship Stability. Elsevier.

Wybo, J.L., 2006. Mastering risks of damage and risks of crisis: the role of organizational learning, Sophia-Antipolis. Int. J. Emerg. Manage. 2 (1/2), 22–34.

Index

Note: Page numbers followed by "*f*" and "*t*" refer to figures and tables, respectively.

A

Abandonment points, 148, 153–154
Absolutely necessary risk, 6–7, 24
Acceptance criteria, 5–6
Accident(s)
  with cryogenic products, 254–265
  in facilities with hydrocarbon inventories and survival, 142–143
  investigations, 441–445
    causal tree or fault tree technique, 444–445
    5 whys technique, 443–444
    SCAT, 442–443
    SHELL method, 444
  in oil and gas industry, 60–62
Accidental scenarios, 113, 367–368
  formation, 27
  selecting and identifying, 268–276
    beyond design accident concept, 275–276
    design basis accident, 271
    external origin accidents, 271–274
Adapt Human to Work approach (AHW approach), 391
Adapt Work to Human approach (AWH approach), 391
Adequately ventilated environment, 193
Adiabatic compression, 192
Age of agents on board, 362
Agents evacuation, 140*f*
  accidents in facilities with hydrocarbon inventories and survival, 142–143
  crisis scenario simulator, 169–170
  escape and abandonment operation, 145–149
  general instructions, 170–171
  human–system interaction during escape and abandonment, 143–145
  instructions about scenario evolution, 171–173
  sea survival equipment, 154–163
  SOS, 163–169
  systems of escape and abandonment, 141
  technical recommendations for escape and abandonment system, 149–154
Agents removal, 39–41
Air conditioning systems, 196–198
Alarm storm, 399
Aluminum alloys, 192
American Bureau of Shipping (ABS), 356
American Petroleum Institute (API), 189, 416
Analytical codes, 83
Anthropometric project, inclusion of, 23
Apparent ignition temperature, 187
Applicable technical standards, 469–471
Artificial Intelligence, 467
Artificial ventilation, environment with, 194
As low as reasonably practicable (ALARP), 315–319
Asphyxiating gas (CO_2) monitoring, 207–208
  post-CO_2 gas confirmation actions, 208
Atmospheric vent, 118–119, 119*f*, 121*f*
Autoignition temperature, 187
Automatic deluge valves (ADV), 220–221, 221*f*
Automatic emergency shutdown, 125–129
  ESD levels, 125–127
  example of emergency shutdown sequence, 127–128
  shutdown requires caution, 129
Automatic fire-fighting systems, 218–244
  carbon dioxide fire extinguishing system, 239–242
  didactic comparison, 219*f*
  fire-fighting water
    distribution system, 236–239
    pumps, 225–236
  foam-water spray systems, 224–225
  water mist fire suppression system, 242–244

Automatic fire-fighting systems (*Continued*)
water spray fixed systems, 220–223
Automation, 21–22, 25, 87, 464–465
Auxiliary equipment, 180, 246
Avatar for "experts" without operational experience, 102–108

B

Balanced risk management, 55
Basic design, 97–98
Benzene, 87–88, 123–124
Best gas sensor in world, 50–52
"Beyond design accident" concept, 275–276
Beyond design-basis accident, 42–43
Blocked ventilation, environment with, 194
Blocking segmentation technique, 115–117
Blowdown valve (BDV), 115
Blowout, 62
Blowout preventer (BOP), 62, 65–66
Boiling liquid expanding vapor explosion (BLEVE), 85–88, 188, 260
formation of explosive atmospheres by, 286, 287*f*
Booster pumps, 227–228
Brainstorming, 331–332
Bronze, 192
Brute force, 218–220
Bulkheads, 247, 249, 253
Bureau Veritas S.A. (BV), 356

C

Cables, special requirements for, 184–185
Capillarity of concepts and principles of risk management, 447–448
Carbon capture, utilization and storage (CCUS), 456
Carbon dioxide (CO_2), 239–240
fire extinguishing system, 239–242, 243*f*
CO_2 deluge alarm, 241–242
for local manual activation, 241
for mechanical manual activation, 241
Cargo handling system, 64
Catalytic detectors, 203, 204*f*
Causal tree technique, 444–445
Center for Chemical Process Safety (CCPS), 3
Centralization of risk management, 433–434
Centralizing objectives in people, 20
Centrifugal pumps, 227–228
Checklist, 332–334
Chemical engineering, 100
Chemical powder fire extinguishers, 264
Chernobyl power plant accident, 9–10
Civil engineering and architecture, 99
Classical blindness, 33
Classical failures, 279–281
Colors, 398
Combating legalism, 26–27
Combustible liquid, 189
Combustion, 187, 203, 283
speed, 188
Completion, drilling rig and, 64–66
Compliant tower platforms (CT platforms), 76, 109
Computational analysis tools, 145
Computational fluid dynamics (CFD), 50, 83, 200–201, 307–308
Computers, 399–400
simulations, 140–141, 297–299
of explosions, 133
of human behavior in emergencies, 309
Conception of redundancies and safety systems, 277–281
"Conceptual project" solicitation, 96–97
Conciseness of rules, 26
Confined equipment, protection systems for, 254
Confinement, 288
Consequence-based decision, 95
Contaminants, 77
Controlled burning and dispersion, 118–124
Controls, 398–399, 405
Conventional fixed-type platforms, 67
Conventional lifeboats, 156–157, 156*f*
Conventional risk analysis studies, 349–350
Correction of conceptual error, 289–292
Crisis management, 147–148, 266–268

Crisis scenario simulator, 169–170
Critical safety functions, 43, 275
Crude oil, 77
Cryogenic products, accidents with, 254–265
Cut set, 444–445
Cyclohexane, 123–124

D

Data acquisition, 435–436
Data Lake, 466–467
Davits, 155–156
  of conventional lifeboats, 159
Decentralization of risk management, 433–434
Decision making, 266–268
Deep Search, 466
Deepwaters, 80
Deflagration, 188, 283–284
Delayed Recovery Scenario (DRS), 452
Deluge system, 220–223, 292–293
Depressurization, 118
Design adaptation to humans, 20–21
Design basis accident, 42, 271
  beyond design-basis accident, 42–43
Designers, 48, 227, 293
Det Norske Veritas (DNV), 356
Detailed design, 98
Detonation, 188, 283
Diesel-engine generators, 181
Digital HAZOP, 467
Digital twin, 465
Digitalization, 464–465
Dispersion, controlled burning and, 118–124
Downstream facilities, 76–81. *See also* Upstream facilities
  marine terminals, 81
  refining facilities and petrochemical plants, 77–79
  transportation and distribution, 79–81
Drilling
  fluid circulating system, 65
  rig and completion, 64–66
    onshore drilling rig, 65*f*
Dual-chamber ionization detectors, 215
Dust explosion, 188

E

Earthquakes, 7–8
Effective evacuation of agents, 139
Efficiency in risk management, 23–29
  combating legalism, 26–27
  conciseness of rules, 26
  fighting heroism, 27–28
  humility, 28–29
  rejecting unnecessary risks, 24
  respect for natural laws, 24–25
  seven principles of, 24–29
  simplicity, 25
Electric actuation, 223
Electric ignition sources, 189–192
Electrical engineering, 100
Electronic engineering, 100
Electrostatic discharge, 190–191
Emergency control, 2, 41–42, 179–180, 179*f*, 479
  accidents with cryogenic products, 254–265
  automatic fire-fighting systems, 218–244
  conception of redundancies and safety systems, 277–281
  crisis management and decision making, 266–268
  explosion phenomena, 281–289
  fire brigade and rescue crew performance, 265–266
  fire detection systems, 209–218
  fire protection systems, additional, 244–246
  flushing, purging, and inerting systems, 199
  gas detection system, 200–208
  heating, ventilation, and air conditioning systems, 196–198
  lessons learned, 289–299
    correction of conceptual error, 289–292
    criteria and results, 297–299
    interactivity, arrangement, and risk management, 293–297
    strategy, 292–293
  and liquefied natural gas cryogenics, 260–264
  passive fire protection, 246–254

Emergency control (*Continued*)
power generation systems, 180–195
protection systems for confined equipment, 254
selecting and identifying accidental scenarios, 268–276
special safety strategies applied to automation, 276–277
subsea safety equipment, 265
Emergency in FPSO, 163–169
Emergency power generators, 180
Emergency scenarios, 147–148
Emergency shutdown system (ESD system), 1–2, 120, 182
Engineering fields, 95
Environment, 401, 404–405
with artificial ventilation, 194
with blocked ventilation, 194
with natural ventilation, 194
Environment Editor module (EVE) software, 354
Environmental damage, 113
Equipment, 398, 405–406
Ergonomics, 101–102, 389–390
Error-inducing environment, human factors and, 18–23
Escape and abandonment
human–system interaction during, 143–145
operation, 145–149
simulation software, 145
systems, 139–141, 343–346
technical recommendations for system, 149–154
applicable materials in escape and abandonment systems, 153
basic dimensions and recommendations for escape routes, 150–152
evacuation, escape, and rescue analysis, 152
meeting points and abandonment points, 153–154
possible operational sequences, 149–150
spaces with limited access and machine rooms, 152–153
Essential consumers, 182–184
common to fixed and floating platforms, 182–183
on FPSO/FSO floating platforms, 183–184
on semisubmersible floating platforms, 183
Evacuability Index module (EVI) software, 353, 354*f*, 355*f*, 363–364
Evacuation, escape, and rescue analysis (EERA), 152
Executive project, 98
Explosion, 188
without flame, 264–265
formation of explosive atmospheres in
BLEVE, 286
closed space, 283–286
open space, 282–283
phenomena, 281–289
shock waves and factors that influence explosions, 287–289
study, 342–343
types of explosion involving flammable products, 282
Explosive atmosphere, 186–187, 189–190
Explosive burning combustion, 84–86
External origin accidents, 271–274
accident classification, 272–274
example of protection against, 274
Extreme sport, 8–9

F

Failure modes and effects analysis (FMEA), 334–335
False safety improvement plans, 471–473
Fast phase transition, 264–265
Fault tree technique, 444–445
Fear, 397
Federal Emergency Management Agency (FEMA), 3
Field experience, 90–96
Fighting heroism, 27–28
Fire and gas detection system, 126–127
Fire brigade and rescue crew performance, 265–266
Fire detection systems, 209–218
fixed temperature heat detection, 216
flame detection, 210–212

heat detection, 212–214
smoke detection, 214–215
thermovelocimetric detection, 216
Fire detectors, specification and positioning of, 216–218, 217*t*
Fire extinguishers, 246
Fire hydrants, 245
Fire prevention engineering, 102
Fire propagation study, 340–341
Fire protection systems, additional, 244–246
auxiliary equipment, 246
fire extinguishers, 246
fire hydrants, 245
fire-fighting monitor cannons, 246
mobile foam generating equipment, 245–246
Fire ring, 237–239
Fire scenarios, 270, 368–369
Fire-fighting monitor cannons, 246
Fire-fighting systems, 141
Fire-fighting water distribution system, 236–239
Fire-fighting water pumps (FWP), 218–220, 225–236
diesel hydraulic, 228*f*, 229*f*
fire-fighting water demand, 231–236, 235*t*
safety requirements, 232*t*
types, 227–231
Fireball, 85
Firefighting measures, 89–90
First-aid measures, 89
5 What technique (5 W technique), 443–444
5 Whys technique, 443–444
5W1H technique, 443–444
Fixed offshore
platforms, 67–70
rigs, 158–159
Fixed rig design, 67, 67*f*, 69*f*
Fixed temperature heat detection, 216
Flame detection, 210–212. *See also* Gas detection system
infrared detector, 211, 212*f*
infrared/ultraviolet detector, 211–212, 213*f*
ultraviolet detector, 210–211, 211*f*
Flammability limits, 189
Flammable clouds, 283
Flammable gas detection, 200–205
post-CH_4 methane gas confirmation actions, 204–205
posthydrogen gas confirmation actions, 205
Flammable gas detectors, 203–204
Flammable liquid, 188–189
Flare system, 120, 123*f*
Flash fire, VCE and, 86
Flash point, 84–85, 188
Floating production storage and offloading (FPSO), 355–356, 355*f*
adaptation of FPSO hull internal areas, 358–359
agent positioning in, 365*f*
emergency in, 163–169
platforms, 71–74
Flow reduction, 83
Flushing, 199
Foam fire-fighting fixed system, 224, 224*f*
Foam flood system test, 260–262, 261*f*
Foam generation system, 224
Foam-generating liquid drums (FGL drums), 245
Foam-water spray systems, 224–225
Free-fall lifeboat davits, 156–159, 157*f*
Fukushima nuclear power plant, 7–8
Full safety analysis (FSA), 2, 344, 352–372
accidental scenarios, 367–368
adaptation of
FPSO hull internal areas, 358–359
process plant area, 357–358
superstructure internal area, 359–360
age of agents on board, 362
agents on board and behavioral parameters, 360
building 3D model, 357
features of analysis of offshore rig, 355–356
fire scenarios, 368–369
gender of agents on board, 361–362
importing documents to build 3D model, 356
measuring the effects of emergency on people's integrity, 366–367

Full safety analysis (FSA) (*Continued*)
naval damage condition scenarios, 369–370
operational experience of agents on board, 360–361
people on board, 360
physical positioning of agents in rig, 363–364
reaction times for agents on board, 363
representative simulations for offshore rigs, 371–372
special tasks for specific agents during emergency, 365
standard and gas leakage scenarios, 368
theoretical scenarios for comparative purposes, 371
travel speeds of agents onboard, 362–363
Fusible plug system, 212–214

G

Gas detection system, 200–208
flammable gas detection, 200–205
monitoring gas contamination, 207–208
specification and location of gas detectors, 208
toxic gases detection, 205–207
Gas detectors
and analyzers, 200
specification and location of, 208, 209*t*
Gas dispersion studies, 341–342
Gas industry, accidents in, 60–62
Gas leaks, 82–84
Gas productive chain components, 58
Gaseous hydrocarbons, 200
relief, 118
Gender of agents on board, 361–362
General accident databases, 158
General safety requirements, 158–160
Global energy matrix, 451
Global positioning system (GPS), 70, 422–423
audio information, 424–425
driver's GPS knowledge, 425
human-system interface, 426
influence on driver, 423
position in dashboard, 423–425
software and configurations, 424–425
visual information, 424–425
Guardrails, 152

H

Hazard, 35–36, 111
"Hazard escape and scenario abandonment system", 139
Hazard identification (HAZID), 2, 88, 319
Hazardous areas, 114, 184–185
Hazards and operability analysis (HAZOP analysis), 2, 322–330, 325*t*
category and failure examples, 327*t*
chaos, 376–378
control failures, 328*t*
establishing premises for, 328–330
examples of division by nodes, 330, 330*f*, 331*f*
Hazards reduction, 1–2, 35–39, 111–112, 179–180, 479
automatic emergency shutdown, 125–129
exercises, 134–135
hydrocarbon inventory disposal during emergency, 117–124
lessons learned from Piper Alpha, 132–134
Piper Alpha hazards reduction failure, 129–132
reduction of hazardous scenario, 36–39
segmentation of hydrocarbon inventory, 113–117
Health, safety and environment (HSE), 93
Heat detection, 212–214
Heat shielding systems, 122–123, 124*f*
Heating, 196–198
Helicopter, 155–156
Heroism, 17
fighting, 27–28
Human decision superiority, 21–22
Human error, 17, 386–388
protection against, 21
Human factors, 141, 388–416, 422–423, 479
analysis, 402–408
behavior, 413
commissioning and startup errors, 412

company or organization, 400–401
computers, 399–400
construction and assembly errors, 411–412
controls, 398–399, 405
design, 409–410
approach, 143
errors, 411
elements related to, 427
emergency response error, 413–414
engineering, 101–102
environment, 401, 404–405
equipment, 398, 405–406
and error-inducing environment, 18–23
fear, 397
inclusion of human factors project, 17
intelligent identification of systems and equipment, 415–416
in life cycle of technological enterprises, 410–415
main influences related to human factors, 393–401
maintenance errors, 412–413
management systems, 406–408
methodologies developed in area of, 390
people, 393–394, 406
procedures, 394–396
programs for consideration of, 408–410
RHFE, 410
safety culture, 401, 404–405
seven principles
adaptation of design to humans, 20–21
centralizing objectives in people, 20
control of human–system interaction, 21
human decision superiority, 21–22
inclusion of anthropometric and psychological project, 23
nonmechanization of human labor, 22
protection against human error, 21
shutdown, decommissioning, and demolition errors, 414–415
stress, 396–397
tasks, 400
workstation, 400
Human Factors and Ergonomics Society (HFES), 390
Human Factors Engineering Review (HFER), 402
Human Factors Society (HFS), 390
Human labor, nonmechanization of, 22
Human reliability analysis (HRA), 391, 416–418
Human-system interaction, 2, 385, 397, 402
control of, 21
during escape and abandonment, 143–145
lessons learned, 422–427
audio information, 424–425
driver's GPS knowledge, 425
GPS position in dashboard, 423–425
human-system interface, 426
influence of GPS on driver, 423
software and configurations, 424–425
visual information, 424–425
limitations of quantification techniques related to human reliability, 416–418
rapid entire body assessment, 418–422
Human–machine interaction systems, 22
Humility, 28–29
Hydrocarbons, 53, 60, 62, 84–85
accidents in facilities with, 142–143
disposal during emergency, 117–124
controlled burning and dispersion, 118–124
pressure relief and depressurization, 118
inventories, 111
safety in physical and chemical operations with, 86–90
composition/information on ingredients, 88
firefighting measures, 89–90
first-aid measures, 89
identification of hazards, 88
segmentation of, 113–117
blocking segmentation technique, 115–117
layout techniques, 113–115
Hydrogen gas (H_2), 203
Hydrogen sulfide (H_2S), 205–206

I

Ignition
  energy, 288
  temperature, 187
Inclusion of human factors project, 17
Industrial engineering, 99
Inerting systems, 199
Infrared (IR), 190–191
  detector, 211, 212*f*
  infrared/ultraviolet detector, 211–212, 213*f*
  point gas detector, 201–202, 202*f*
  sensors, 201–202
International Atomic Energy Agency (IAEA), 11
International Electrotechnical Commission (IEC), 189
International Energy Agency (IEA), 450
International Maritime Organization (IMO), 438
International Nuclear Safety Advisory Group, 11–12
International Organization for Standardization (ISO), 139
Ionization detectors, 215
ISO 13702 standard, 139, 152

J

JET fire, 85
John Davison Rockefeller and risk management, 56–58, 59*f*

K

Kerosene, 57
Kick, 62
Knock Out Drum (KOD), 120
Kraftwerk-Kennzeichen-System (KKS), 415

L

Lagrangian transport model, 348
Layer of protection analysis (LOPA), 336–338
Layout techniques, 113–115
Legalism, 17
Life rafts, 161–162, 161*f*, 168–169
Lifeboats, 155–160
  general safety requirements, 158–160
Lighting, special requirements for, 184–185
Limited ventilation environment, 194
Liquefied natural gas (LNG), 81, 82*f*, 85–88, 224–225, 254–265, 350
  comparison of boiling points between liquefied natural gas and materials, 256*t*
  cryogenics
    basic accidental scenarios and, 257–260
    characteristics, 255–257
    emergency control and, 260–264
    rapid phase transition and, 264–265
  evaporation control mechanism, 262*f*
  pool, 224–225
  relative characteristics, 256*t*
  spill burn control mechanism, 263*f*
  vapor cloud, 257
Liquefied petroleum gas (LPG), 77, 88
Liquid containment and environmental control
  analysis of loss of, 346–351
  practical results, 350–351
Liquid leaks, 82–84
Living quarters. *See* Superstructure areas
Lloyd's Register Group, 356
Loss of coolant accident (LOCA), 43
Low probability scenario, 448–449
Lower flammability limit (LFL), 189

M

Machine Learning, 467
Malfunctioning, 123–124
Manophobia, 267–268
Marine salvage resources infrastructure, 114
Marine terminals, 81
Maritime Safety Committee, 361–362
Maritime transportation, 79–80
Material Safety Data Sheet (MSDS), 87–88, 90
Mechanical engineering, 100
Mechanical manual actuation, 223
Mechanical sparks, 192
Meeting points, 153–154

Membrane diagram, 92–94, 92*f*
Methane gas (CH_4), 83, 87, 201–202, 207
  detectors, 201
Midstream, 58, 108
Minimum cutting set, 444–445
Mobile foam generating equipment, 245–246
Multidisciplinarity, 14–16
Multispectral IR detectors, 211
Muster stations, 148

N

National Electrical Code (NEC), 189
Natural gas, 87
Natural laws, 27
  respect for, 24–25
Natural ventilation, environment with, 194
Naval damage
  condition, 351–352
    scenarios, 369–370
  in offshore rigs, 271
Net-Zero Emissions by 2050 case (NZE2050), 452
Node, 330
Nonelectric ignition sources, 189–192
Nonfloating offshore rigs. *See* Fixed offshore, rigs
Nonhazardous area, 114
Nonmechanization of human labor, 22
Nonmetallic materials, 192
Nonquantifiable risk, 7–9
  surfing, 9*f*
Nonsparking metal, 192
Normally inhabited platform, 68
Nuclear accident at Chernobyl power plant, 9–10
Nuclear energy sector, 10
Nuclear safety, 11

O

Obstruction, 287–288
Occasionally inhabited platform, 68
Occupational safety, 436–438
Offshore hydraulic diesel pump, 228–229
Offshore platforms
  fixed, 67–70
  semisubmersible, 70–71
  special
    CT platforms, 76
    self-elevating platforms, 75
    submersible platforms, 74
    tension leg, 75–76
Offshore rigs, 155–156
  features of analysis of, 355–356
  representative simulations for, 371–372
Oil
  industry, 56–62
    accidents in, 60–62
    components of oil and gas productive chain, 58
    John Davison Rockefeller and risk management, 56–58
    onshore and offshore facilities, 58–60
  productive chain components, 58
On-land transportation, 79–80
Open path infrared gas detector, 201–202, 202*f*
Operational experience of agents on board, 360–361
Operational flare failures, 123–124
Operational practice, 90–96
  professional work in operational activities and in field, 94–96
  safety barrier, 94
Operational technical knowledge, 58
Organizations, 400–401, 408

P

Parallel system, 278
Passive fire protection (PFP), 246–254
  determination of type of partitions, 249–250
  interference
    of classified bulkhead and penetrations, 253
    between classified bulkheads and doors and windows, 253
  materials for, 254
  observations, 250–253
  structural protection, 253–254
Passive protection, 246–247
Pellistor, 203
People, 393–394, 406, 409

People on board (POB), 154, 351–352, 360
Permit to work (PW), 95, 164–165
Personal protection equipment (PPE), 434
Petrochemical plants, refining facilities and, 77–79
Petroleum coke, 78
Photoelectric smoke detectors, 214
Physical barrier, 94
Physical positioning of agents in rig, 363–364
Pipelines, 113
Piper Alpha, 69–70
  Cullen Report, 132
  hazards reduction failure, 129–132
  lessons learned from, 132–134
Piping and Instrumentation Diagram/Drawing (P&ID), 466
Pneumatic actuation, 223
Pool fire, 85
Post-CH_4 methane gas confirmation actions, 204–205
Post-CO_2 gas confirmation actions, 208
Post-H_2S gas confirmation actions, 207
Posthydrogen gas confirmation actions, 205
Potassium bicarbonate, 264
Power generation, 180–195
  area classification, 185–195
    American and international standards, 189
    concepts, physical, and chemical phenomena, 187–189
    degree of risk source, 192–193
    electric and nonelectric ignition sources, 189–192
    group and zone classification, 194–195
    ventilation, 193–194
  essential consumers, 182–184
    common to fixed and floating platforms, 182–183
    on FPSO/FSO floating platforms, 183–184
    on semisubmersible floating platforms, 183
  safety consumers, 184
  special requirements for cables and lighting, 184–185
  and transmission system, 64
Preliminary hazard analysis (PHA), 2, 314–315, 319–322
Preliminary risk analysis (PRA), 314–319, 440
  description table of level of control, 319*t*
  example of
    risk classification matrix used in, 318*t*
    spreadsheet used in application of, 316*t*
Pressure relief, 118
Pressure waves, 287, 288*f*
Primary processing equipment, 66–67
Primary risk source, 193
Prioritization, 16
Probabilistic analysis, 349
Procedures, 394–396, 407
Process safety, 82–90
  loss of containment, 82–84
  safety in physical and chemical operations with hydrocarbons, 86–90
  stable or explosive burning combustion, 84–86
    fireball, 85
    flash point, 84–85
    JET fire, 85
    pool fire, 85
    VCE and flash fire, 86
Production Excellence Plan, 472
Professional work
  in operational activities and in field, 94–96
  in project activities, 98–102
    chemical engineering, 100
    civil engineering and architecture, 99
    electrical engineering, 100
    electronic engineering, 100
    fire prevention engineering, 102
    human factor engineering and ergonomics, 101–102
    industrial engineering, 99
    mechanical engineering, 100
    risk and safety management engineering, 100–101
Project routine, 96–102
  professional work in project activities, 98–102
  safety systems design documents, 102

Protection against human error, 21
Protection system, 276
  for confined equipment, 254
Protective distance, 288–289
Psychological project, inclusion of, 23
Purging, 199

Q

Qualitative risk analyses, 310, 312–314
Quality HSE (QHSE), 93
Quantitative risk analyses, 7, 44, 310–314

R

Radiation energy, 190–191
Rapid entire body assessment (REBA), 418–422
  anthropometry, 420–421
  evaluation, 422
  example of application of, 419
  human body mechanics during execution of tasks, 419–420
  recommendations, 422
  repetitive work, cumulative trauma, and use of hand tools, 421
  static work, 421
  website offers step-by-step execution, 420*f*
Reaction times for agents on board, 363
Reduction of unpredictability, 307
  full safety analysis, 352–372
  lessons learned, 372–378
    HAZOP chaos, 376–378
    risk analysis and team work, 372–376
  risk
    analysis techniques, 310–338
    management strategic line, 308–309, 308*f*
  studies and consequence analyses, 338–352
Redundant configurations, 278–279
Reference standards, 470–471
Refining facilities and petrochemical plants, 77–79
Reliability, 277–278
Remotely operated vehicle (ROV), 164–168
Removal of agents (people), 2, 39–41, 479
Rescue boat, 148–149, 162–163, 162*f*
Rescue equipment, 154
Rescue study, 343–346
Residual risk, 311
Review of Human Factors Engineering (RHFE), 409–410
Right attention, 16–17
Risk, 5–6, 36
  acceptance, 9–17
  classification matrices, 315
  degree of risk source, 192–193
  rejection, 14
Risk analysis
  brainstorming, 331–332
  checklist, 332–334
  example of criterion for applying, 313*t*
  FMEA, 334–335
  HAZOP analysis, 322–330, 325*t*
  human behavior, 467–469
  LOPA, 336–338
  preliminary hazard analysis, 319–322
  preliminary risk analysis, 314–319
  quantitative and qualitative, 312–314
  questions and limitations, 312*t*
  and team work, 372–376
  techniques, 310–338
  what-if technique, 335–336
Risk management, 1–2, 5–6, 8–9, 55, 71, 98, 111, 149, 386, 431, 479
  accident investigations, 441–445
  applicable technical standards, 469–471
  association of technical fields, 434–435
  best gas sensor in world, 50–52
  capillarity of concepts and principles of risk management, 447–448
  centralization and decentralization, 433–434
  in corporate environment, 432–433
  design-basis accident, 42
  beyond design-basis accident, 42–43
  efficiency, 23–29
  emergency control, 41–42
  in energy industry postpandemic COVID-19, 448–464

Risk management (*Continued*)
engineering, 100–101, 480
exercise, 52–54
experts, 311
hazard reduction, 35–39
historical data records and management by indicators, 435–436
human factors and error-inducing environment, 18–23
John Davison Rockefeller and, 56–58
lessons learned, 471–473
nonquantifiable risk, 7–9
occupational safety and safety engineering, 436–438
in postpandemic world energy future, 461–464
and potential of new digital tools, 464–469
new risk analysis methods, 465–467
professionals, 96–97
reducing unpredictability, 43–46
removal of agents, 39–41
risk management strategic line, 29–31
safety culture and risk acceptance, 9–17
safety peer review, 439–441
strategic line, 24, 29–31, 111–112, 308–309, 308*f*, 479
surveillance system, 446–447
technical and operational knowledge, 31–35
theory specialist, 46–49
Risk-based design (RBD), 438–439
Rotary drilling rigs, 64–66
Rotation system, 65

S

Safe temporary refuge (STR), 133
Safety
barrier, 94
consumers, 184
engineer, 33–34
engineering, 34, 436–438
pendulum, 14
in physical and chemical operations with hydrocarbons, 86–90
safety-related factor, 69
safety-related flare system, 122–123
studies, 307, 309
systems, 93, 167
design documents, 102
Safety Case, 132
Safety culture, 9–17, 145, 401, 404–405, 479
right attention at right time, 12–14
safety pendulum, 14
seven principles of, 14–17
inclusion of human factors project, 17
multidisciplinarity, 14–16
prioritization, 16
right attention, 16–17
right time, 17
subjectivity, 16
technical intelligence, 17
Safety instrumented function (SIF), 336–337
Safety instrumented level (SIL), 336–337
Safety instrumented system (SIS), 336–337
Safety management, 1. *See also* Risk management
in energy industry postpandemic COVID-19, 448–464
engineering, 100–101
Safety peer review (SPR), 439–441
Salvage equipment, 163
Sea survival equipment, 154–163
life rafts, 161–162
lifeboats, 155–160
rescue boat, 162–163
salvage equipment, 163
Secondary grade risk source, 193
Segmentation of hydrocarbon inventory, 113–117
Self-elevating platforms, 75, 75*f*
Self-licensing, 13–14
Semisubmersible (SS), 184–185
offshore rig project, 152, 377
platforms, 70–71, 70*f*
rigs, 184–185
Series system, 278
Series-parallel system, 278
75-INSAG-4 (Safety Series number), 11
Shock waves, 192
and factors that influence explosions, 287–289

Shut-off valves (SDV), 113, 120, 126
Shutdown, 125–126
Shutdown valve (SDV), 115, 116*f*
Simplicity, 25
Smoke
detection, 214–215
dispersion analysis, 341–342
Single Point Anchor Reservoir (SPAR) platforms, 75–76
Software, Hardware, Environment, Liveware 1
Liveware 2 method (SHELL method), 444
SOS, 163–169
Space classification strategy, 247–249
Spread mooring, 72–73
Stability studies, 351–352
Stable burning combustion, 84–86
Standard and gas leakage scenarios, 368
Startup failure, 279
State-of-the-art FPSO rig, 163–164
Stated Policies Scenario (STEPS), 451, 453
Static electricity, 86–87
Stress, 396–397
Structured what-if technique (Swift technique), 335–336
Subjectivity, 16
Submersible platforms, 74
Subsea safety equipment, 265
Sulfidic acid, 205–206
Superstructure areas, 359
Support system, 64
Surveillance software, 446
Surveillance system (SS), 446–447
Survival conditions, accidents in facilities with, 142–143
Sustainable Development Scenario (SDS), 452
Swiss cheese model, 92–93, 109
Systematic cause analysis technique (SCAT), 442–443

T

Tasks, 400
Technical and operational knowledge, 1, 31–35, 47, 55, 111–112, 139, 479
avatar for "experts" without operational experience, 102–108
downstream facilities, 76–81
exercises, 108
knowing operational practice, 90–96
knowing process safety, 82–90
knowing project routine, 96–102
oil industry, 56–62
upstream facilities, 62–76
Technical intelligence, 17
Technological enterprises, 20–22, 28–29
Technological management, 431
Technology, defined, 5
Tension leg platforms (TLP), 75–76, 109
Theory specialist, 46–49
Thermal detectors, 216
Thermocouple, 216
Thermovelocimetric detection, 216
Totally enclosed motor propelled survival craft (TEMPSC), 155
Toxic gases detection, 205–207
Transportation and distribution, 79–81
Travel speeds of agents onboard, 362–363
Triple IR sensor detectors (IR3 detectors), 211
Tsunamis, 7–8
Tubular fluorescent lamp, fracture of, 192
Turbo generators, 181
Twin Design, 465

U

Ultraviolet detector (UV detector), 190–191, 210–211, 211*f*
Unavailability failure, 279
Uninterruptible power supply (UPS), 182
Unnecessary risks, rejecting, 24
Unpredictability reduction, 43–46, 479
Unquantifiable risk, 7
Upper flammability limit (UFL), 189
Upstream facilities, 62–76, 108. *See also* Downstream facilities
drilling rig and completion, 64–66
fixed offshore platforms, 67–70
FPSO platforms, 71–74
primary processing equipment, 66–67
semisubmersible offshore platforms, 70–71
special offshore platforms, 74–76

V

Vacuum distillation process, 78
Vapor cloud explosion (VCE), 86, 256–257
  and flash fire, 86
Vapor pressure, 188
Ventilation, 193–194, 196–198
Very large tanker crude carrier (VLCC), 355–356

W

Water mist fire suppression system, 242–244
Water spray fixed systems, 220–223, 222*f*
Well safety system, 65–66
Well-drilling process, 62
Wellhead Safety Equipment, 65–66
What-if technique, 335–336
Workstation, 400
World energy outlook, 450–460

Kolibri 360 GmbH
Ferdinand-Jühlke-Straße 7
99095 Erfurt, Deutschland
produktsicherheit@kolibri360.de